Spatial data analysis in the social and environmental sciences

Spatial data analysis in the social and environmental sciences

ROBERT HAINING

University of Sheffield

CAMBRIDGE
UNIVERSITY PRESS

PUBLISHED BY THE PRESS SYNDICATE OF THE UNIVERSITY OF CAMBRIDGE
The Pitt Building, Trumpington Street, Cambridge CB2 1RP, United Kingdom

CAMBRIDGE UNIVERSITY PRESS
The Edinburgh Building, Cambridge CB2 2RU, United Kingdom
40 West 20th Street, New York, NY 10011–4211, USA
10 Stamford Road, Oakleigh, Melbourne 3166, Australia

© Cambridge University Press 1990

First published 1990
First paperback edition 1993
Reprinted 1997

A catalogue record for this book is available from the British Library
Haining, Robert P.
Spatial data analysis in the social and environmental sciences.
1. Social sciences. Spatial analysis.
I. Title
300.1

Library of Congress Cataloguing in Publication data
Haining, Robert P.
Spatial data analysis in the social and environmental sciences
Robert Haining.
    p. cm.
Includes bibliographical references.
ISBN 0-521-38416-8 (hbk)   ISBN 0-521-44866-2 (pbk)
1. Spatial analysis (Statistics) 2. Population geography –
Statistical methods. I. Title
HA30.6.H35   1990
300′.15192-dc20   90-1361 CIP

ISBN 0 521 38416 8   hardback
ISBN 0 521 44866 2   paperback

Transferred to digital printing 2000

*This book is dedicated to my wife Rachel
and to our children Celia, Sarah and Mark*

Contents

List of tables and displays   xiii
Preface   xviii
Acknowledgements   xx

PART A **Introduction to issues in the analysis of spatially
referenced data**

1	**Introduction**	3
	Notes	10

2	**Issues in analysing spatial data**	12
2.1	Spatial data: sources, forms and storage	13
2.1.1	Sources: quality and quantity	13
2.1.2	Forms and attributes	17
2.1.3	Data storage	18
2.2	Spatial data analysis	21
2.2.1	The importance of space in the social and environmental sciences	21
2.2.1	(a) measurement error	21
2.2.1	(b) continuity effects and spatial heterogeneity	22
2.2.1	(c) spatial processes	24
2.2.2	Types of analytical problems	26
2.3	Problems in spatial data analysis	32

Sections starred (*) could be omitted at a first reading. Details in section 3.2 could be omitted as well.

Contents

2.3.1   Conceptual models and inference frameworks for
spatial data   32
2.3.2   Modelling spatial variation   37
2.3.3   Statistical modelling of spatial data   .   40
2.3.3   (a) dependency in spatial data   40
2.3.3   (b) spatial heterogeneity: regional subdivisions and
parameter variation   43
2.3.3   (c) spatial distribution of data points and boundary effects   44
2.3.3   (d) assessing model fit   45
2.3.3   (e) distributions   46
2.3.3   (f) extreme data values   46
2.3.3   (g) model sensitivity to the areal system   47
2.3.3   (h) size–variance relationships in homogeneous aggregates   49

2.4   A statistical framework for spatial data analysis   50
2.4.1   Data adaptive modelling   50
2.4.2   Robust and resistant parameter estimation   54
2.4.2   (a) robust estimation of the centre of a symmetric
distribution   55
2.4.2   (b) robust estimation of regression parameters   56

   Notes   61

PART B   **Parametric models for spatial variation**

3   Statistical models for spatial populations   65

3.1   Models for spatial populations: preliminary considerations   66
3.1.1   Spatial stationarity and isotropy   66
3.1.1   (a) second order (weak) stationarity and isotropy   66
3.1.1   (b) second order (weak) stationarity and isotropy of
differences from the mean   67
3.1.1   (c) second order (weak) stationarity and isotropy of
increments   67
3.1.2   Order relationships in one and two dimensions   69

3.2   Population models for continuous random variables   75
3.2.1   Models for the mean of a spatial population   75
3.2.1   (a) trend surface models   75
3.2.1   (b) regression model   76
3.2.2   Models for second order or stochastic variation of a spatial
population   80

3.2.2	(a) interaction models for **V** of a MVN distribution	80
*3.2.2	(b) interaction models for other multivariate distributions	90
3.2.2	(c) direct specification of **V**	90
3.2.2	(d) intrinsic random functions	94
*3.3	Population models for discrete random variables	99
*3.4	Boundary models for spatial populations	101
3.5	Edge structures, weighting schemes and the dispersion matrix	110
3.6	Conclusions: issues in representing spatial variation	113
	Notes	115
	* Appendix: simulating spatial models	116
4	**Statistical analysis of spatial populations**	118
4.1	Model selection	118
4.2	Statistical inference with interaction schemes	123
4.2.1	Parameter estimation: maximum likelihood (ML) methods	123
4.2.1	(a) μ unknown; **V** known	123
4.2.1	(b) μ known; **V** unknown	124
4.2.1	(c) μ and **V** unknown	127
*4.2.1	(d) models with non-constant variance	129
4.2.2	Parameter estimation: other methods	130
4.2.2	(a) ordinary least squares and pseudo-likelihood estimators	130
4.2.2	(b) coding estimators	131
4.2.2	(c) moment estimators	133
*4.2.3	Parameter estimation: discrete valued interaction models	134
*4.2.4	Properties of ML estimators	134
4.2.4	(a) large sample properties	134
4.2.4	(b) small sample properties	135
4.2.4	(c) a note on boundary effects	137
*4.2.5	Hypothesis testing for interaction schemes	142
4.2.5	(a) Likelihood ratio tests	142
4.2.5	(b) Lagrange multiplier tests	145
4.3	Statistical inference with covariance functions and intrinsic random functions	147
4.3.1	Parameter estimation: maximum likelihood methods	150

Contents

*4.3.2	Parameter estimation: other methods	151
*4.3.3	Properties of estimators and hypothesis testing	154
4.4	Validation in spatial models	158
4.5	The consequences of ignoring spatial correlation in estimating the mean	161
	Notes	166

PART C   Spatial data collection and preliminary analysis

5	Sampling spatial populations	171
5.1	Introduction	171
5.2	Spatial sampling designs	175
5.2.1	Point sampling	175
5.2.2.	Quadrat and area sampling	177
5.3	Sampling spatial surfaces: estimating the mean	177
5.3.1	Fixed populations with trend or periodicity	178
5.3.2	Populations with second order variation	178
5.3.2	(a) results for one-dimensional series	180
5.3.2	(b) results for two-dimensional surfaces	181
5.3.3	Standard errors for confidence intervals and selecting sample size	183
5.4	Sampling spatial surfaces: second order variation	186
5.4.1	Kriging	186
5.4.2	Scales of variation	189
5.5	Sampling applications	191
5.6	Concluding comments	195
6	Preliminary analysis of spatial data	197
6.1	Preliminary data analysis: distributional properties and spatial arrangement	198
6.1.1	Univariate data analysis	198
6.1.1	(a) General distributional properties	200
6.1.1	(b) Spatial outliers	214
6.1.1	(c) Spatial trends	215
6.1.1	(d) Second order non-stationarity	222
6.1.1	(e) Regional subdivisions	223

6.1.2	Multivariate data analysis	223
6.1.3	Data transformations	227
6.2	Preliminary data analysis: detecting spatial pattern, testing for spatial autocorrelation	228
6.2.1	Available test statistics	228
6.2.2	Constructing a test	231
6.2.3	Interpretation	234
6.2.4	Choosing a test	237
*6.3	Describing spatial variation: robust estimation of spatial variation	239
6.3.1	Robust estimators of the semi-variogram	241
6.3.2	Robust estimation of covariances	244
6.4	Concluding remarks	244
	Notes	245

PART D   **Modelling spatial data**

7	**Analysing univariate data sets**	249
7.1	Describing spatial variation	250
7.1.1	Non-stationary mean, stationary second order variation: trend surface models with correlated errors	251
7.1.2	Non-stationary mean, stationary increments: semi-variogram models and polynomial generalised covariance functions	282
*7.1.3	Discrete data	288
7.2	Interpolation and estimating missing values	291
7.2.1	*Ad hoc* and cartographic techniques	293
*7.2.2	Distribution based techniques	296
7.2.2	(a) Sequential approaches (sampling a continuous surface)	297
7.2.2	(b) Simultaneous approaches	304
7.2.3	Extensions	307
7.2.3	(a) Obtaining areal properties	307
7.2.3	(b) Reconciling data sets on different areal frameworks	309
7.2.3	(c) Categorical data	310
7.2.3	(d) Other information for interpolation	310
	Notes	311

Contents

8   Analysing multivariate data sets   313

8.1   Measures of spatial correlation and spatial association   313
8.1.1   Correlation measures   313
8.1.2   Measures of association   324

8.2   Regression modelling   330
8.2.1   Problems due to the assumptions of least squares not
        being satisfied   334
8.2.2   Problems of model specification and analysis   339
8.2.2   (a) Model discrimination   341
8.2.2   (b) Specifying **W**   341
8.2.2   (c) Parameter estimation and inference   344
8.2.2   (d) Model evaluation   347
8.2.3   Interpretation problems   348
8.2.4   Problems due to data characteristics   348
8.2.5   Numerical problems   349

8.3   Regression applications
        Example 8.1: Model diagnostics and model revision
                     (a) new explanatory variables   350
        Example 8.2: Model diagnostics and model revision
                     (b) developing a spatial regression model   354
        Example 8.3: Regression modelling with census variables:
                     Glasgow health data   365
        Example 8.4: Identifying spatial interaction and
                     heterogeneity: Sheffield petrol price data   372

        Notes   383

*Appendix: Robust estimation of the parameters of
            interaction schemes   384

        Postscript   386

        Glossary   389
        References   391
        Index   406

List of tables and displays

Tables

2.1	A general framework for statistical analysis	13
2.2	Considerations in defining the spatial attributes of a region and regional data prior to data analysis	39
2.3	Robust estimators: weighting functions	58
3.1	Spatial correlations for the stationary form of the single parameter CAR model	104
3.2	Spatial correlations for the stationary form of the two parameter CAR model	105
3.3	Spatial correlations for the stationary form of the single parameter SAR model	106
3.4	Lag one $(R(0,1))$ correlations for finite lattice schemes	107
3.5	Modelling strategies for different forms of spatial stationarity and non-stationarity	114
4.1	Arithmetic mean, variance and mean square error of the exact ML estimator of the parameter of a first order SAR model	137
4.2	Arithmetic mean and standard deviation of the exact ML estimators of the parameter of a first order CAR model and the surface mean	139
4.3	Arithmetic mean and standard deviation of the Whittle estimator of the parameter of a first order stationary CAR model	140
4.4	5% critical values for $-2\ln\lambda$ obtained by simulation methods	144
4.5	Percent power for likelihood ratio (LR) and Moran tests for a first order SAR model	147
4.6	Standard error of $\hat{\mu}$ (4.4) and $\tilde{\mu}$ (4.7)	162

4.7	Information loss as a percentage of total information for each directional parameter of a first order trend surface	165
6.1	Glasgow community medicine area data: standardised mortality rates	199–200
6.2	Petrol price data: petrol stations in southwest Sheffield	208–9
6.3	Application of median polish to remotely sensed data: area 1	217
6.4	Application of median polish to remotely sensed data: area 2	219
6.5	Application of median polish to remotely sensed data: area 3	220
6.6	Row and column D statistics for the residual values for area 1	220
7.1	Ordinary least squares estimates for the zero order trend surface model: pollution data	255
7.2	Lag spatial correlations for the least squares residuals from the zero order trend surface model: pollution data	255
7.3	Ordinary least squares estimates for the first order (linear) trend surface model: pollution data	256
7.4	Lag spatial correlations for the least squares residuals from the first order (linear) trend surface model: pollution data	256
7.5	Estimates using an SAR model for V in (7.2): order 0 model on pollution data	256
7.6	Estimates using an SAR model for V in (7.2): first order model on pollution data	257
7.7	Maximum likelihood estimates for (7.2): zero order model on pollution data	257
7.8	Maximum likelihood estimates for (7.2): first order model on pollution data	257
7.9	Maximum likelihood estimates using an SAR model for V in (7.2): second order model on pollution data	258
7.10	Glasgow health data: (a) contiguous CMAs	266–7
	(b) digitised co-ordinates for each CMA	268–9
7.11	Ordinary least squares estimates for the second order trend surface model (7.1): cancer data	270
7.12	Maximum likelihood estimates for the second order trend surface model (7.2): cancer data	276
7.13	Leverage values for CMAs	276

7.14 Population weighted second order trend surface model (7.2): cancer data — 279

7.15 Possible outliers: area 1 pollution data — 281

7.16 Possible outliers: area 1 pollution data after adjusting for initial outliers — 283

7.17 Possible outliers: area 1 pollution data (*a*) after square root transformation, (*b*) after square root transformation of outlier adjusted data — 283

7.18 Model fits: pollution data (area 1) — 285

7.19 Fitting orders of logistic and autologistic model to TB control adoption data — 291

7.20 Estimated population density based on all other states (worst cases only) — 296

7.21 Henley's example — 303

7.22 Missing value estimates: different forms for V. Case (a) CAR model; case (b) Estimated empirical covariances — 308

8.1 (*a*) Asymptotic value of n Var(r) when Y_1 and Y_2 are first order CAR models — 315
(*b*) Asymptotic value of n Var(r) when Y_1 and Y_2 are first order SAR models — 316

8.2 Type I errors for a 5% test on the Pearson product moment correlation coefficient based on 300 simulations of first order SAR processes. The upper figure is the unadjusted test and the lower figure is the test based on the Clifford and Richardson (1985) adjustment (*a*) 11×11, (*b*) 7×7 — 319

8.3 Properties of N' for different levels of spatial dependence in Y_1 and Y_2. The Pearson product moment correlation coefficient   (1) 11×11 ($n-121$)   (2) 7×7 ($n-49$) — 320

8.4 Type I errors for a 5% test on the Spearman rank correlation coefficient based on 300 simulations of first order SAR processes converted to rank order values. The upper figure is the unadjusted test and the lower figure is the test based on the Clifford and Richardson (1985) adjustment.   (*a*) 11×11 (*b*) 7×7 — 322

8.5 Properties of N' for different levels of spatial dependence in Y_1 and Y_2. The Spearman rank correlation coefficient   (*a*) 11×11 ($n=121$)   (*b*) 7×7 ($n=49$) — 323

8.6 Spearman's rank correlation measure for the 1980 and 1981 burglary data adjusted for the spatial structure of the data — 330

8.7   Problems that may arise in fitting the regression model
        using ordinary least squares: (*a*) Problems due to
        assumptions of least squares   (*b*) Problems due to the
        nature of the data   332–3
8.8   Population density data (1970) for the 48 states in
        Weisberg (1985) table 2.1   354
8.9   Ordinary least squares analysis of Irish data   (*a*) Lag
        correlations computed for the residuals
        (*b*) Autocorrelation tests on the residuals   356
8.10   Fitting different spatial regression models to Irish
        data   (*a*) Lagged explanatory variable regression
        model   (*b*) Lagged response variable regression model   358
8.11   Fitting model (8.12) to the Irish data   (*a*) SAR
        errors   (*b*) Other spatial error models   360
8.12   Leverage values (as % of total leverage) for model
        (8.11) and model (8.12) with $V = V_{SAR}$, W the
        weighted form   362
8.13   Robust estimation of (8.12) for β using Tukey's bi-
        weight   365
8.14   Glasgow CMA data   367
8.15   Diagnostics for the fit in Display 8.4   368
8.16   Testing for spatial heterogeneity in the petrol price data   382–3

Displays

6.1   First two moments of autocorrelation statistics   233
8.1   Analysis of US fuel consumption data   (*a*) Model
        fit   (*b*) Diagnostics   (*c*) Revised model   351
8.2   Evaluating the effects on parameter estimates of
        deleting two counties (Donegal and Mayo) with high
        leverages   362
8.3   Exploratory fit of census variables to Glasgow health
        data: least squares fit   366
8.4   Fitting models to the cancer SMR data   (1) Fitting
        with additional quadratic trend: least squares
        fit.   (2) Fitting with non-constant error variance
        (weighted least squares).   (3) Fitting with additional
        quadratic trend: resistant R-regression.   (4) Stepwise
        Poisson log linear regression   370–1
8.5   Analysis of petrol price data: temporal model I   376
8.6   Analysis of petrol price data: temporal model II   378

8.7   Analysis of petrol price data: spatial models   (*a*) Spatial
       error model ($V = V_{SAR}$)   (*b*) Lagged response variable
       model   379

8.8   Robust estimation of spatial error model with $V = V_{SAR}$
       (*a*) Robust estimation of β only   (*b*) Robust estimation
       of β and ρ 379

Preface

This book is about methods of analysing spatially referenced data where quantitative observations are associated with fixed points or areas on a map. My feeling was that there was a place for a book that tried to identify the main approaches to analysing spatial data and brought them within a coherent framework. Specifically, I wanted to examine the substantive justification for certain modes of analysis, to try to lay out some of the more important parts of statistical theory as it applied to spatial data analysis and consider important problems and discuss applications. The aim was to produce a text that proceeded from data collection to preliminary or exploratory analysis to uni-variate and multi-variate data analysis.

The purpose of Chapter 2 is to try and answer questions about what one is doing in spatial data analysis. As the introduction makes clear, I am not however advocating any form of separatism and in fact I see many of the problems associated with spatial data as amenable to treatment by currently available methods widely used in other branches of data analysis.

Chapters 3 and 4 lay out those areas of theoretical statistics which I think are relevant to the problems of representing spatial variation in the social and environmental sciences. No doubt there will be those who feel that the models are far too simple and the techniques far too difficult to justify application. I am not wholly unsympathetic to this view but also feel (and seek to demonstrate) that the methods are useful and as far as I am aware they are the best we have got. I suspect that the problems lie not so much in the models and techniques themselves but how they are used to develop adequate models for spatial data.

Chapters 5 to 8 are mainly about method and application. I make no apology for reworking some familiar data sets, nor, in reworking some of my own data, for coming to slightly different conclusions than I had previously come to. I have made available to the reader all the data sets I have analysed.

The book is aimed at the graduate and research level as a guide to methods of data analysis in the two fields. Some parts are, I think, quite difficult and I have starred sections that I feel could probably be omitted in a first reading. It should be possible, for example, to just dip into Part B on a first reading, returning to it later for reference and technical detail. I have deliberately concentrated technical material into Part B in order to 'free' the rest of the book, but the price for this has been that this section does make for rather heavy reading. Parts of the book could be used to introduce spatial data analysis to an undergraduate audience. The sections I have in mind are Chapters 1 and 2, sections 3.1.2, 3.2.1, 3.2.2(a) (examples), 4.1, 4.2, 4.4, 5.2, 5.5, Chapter 6, sections 7.1.1, 7.2.1 and Chapter 8. It would be relatively easy to put together a short course using MINITAB software in which students explore the data, provide data descriptions, try out graphical tests of hypotheses and fit simple models to small data sets. Following this route and analysing some of the data sets provided seems to me a good way to get started.

Robert Haining
Sheffield

Acknowledgements

This book was written and revised over a period of two years from the summer of 1987. During the first year I was in receipt of a Nuffield Social Sciences Research Fellowship which enabled me to concentrate full time on developing and writing the book. Academics, who must make time for research in conjunction with other duties, will know what a luxury it is to have such a concentration of time and I am most grateful to the Nuffield Foundation for their support.

The roots of this book go back over many years. My early interest in geography was stimulated by Ron Dolby and the late Charles Larkinson at Wallington CGS and developed under the amiable but sharp eye of A.A.L. ('Gus') Caesar at St Catharines College, Cambridge. My interest in the topics of this book first took shape as a graduate student at Northwestern University in the early 1970s. I was fortunate indeed to be a student there at a time when so many talented geographers were on the staff. I owe a special debt of gratitude to Mike Dacey and Walter Fisher for their encouragement and support in my early endeavours.

North American geography has had a strong influence on my work. Contacts with friends and colleagues over the last ten years at, amongst other places, Buffalo, McMaster, Wilfred Laurier, Santa Barbara, Los Angeles, Urbana and Bloomington have helped to shape this work. My special thanks to Bob Bennett and Dan Griffith with whom I have shared an extended period of research. Both have encouraged me over several years to write a book in this area. Thanks also to Luc Anselin, Noel Cressie, Robin Flowerdew, Art Getis, Reg Golledge and Richard Martin for advance copies of their work. Thanks also to the anonymous referees who took time to comment on the original manuscript. Of course they are in no way responsible for errors that remain and I would welcome criticism (preferably constructive) on all or any part of this.

This book is dedicated to my wife Rachel not least for her help in typing it,

xx

a task which compared to the joys of raising three children and teaching French and Italian comes a poor third. This book is also for my parents for their love and support.

Maps and diagrams have been drawn by Paul Coles and at various times Peter Bragg and Steve Black have helped with data collection. I am grateful to John Womersley for allowing me to use the Glasgow health data. The Glasgow Health Authority Information Services Unit was responsible for defining the Community Medicine Areas and preparing the data. The remotely sensed data arise from an aerial survey by the Southern Water Authority monitoring pollution levels in an area off the South Coast of England. Areas 1 and 2 are at an equal distance from the discharging pipe but area 1 is closer to an exit point from the pipe. Area 3 is further offshore than 1 and 2.

I gratefully acknowledge the following sources for permission to reproduce figures in the text: American Geographical Society (Figure 8.8a), American Geophysical Union (Figures 5.4, 5.5, and 5.10), American Statistical Association (Figures 7.1 and 7.2), Biometrika Trustees (Figure 4.9), British Society of Soil Science (Figures 5.6, 5.7, 7.26 and 7.27), Cambridge University Press (Figures 3.4c, 8.6 and 8.8b), David Fulton Publishers (Figure 8.1), Institute of British Geographers (Figures 3.18, 4.7, 4.8, 7.20 and 7.22), Institute of Electrical and Electronic Engineers (Figure 3.16), John Wiley and Sons (Figures 7.18(*a*),(*d*),(*e*), 7.19 and 8.10), Journal of Applied Probability (Figure 3.15), Kluwer Academic Publishers (Figures 6.17 and 6.18), Marcel Dekker Publishing (Figures 3.6, 3.8, 4.11 and 4.12), Martinus Nijhoff Publishing (Figure 8.9), Ohio State University Press (Figures 3.4(*a*), 5.9, 8.2, 8.3 and 8.4), Pion Limited (Figures 4.4, 4.5, 8.5 and 8.8(*a*)), Springer Verlag (Figures 7.18(*b*) and (*c*)).

PART A

Introduction to issues in the analysis of spatially referenced data

1

Introduction

A spatial data set consists of a collection of measurements or observations on one or more attributes taken at specified locations. Data sites are referenced so that the relative positions of sites are recorded, for the spatial organisation of the data is important whether the purpose of data analysis is to build a model for the data or to assess the relative merits of different hypotheses concerning some arrangement property of the data or some other (non spatial) characteristic of the data.

The principal purpose of this book is to describe and evaluate methods for spatial data analysis in order to show what is available, how the different techniques relate to one another and what can be achieved – in short, to contribute to the development of a sound inductive methodology for research areas that deal with data in their spatial context. In doing so, the book is aimed primarily at the social and environmental sciences and most of the examples are drawn from those areas. Apart from the fact that there are important links between social and environmental systems so that the study of one may draw in theory and data from the other, there are two other reasons for a methodological book that takes in both areas of research. First, both deal with observational rather than experimental data. In experimental situations the values of 'treatment' or explanatory variables are under experimental control and, moreover, the significance of these variables in influencing a response variable can be assessed by repeated experimentation. In the case of observational data these variables cannot be controlled: only events observed in the data can be modelled or predicted (confounding effects between variables may be present and raise serious problems for interpretation of findings); measurement error problems may be more pervasive particularly in the explanatory variables (since their levels are not controlled); and the analyst is particularly uncertain about the underlying situation influencing the response data since no formal experiment defines the situation. Analysis is restricted to what the real world offers. A consequence of this is that both fields are faced by similar sorts of problems in analysing and drawing inferences from such data and often considerable effort has to be expended in identifying a suitable model for the data.[1] Second, social and environmental analyses are often directed at similar spatial scales and data structures (the spatial arrangement of the sites or areas are often highly irregular) so that many of the types of data analytical methods relevant in one for describing spatial variation may be applicable in the other. The methods described in this book are already applied in branches of geography, economics, sociology, regional science and demography, and in areas of the environmental sciences, including ecology, soil science, resource management, hydrology, environmental epidemiology and remote sensing.

Data analysis must confront the following types of problem: those that arise when statistical assumptions are not met (how severe do departures have to be before they matter, how can departures be detected, what action should be taken?); those that arise from the particular data set; and those that arise from the introduction of modelling assumptions that may often stem from the particular subject matter theory underlying the research. An aim of this book is to present spatial data analysis as a part of general data analysis; to promote unity (rather than separatism) whilst alerting the reader to the special difficulties that spatial data may create from sampling, through preliminary data analysis, modelling, inference and evaluation. In particular the role of exploratory data analysis (EDA) and robust and resistant estimation and fitting procedures in spatial data analysis are examined.

EDA is concerned with data description, identification of statistical properties and the preliminary identification of data structure, with the objective of encouraging hypothesis formulation from the data. The methods of EDA are also useful in model evaluation. Many of the methods of EDA derive from conventional descriptive statistics; where they differ is in emphasising the utilisation of resistant methods (methods that are not sensitive to extreme data values) and in presenting data summaries in numerous graphical or pictorial forms and other ways that 'match the information handling capabilities of the brain and facilitate the detection of structure' (Good, 1983).[2] It has been suggested that the methods of EDA, and particularly those aspects concerned with encouraging hypothesis formulation, are particularly appropriate to the areas of 'uncomfortable science' where observational data are not obtained by means of any formal experimental design, are not always very accurate or at a high level of measurement, and where real repetition is not feasible or practical, so that scientific thinking depends on qualitative assessments of what constitute parallel situations.

Resistant estimation procedures are designed to deal with situations where, for example, the set of measurements follow a symmetric distribution but contain a small number of extreme values (outliers) which might have a disproportionate influence on estimates (such as the centre of the distribution or its spread) or the fit of a model. Values may not necessarily be 'wrong', merely different enough from the mass of data so that the analyst is concerned about how they might be influencing conclusions. In regression analysis, in addition to extreme values of the response variable, there may also be problems associated with observations on the explanatory variables. A small number of these may have a disproportionate influence (leverage) on the fit of the model if they are extreme with respect to the mass of observations. Resistant methods are often informal and repeated many times with different levels of resistance. They often form part of a sensitivity analysis concerned with assessing model 'fragility' to data attributes and may be used to help construct a better model by providing the analyst with new sets of residuals. Robust methods, on the other hand, aim to produce a single set of parameter estimates and confidence intervals which are efficient in situations where the data are believed to follow some 'heavy tailed' rather than symmetric distribution. Whereas resistant methods are most closely associated with exploratory stages of analysis, robust methods are more closely associated with the confirmatory stage.

These developments raise issues for spatial data analysis. How should the methods of EDA be used for detecting *spatial* structure in data sets? Methods are needed for analysing spatial data that can detect and deal with not only the data problems mentioned above but also problems that are associated with the spatial organisation of the data and which may affect the fitting of spatial models. Such problems include local spatial outliers (isolated values or clusters that are abnormal with respect to the local configuration of values and which might give rise to outlier or leverage effects in estimating spatial parameters or lead to the selection of an inappropriate model), uneven spatial coverage and the intrusive effects of an areal partition. Spatial dependence is a common characteristic of spatial data that is of intrinsic interest as well as creating problems for the application of certain statistical procedures. Methods are needed for detecting and representing this attribute. If the aim of modelling is to move towards better fitting models with better behaved residuals, what are the criteria that should be used to decide when a model is satisfactorily representing spatially referenced data? It is important to assess the sensitivity of results to modelling assumptions, particularly when there are several plausible alternatives. Certain spatial modelling assumptions (such as how spatial relationships between observations are represented and what conditions are assumed at the boundary of the study region) could be of considerable importance in affecting model fit.

A second aim of this book is to examine different approaches to describing spatial variation. Environmental and earth scientists have been much influenced by regionalised variable theory developed by Matheron and colleagues. This has lead to the widespread use, particularly in geostatistics, of the semi-variogram for describing spatial variation. Where slightly stronger assumptions hold, spatial variation can be described using spatial correlations and covariances. A second approach to representing spatial variation uses interaction schemes that specify dependency relationships explicitly in terms of random variables associated with sites or areas. Social scientists have tended to favour a third approach, 'autocorrelation' measures for characterising spatial variation, and modelling follows econometric principles using a spatial analogue of the temporal autoregressive model in order to describe spatial variation. The development of these different routes is due partly to the nature of the space, the types of data analysed and the types of problems tackled.[3] Nonetheless, this seems to be a useful area for some cross fertilisation (particularly between the first and second and the second and third approaches) for model specification, estimation and validation procedures.

A third theme in this book concerns subject matter theory and the role it should play in data analysis. This is an area of controversy concerning the *extent* to which subject matter theory should influence data analysis, *when* (at what stage) the two types of information should be mixed and *how* they should be mixed. One controversial aspect of the EDA literature, for example, is the extent to which it emphasises the importance of data dependent properties in hypothesis formulation and modelling. Concessions appear to be made to areas of the physical sciences where theory is strong but for many other fields, notably those for which the methods were ostensibly derived where theory is relatively weak, data properties (current experience) are considered of paramount importance and 'preconceived ideas' are not to be allowed to override conflicting evidence in the data.[4] Such a view tends to encourage an approach to modelling in which 'soft' or preliminary models are made firmer by a process of iterative fixing or model adjustment in the light of residual properties (see for example Figure 2.4). On the other hand Mulaik (1985) notes that those who 'may find some use for exploratory statistics in provoking hypotheses should realize that looking at patterns in data is not the only basis for the generation of hypotheses and may often yield ambiguous results' (p. 427). In this book we hold to the opinion that those who would engage in data analysis must observe the rules of the game but that there is more to our understanding and notions of what exists in the world than is given by the current set of data and indeed such understanding may be essential in order that sensible conclusions can be reached from data analysis. We require not only methods that facilitate data exploration and the detection of structure in data sets, but also methods which enable us to confirm or refute the presence of *expected* patterns in data sets. One interpretation of this perspective (widely adopted in spatial econometrics; Anselin, 1988b) is to start by specifying one or more models that include the expected features (such as spatial correlation, specified variate relationships, etc.) and that derive from the substance of the problem. Subsequent analysis seeks to discriminate between these alternative specifications and assess how well the models describe the data. A danger with this in areas of poorly developed theory is that it may focus theoretical attention too soon before all the available evidence (from the data) has been assessed (see for example Leamer, 1983).

There is also the question of *how* to link current and prior information (what Leamer, 1983 classifies as truths, facts, opinions and conventions). There are many recommendations, such as adopting a Bayesian approach (Leamer, 1978); fit 'big' models; adopt a wide horizon and try out many models. The principle followed in this book is to always try to make clear what prior assumptions (or modelling assumptions, as distinct from statistical assumptions which can be checked in the data) are being used as part of statistical data analysis, state their origin and assess the sensitivity of results to them.

The book is organised into four main sections. Chapter 2 outlines a framework for spatial data analysis and discusses the sorts of problems and questions that fall within its boundaries. We describe types of spatial data and the characteristics of spatially referenced data that are likely to raise problems for data analysis. It is at least in part the *mix* of problems surrounding spatial data that gives the field of spatial data analysis its distinctive quality, and it should be noted that the treatments for many of these problems are still in their infancy.

Part B of the book is the most difficult. Chapter 3 defines statistical models for describing spatial variation. The representation of spatial variation occupies an important position in spatial statistics. The data analyst needs to be aware of different models and their properties in representing spatial variation. In Chapter 4 we discuss inference questions: specification, parameter estimation and validation for spatial models. These two chapters summarise results in what might be termed mainstream parametric spatial statistics, which depend on strict distributional assumptions being met. Although these results extend the area of 'classical' or 'standard' statistics into areas where independence no longer holds, they still make strong assumptions (normality is often assumed) and the methods and results of this section should be used with care at the confirmatory stage of analysis.

Part C deals with data collection and preliminary data analysis. Chapter 5 discusses spatial sampling. Chapter 6 is much influenced by developments in exploratory data analysis and resistant estimation. The aim of this chapter is the description of methods that encourage informative presentation of spatial data, and the preliminary identification of data structures with the aim of better description and modelling of data. This chapter includes the extensive work on spatial autocorrelation tests. Some may feel such tests aim at more than just *preliminary* data analysis. However, within the context of this book I feel that they are properly placed here. I also feel that there has been a tendency to over-elaborate this area of spatial analysis to the detriment of wider issues.

Part D deals with practical issues in the further analysis of spatial data. Chapter 7 examines models for describing spatial variation in a single variable (univariate data analysis). Good summary description can be informative and help answer questions relating to comparative spatial structures both through time and between regions. Description is also a precondition for many methods of data interpolation. Although there are many cartographic methods for data interpolation and mapping that rely largely on the geometrical properties of the sample points (they are widely used in automated mapping routines, for example), there are other methods that make deeper use of the statistical properties of the observed data. In addition, these methods can be used to derive prediction errors. The best known example is kriging.

Chapter 8 deals with bi-variate correlation and regression modelling with spatial data. This includes regression models with a spatial error model or regression models in which lagged response and/or explanatory variables appear in the specification.

Throughout the book I have emphasised continuity of argument and stressed properties of method rather than mathematical detail. For those interested in following up specific points the references will, I hope, be adequate. There is a growing literature in this field and I have found particularly useful the earlier texts by Ripley (1981) *Spatial Statistics*; Cliff and Ord (1981) *Spatial Processes: Models and Applications*; Upton and Fingleton (1985) *Spatial Data Analysis by Example*. These books also discuss point pattern analysis which is not included here. Our interest is in the analysis of spatial variation associated with attributes of points or areas which are fixed in location.

I have found the book by Hoaglin, Mosteller and Tukey (1983) *Understanding Robust and Exploratory Data Analysis* and their second volume (1985) *Exploring Data Tables, Trends and Shapes* invaluable in understanding these methods. For some excellent papers in this area applied to spatial data, but specific to the area of kriging, I have also benefitted from the book in the NATO ASI series edited by Verly *et al.* (1984) *Geostatistics for Natural Resources Characterization.* Parts of this book also lean heavily on recent developments in regression modelling particularly relating to sensitivity analysis. I have found the books by Wetherill *et al.* (1986) *Regression Analysis with Applications*, Weisberg (1985) *Applied Linear Regression* and Belsley, Kuh and Welsch (1980) *Regression Diagnostics* very helpful. My source text for standard methods of applied regression analysis is Johnston (1984) *Econometric Methods*. I have found theoretical and applied contributions to spatial statistics in many journals. However, if the past is anything to go by, I would particularly commend the reader interested in keeping abreast of future developments to keep a watch on the *Journal of Soil Science; Soil Science; Water Resources Research; Geographical Analysis; Journal of Regional Science; Review of Economics and Statistics; Environment and Planning (A); Biometrics; Technometrics; Journal for the International Association of Mathematical Geologists; Computers and Geosciences; Journal of Ecology;* and *The International Journal of Remote Sensing.* This is the addition to developments in statistical theory where journals such as *Biometrika, Journal of the Royal Statistical Society (A and B), Journal of the American Statistical Association, Applied Statistics* and *Communications in Statistics* frequently carry papers on spatial statistics.

In concluding these introductory remarks a comment on what has been termed the 'hidden agenda' in this area of research is, I think, in order. It is still not routine to do computerised spatial analysis and the field is definitely underdeveloped relative to, say, time series analysis which is perhaps the fairest comparison. Specialist packages are not yet available and the problems of writing a satisfactory program of one's own and selecting appropriate algorithms for such operations as numerical inversion and function maximisation are further compounded if data sets are large, when many of the fitting procedures discussed in this book may become expensive to implement.[5] There is a need for good, efficient, preferably interactive, software which minimises the 'start up' costs of working in this area where one might have data and a (complicated) map recording the sites or defining the areal boundaries of some county system. Software capability should be able to handle exploratory and confirmatory data analysis with good graphical and mapping facilities. For very large data sets encountered in remote sensing there are at least the benefits of often working on a lattice of observations.

In 1987 the (U.K.) Report of the Committee of Enquiry chaired by Lord Chorley on handling geographic information appeared. 'Geographic Information Systems' is the label attached to the automated storage, manipulation, retrieval and display of spatially referenced data: 'a system as significant to spatial analysis as the invention of the microscope and telescope were to science, the computer to economics and the printing press to information dissemination. It is the biggest step forward in the handling of geographic information since the invention of the map.' (p. 8). It called for the development of statistical methodology relevant to spatial data, and in particular for easy to use software for processing large spatially referenced data sets. This is the additional link in the chain from data storage through manipulation to data retrieval and display. Whilst there are certainly important problems that are specific to analysing spatial data the position taken in this book is that many of these can be tackled within existing statistical methodologies. An important step here is to make these methodologies available within software packages that both draw on the data bases of working geographic information systems (or image processing systems) and interact with the graphics and mapping capabilities of these systems. Such a program may provide a way forward and go someway to meeting both the 'hidden agenda' and the proposals of the Chorley Report. I hope that by the time this book appears, progress will have been made in this direction.

NOTES

1   When spatial data arise from controlled (or even partially controlled) experimental situations, as in the case of agricultural uniformity trials, an important area of interest is to design experiments to assess treatment effects. This includes designing the spatial layout of treatment blocks in order to counter the influence of spatial correlation between the blocks (Kiefer and Wynn, 1983). This route is clearly not open to areas that depend on observational data but sampling schemes can be devised to minimise correlation effects.

2   These developments are due, in part at least, to the growth and availability of computer power, improvements in computer graphics and interactive software. The methods of EDA encourage the analyst to look at the data from many different angles and try out many different analyses.

3   In the environmental sciences the use of regionalised variable theory has largely been descriptive, to meet the problems of efficient interpolation and

10

mapping of continuous surfaces. In the social sciences spatial econometrics is largely concerned with the development of better methods of regression analysis with spatial data recorded for sites or areas. Regression is one of the principal techniques used for explaining, understanding and controlling social systems and considerable effort is expended in model specification (Solow, 1984; Leamer, 1978). In mapping problems finding a good model to describe spatial variation is essential. In regression modelling the description of spatial variation *per se* may be of lesser importance although an adequate representation is still needed.

4   The extent to which the assessment of evidence (in this case data) is influenced by preconceptions is the subject of much research in psychology (the anchoring phenomenon). Indeed, it is not even necessary for the preconditions to be of the form of a theory, for even random start points can affect the outcome of experiments. The tendency for data to be used simply to strengthen belief in pre-existing ideas to a degree unsupported by the characteristics of the data is a problem we shall not develop here, but the interested reader can consult Kahneman, Slovic and Tversky (1982) and Nisbett and Ross (1980) for collected studies. An interesting demonstration of the effects of preconceived ideas in evidence assessment is discussed in Leamer (1983).

5   In some cases it may be possible to modify existing packages to carry out analysis (Griffith, 1988; Upton and Fingleton, 1985). This may not always be an efficient way of doing things nor may the algorithms contained within certain packages be appropriate.

2

Issues in analysing spatial data

This chapter considers problems arising in handling and analysing spatial data. The first section deals with the nature of spatial data, the sources and quality of such data with particular reference to the 'new' data sources and the problems they may create for analysis. Subsequently, we discuss forms of spatial data and the problems of computerised spatial data storage. The second section examines why the spatial referencing of data is important in the social and environmental sciences and how spatial structure in data comes about. This leads to a discussion of the types of problems that are addressed and for which methods of data analysis are required. The third section is concerned explicitly with the distinctive problems of *spatial* data analysis. These are problems associated with the observed values, their spatial configuration and the areal system across which the values are observed.

The final section describes a framework for data analysis which summarises the issues of the first two chapters. Exploratory and confirmatory data analysis provide a framework for statistical analysis (Table 2.1). These two elements are concerned with the identification of structure in general data sets and the development of models for such data. The table distinguishes different approaches to confirmatory data analysis. If strict distributional assumptions are known to hold, parametric methods are very efficient (estimators have small sampling variances and hypothesis tests have high power), but otherwise they may be subject to serious error. Non-parametric methods on the other hand tend to be conservative and wasteful of data, losing much in efficiency relative to parametric methods if distributional assumptions do hold, although power and efficiency losses can sometimes be compensated for by increasing the number of observations. But for spatial data collected over a fixed areal grid, increasing the number of observations may not be possible without changing the nature of the problem. Robust methods occupy a middle ground, offering a potentially

12

Table 2.1. *A general framework for statistical analysis*

Exploratory data analysis	Confirmatory data analysis
Resistant methods of data analysis for detecting data structure, hypothesis formulation, and the specification of 'soft' data models. Methods include descriptive statistics. Also used for model evaluation.	Statistical inference and the development and testing of models. (i) *Parametric approach* Assumes that strict distributional assumptions hold. (ii) *Robust approach* Assumes that data have arisen from within a class or set of possible distributions. (iii) *Non parametric approach* All distributions are equally likely to underlie the generation of the data.

safer and more effective approach in which data are assumed to have arisen from within a *class* of possible distributions including so called contaminated distributions in which deviant data points (outliers) or substructures occur within the data. An introduction to these issues is given in Hampel, Ronchetti, Rousseeuw and Stahel (1986, Chapter 1).

2.1 Spatial data: sources, forms and storage

2.1.1 *Sources: quality and quantity*

Spatial data acquisition over the last twenty or so years has undergone significant changes. To the traditional sources of spatial data, including archival sources (maps, census material, air photographs), field observation (directly observed survey and sample data), and experimental or simulation work (where processes are reproduced and data recorded in a laboratory environment) have been added data collection through remote sensing of the environment by satellite as well as new government and commercial spatially referenced data bases that have been appearing in an ever rising flood (Rhind, 1987). Both these new areas owe their growth to technological developments: the ever increasing power and falling real costs of computers and the creation of software systems (geographic information and image processing systems) that overcome the technical and organisational problems that underlie the collection, storage, manipulation and display of geographic data.

13

For government and commercial data the important questions regarding the handling of such data are: how accessible will the data be, and of what quality, and what types of questions will the organisations collecting the data want answers to? In many areas, data banks are still at the planning stage. However, it is likely that in the future such data banks will become the major sources of social and environmental data and, consequently, familiarity with the handling of large data files and the development of methods that are capable of handling large data files quickly and efficiently and which are robust to their underlying problems will become of paramount importance.

All data have only a limited accuracy, arising from human error and instrument error. There are few studies that attempt to document the seriousness of this problem but Hampel *et al.* (1986), who briefly review the problem of measurement error, suggest that as a matter of routine between 1% and 10% of values will contain gross errors; that is, occasional but powerful errors in addition to the slight distortions (due to rounding, scale reading, etc.) that naturally occur. Rosenthal (1978) in a study of errors in psychology found rates of between 0% and 4% (with an average in a very skewed distribution of 1%) in a study of 15 data sets. Hampel *et al.* (1986) warn that the growing practice of putting large masses of data unscreened into the computer will lead to problems of data error becoming more serious, not less. Where data recording methods are automated the risk of undetected error increases. The key questions are whether the data have been properly collected and properly stored in the computer. 'There seems to be some doubt as to the wisdom of collecting large quantities of badly handled data, when only a small proportion of it may ever get analysed. Perhaps the philosophy "The greater the amount of information the less you know" is not completely out of place here.' (Leonard, 1983, p. 18). Even data that are not usually subject to such criticism and which are usually checked carefully can create problems. Coale and Stephan (1962) report the occurrence of data errors in the 1950 U.S. Census of population despite careful checking, in which subtle errors led to excessive widowhood and divorce among native Americans and middle aged males becoming teenagers! Fuller (1975) mentions measurement errors associated with the 1972 U.S. Census. Nor are such problems restricted to social data. Remotely sensed data may display very serious distortions over large areas of the monitored surface due, for example, to topographical or meteorological conditions. Further errors may be introduced by the procedures used to pre-process the data or to classify it into, for example, land cover types.

Another possibly serious data problem is that of incompleteness. The problem may be only temporary in some cases (Rhind, 1981), in other

14

instances permanent due to instrument failure (Bernstein *et al.*, 1984) or, in the case of census data, suppression for reasons of confidentiality.

Spatial data acquired by sampling may raise special problems. Haggett (1981) gives an example where measured rates for the incidence of a blood disease in 36 cities in El Salvador were based on a sample of about 5000 people. However, different sized samples were drawn from each city. The patterns of highs and lows, he suggests, may not therefore be due to real variations in the disease across the cities but rather an artifact of the sampling process. He uses a James–Stein estimator to adjust the estimated values the result of which does not even preserve the order properties of the original data set.

In addition to measurement error problems associated with attribute values, errors in spatial data may also be associated with recording or representing the areal framework. The locations of point sites and areal boundaries on a map are often inaccurate because of scale effects for example. Area boundaries and line features (such as roads) are stored in geographic information systems (GIS) in terms of points which are joined by straight lines in order to represent the feature. This introduces inaccuracy which is a function of the complexity of the line and the density of points used to represent it. There may also be operator errors in digitising the points from the source map. The source map may also contain mapping errors as well as those arising from expressing a curved surface on a flat piece of paper and shrinkage and distortion effects associated with the paper itself. The errors may be compounded in different types of GIS operations such as map overlay (Burrough, 1986; Goodchild and Gopal, 1989).

A major difficulty arises when data are collected over different (and incompatible) areal partitions. The analyst may be interested in analysing relationships between variables that have been collected by different agencies that have used different areal frameworks. (Social and economic data may be recorded for enumeration districts while health data are recorded for postcode sectors.) Solutions to this problem range from ignoring the incompatibility to attempting some form of adjustment such as areal weighting or aggregation. With areal weighting solutions a new set of (target) zones are defined and values assigned from the original (source) zones based on the degree of overlap between the two sets. The accuracy of such methods rest on assumptions of intra-unit uniformity. Where zonal incompatibility or areal non-uniformity is severe the analyst will have to decide whether any *ad hoc* approach is likely to be worthwhile.

What are the special problems for data analysis that are created by large data sets? Is it possible for example to obtain convenient summaries of large data sets for exploratory data searches? What are the problems of modelling large data sets, in particular evaluating and comparing the adequacy of different models? What effect will large sample size have on exploratory model fitting and hypothesis testing procedures?

As an example of this last issue consider the researcher who fits the standard regression model:

$$y = X\beta + e$$

where y is an $(n \times 1)$ vector of observations on the response variable, X is an $(n \times (k+1))$ matrix of observations on the explanatory variables (with the first column consisting of 1s), $\beta = (\beta_0, \ldots, \beta_k)^T$ is a vector of parameters and e is an $(n \times 1)$ vector of unobservable disturbances which are normally distributed with mean zero and variance covariance matrix $\sigma^2 I$. Suppose we wish to test a null hypothesis

$$H_0: \beta_1 = \beta_2 = \ldots = \beta_k = 0$$

against an alternative

$$H_1: \beta_i \neq 0 \text{ for at least one } i \ (i = 1, \ldots, k)$$

The F test is given by

$$F = [R^2/(1 - R^2)][(n - k - 1)/k]$$

where R^2 is the coefficient of determination and H_0 is, under classical hypothesis testing, rejected if $F > F_{(k, n-k-1)}$ where $F_{(k, n-k-1)}$ is obtained from the F tables at the chosen significance level. It is a decreasing function of sample size. It is evident from the formula for F that very small values of R^2 can give arbitrarily large values of F if the degrees of freedom $(n - k - 1)$ are large enough. The implication is that with large samples almost any null hypothesis is going to be rejected. Leamer (1978) poses the question: 'Is classical hypothesis testing at fixed levels of significance a "good" way to summarize the evidence in favour of or against hypotheses' of the type given above? He argues that significance levels need to be a decreasing function of sample size so that much larger computed F values are required to reject H_0 if n is large.

At the other end of the quantitative spectrum are those data sets, typically under 100 observations, where the problems may not be so much associated with data accuracy (although a few aberrant values may have a more serious distorting effect when the data set is small than when it is large) as whether we have enough information to undertake meaningful data analysis, whether our analytical procedures are safe in small sample situations and whether we have enough data to discriminate between competing models.

2.1.2 Forms and attributes

Two main forms of spatial data are point and area referenced data. Point referenced data (when observations are taken at a fixed number of point sites) may be all such sites in a region: for example prices charged for a particular good at all retail sites of a given type in a city. Alternatively, point site measurements might be samples from a surface (e.g. soil pH readings across a field, temperature or precipitation measurements) or a quasi-surface (e.g. households in an urban area). The pattern and intensity of sampling will be partly a function of the nature of the underlying spatial variation (Chapter 5).

For area referenced data, one important issue concerns the type of areal framework that should be adopted for long term, multi-purpose data collection (e.g. within the context of building up a data base for a geographic information system); another concerns the appropriate framework for any specific application. Ideally, areas should be small and uniform (with respect to attributes), for the aim in designing a data collection system is to have a basis for flexible aggregation of areal units (to meet a wide range of user needs) that facilitate merging data from different sources and which can be modified to meet changing circumstances. In the case of any specific application, the choice of areal framework is a function of the spatial scale of the problem, the objectives of the study (including, where appropriate, requirements for good statistical analysis) and the specific attributes of the study (Visvalingam, 1988).

Area referenced data (where a region has been partitioned and observations are available for each area) derive from aggregations of primary units and form either a regular grid or an irregular mosaic. Much social and economic data refer to aggregations of households (as the primary units). The areal partitions are usually irregular (postcode sectors, enumeration districts, electoral wards, administrative regions, counties). The 1971 U.K. census provided statistics for 100 metre and 1 kilometre squares. The Chorley report (1987) has recommended the use of postcode sectors for reporting personal data. They are convenient to use and because of their small size offer wide scope for linking different data sets together and flexibility in spatial aggregation. Visvalingam (1988), however, comments critically on the adoption of these areal units and argues that only tessellations provide a stable framework for analysing change and for comparative studies. Environmental data deriving from multi-spectral scanners represent integrations across surfaces. The areal partition produces square grided (raster) data sets. Landsat and SPOT satellites, for example, generate grid square data at various resolutions. A high

17

resolution is usually desirable for environmental monitoring when surfaces change continuously and boundaries are indistinct. Discontinuous change (e.g. across land parcels and property registers) usually requires a lower resolution.

The observation for an area may be descriptive of any primary unit within the area, as in the case of a categorical (presence/absence) geographic attribute. Observations can be area dependent however. This arises in one of two distinct ways. In one case observations refer to attributes of primary units but are averages over all the primary units within each areal unit (e.g. average age, unemployment rate, reflectance levels on a remotely sensed image). Alternatively, observations may have no meaning at the level of a primary unit (e.g. population density where the primary unit is a unit of land). In both cases though the value assigned to an area ought to be representative of any sub unit of the area.

In addition to attributes that relate to primary units or areas, there are attributes that refer to relationships: between primary units (e.g. contact frequencies between households); between areal units (e.g. migration rates); and between either of these levels and higher levels which constitute a broader spatial context (such as national or supra-national influences). These relational attributes may be responsible for the spatial variation observed in area and primary unit attributes and such relational inform-ation may be extremely important in helping to model this variation, as will be discussed in Chapter 3.

2.1.3 Data storage

Many analytical methods require a knowledge of the spatial referencing of data values, so it is not enough just to label point sites or areas it is also necessary to know how they are located relative to each other. Point data can be digitised and stored in a data file as a triple $\{(x_i,y_i),z_i\}$, where (x_i,y_i) references the location of point i with respect to some appropriate co-ordinate system and z_i is the measured attribute at point site i. If several variables are measured then z_i becomes a vector. A data format of this type lends itself readily to spatial statistical analysis and the commonly used methods of spatial data description. Data relating to areas can also be stored in this way by representing each area by, for example, its visual or exact centroid (Visvalingam, 1988).

There are two basic models of data storage for areas, although hybrid forms have been developed which preserve relational information. The two models are the vector and tessellation models. In the vector model the basic logical unit is the line. Areas are modelled as polygons and represented by the set of lines that form their boundaries. Associated with each area is a collection of attributes. The topological model is a version of the vector model in which the network of lines partitioning the map are represented as a planar graph. Line segments correspond to arcs in the graph and their endpoints to nodes.

The tessellation model for areal data storage consists of an aggregate of cells that partition the plane. The basic logical unit is a unit of space and attached to each unit is a collection of attributes. Spatial relations between logical units are implicit in the tessellation. Since spatial variation may not be constant across an area it is useful to be able to adapt unit size in order to optimise storage. If properties of polygonal areas are required these can be assembled by aggregating the constituent basic units. There are a number of alternative systems for addressing the basic units and these are based on different linear orderings. The implications of the choice of ordering for the efficiency with which map operations or spatial functions (addition, subtraction, intersection, translation, rotation, etc.) can be performed by a computer are fundamental to the performance of geographic information system software.

Two properties are required of planar tessellations: first, they should result in an infinitely repetitive pattern in the plane and second, they should be infinitely recursively decomposable into similar patterns of smaller size (Smith *et al.*, 1987). Property one allows data bases of any size to be represented and property two makes it possible to use hierarchical data structures. Only the square and triangle satisfy both properties and the most widely used is the square, which corresponds to the raster data structure. Multi-spectral scanners produce raster data as do optical scanners for digitising maps.

Raster schemes have some highly desirable properties facilitating data search, data compaction, data browsing and accessing data by areas. Properties of this scheme and the other ways of structuring spatial data are discussed in Smith *et al.* (1987) and the reader should refer to papers in the *International Journal of Remote Sensing* that document specific examples of data storage. Figure 2.1 shows how vector and raster storage might work on map data with irregular attribute boundaries. The quality of the raster representation depends on the size of the basic units (square cells) while that of the vector representation depends on the length of the line segments in relation to the complexity of the boundaries on the surface. Optical scanners are currently expensive but may reduce the problems of translating a map or image into raster format. When data storage must proceed by digitising map data the effort involved should be carefully assessed in relation to the volumes of data that are to be fed into the structure and the operations to be performed. With such an activity the possibilities for excessive numbers of points, erroneous ('sliver') polygons and closure errors (where line segments don't meet) are strong. Geographical information system software often facilitate this stage of data preparation by checking for and correcting these types of errors. Green *et al.* (1985) discuss a system which makes these adjustments.

The tessellation models are usually considered most appropriate where map data is continuous and the scale of the grid is small relative to the scale of surface variation. It is sensible then to analyse these data files using the attribute data attached to each logical unit. Alternatively data might be sampled by selecting, according to a spatial design, a subset of logical units. The storage and manipulation of tessellated data is simpler than for vector data and the regular areal structure underlying this data file, as well as the ease with which blocks of logical units can be aggregated, makes this a particularly suitable type of data file for subsequent statistical analysis. This will become evident in later chapters. Overlay analysis of categorical data is easier using this data structure.

As noted above, particularly in the social sciences, spatial data are area dependent. Data values may be meaningless at small spatial scales or with

Figure 2.1. Raster and vector storage of spatial data on irregular areal units.

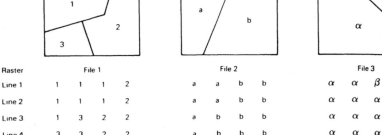

Map 1	Map 2	Map 3

Raster	File 1				File 2				File 3			
Line 1	1	1	1	2	a	a	b	b	α	α	β	β
Line 2	1	1	1	2	a	a	b	b	α	α	α	β
Line 3	1	3	2	2	a	b	b	b	α	α	α	β
Line 4	3	3	2	2	a	b	b	b	α	α	α	α

Vector

Node file
1 = (0.0)
2 = (1.5,0)
3 = (3.0, 0) etc.

Region file
A = (1,2,6,5)
B = (2,8,7,6)
C = (2,3,8) etc.

Attribute file
a = A,E
1 = A,B,C etc.

respect to some arbitrary tessellation. Attribute measures may not even carry down to the level of arbitrary logical (spatial) units. In such cases statistical analysis is with respect to the attributes associated with areas not the set (or subset) of logical units into which that region could, formally, be decomposed. However, it may still be more convenient to store data using the tessellation model and then aggregate as necessary.

Relational data bases (which preserve spatial relationships between data) and the mapping and graphics facilities within geographic information system software provide a potentially valuable structure for developing automated spatial statistical analysis.

2.2   Spatial data analysis

2.2.1   *The importance of space in the social and environmental sciences*

In this section we examine why the spatial referencing of data may be important by considering how spatial structure can arise in social and environmental data. The answer to this question has implications for the statistical analysis of spatial data – indicating when to anticipate spatial data patterns and the types of models likely to represent the patterns. Spatial structure in a measured attribute arises most often from measurement error, continuity effects including spatial heterogeneity, or from the operation of some form of space-dependent process or mechanism.

2.2.1   *(a) Measurement error*

Suppose a region is to be surveyed. The region is large and several operators are required to visit sites or areas within the region. If the region is divided up by assigning each operator to one contiguous block of sites or areas then, if there is any operator-dependent bias in the recording process, dependency effects between the sites or areas could be generated, with (mild) discontinuity at the block boundaries, because of the way the operators have been assigned to cover the region. Milne (1959) alludes to a problem of this kind in the results of a forest survey.

Remotely sensed data suffer from instrument induced spatial dependency effects. The problem arises because sensors measure light reflectance from the earth's surface, but this light is scattered so that reflectances from one small area of ground can be distributed over several contiguous pixels (picture elements) on the image (Forster, 1980). Craig and Labovitz (1980) discuss the effects of the type and age of the hardware employed and natural conditions (sun angle, cloud cover, geographic location and season) on the measurement process and induced spatial correlation. This type of error appears to be most serious along scan lines so that measurement induced dependency may take the form of a serial structure (Craig, 1979; Labovitz and Masuoka, 1984). The errors associated with sets of contiguous Landsat pixels are often modelled as normally distributed random variables with low order spatial correlation.

2.2.1   (b)   *Continuity effects and spatial heterogeneity*

In environmental studies spatial continuity of form is due to the spatial continuity of events that occur 'in space' and which are responsible for form. These events are responsible both for the alternation of surface forms and the scales of that alternation. They are also responsible for the form of alternation; whether there are relatively sharp boundaries or smooth continuous variation. Webster (1985) describes the different scales of events responsible for soil variation (from major land forming processes to the operation of earthworms) and Mejia and Rodriguez-Iturbe (1974) describe how spatial variation in precipitation levels may reflect the type of rainfall event (frontal or convection rainfall). In the social sciences, spatial continuity in social and economic variables is a consequence of the social instincts of individuals as well as of patterns of behaviour and economic constraints which taken together help bind social space into recognisable structures. Further, the operations and activities of public institutions and private corporations which deal with space in terms of areas, not points, also contribute to the structuring of social and economic space.

In both fields, there has been a long tradition of representing spatial variation in terms of regions: a mosaic of homogeneous (or nearly homogeneous) areas in which each patch of the mosaic is demarcated from its neighbours in terms of attribute levels. As a model for spatial variation, the regional concept encourages the analyst to view individual primary units or tiles (households, soil units, etc.) within each region as essentially identical members of the same population. Relationships between these primary units are often left unspecified or assumed to be independent. Since regions are distinguished by sharp boundaries (spatial heterogeneity), observations on different regions are also independent.

As a model for spatial variation, for purposes of spatial data analysis, the regional concept suffers from two problems: sharp regional delimitation and internal uniformity are difficult to sustain; and both these properties become particularly difficult to justify at small spatial scales and as more attributes are included in the classification. Attribute variation in space is usually subject to continuous change: it is an amalgamation of different components of variation or scales of continuous or quasi-continuous variation possibly associated with different processes. So individual primary units are not independent and units close together are often more closely related than those further apart. When areal aggregates are formed that are smaller than the scale at which surface forms vary then attribute values for these too are likely to be spatially correlated.

This change in emphasis does not imply that the region as a model for spatial variation is redundant. Particularly in the social sciences, regional delimitation may be an important element of spatial variation. The institutions of government manage and control society through areas and public institutions impose policies at areal and regional scales. Private corporations service markets that are spatial and introduce pricing policies that may distinguish sharply between different regions and submarkets. The built form of a city is in part an expression of the policies and activities of developers, estate agents and financial institutions dealing with space in terms of areas. A further consequence of these processes may be a partitioning of space in terms of response (spatial heterogeneity of response). Sub-regions respond differently to identical conditions. The housing market of an urban area is spatially heterogeneous (measured by price differences for identical housing units or components) because supply and demand conditions vary between areas. Supply constraints operating differentially over an urban area can underlie submarket delimitations whilst demand constraints operate because 'prospective owners and renters often examine housing in only a limited area because of search costs, racial discrimination, or desired proximity to friends or workplace' (Goodman, 1981, p. 175). Brueckner (1981) develops a dynamic model of urban growth in which the demolition and development cycle inherent in the growth process generates population density gradients with distance from the city centre, that show marked discontinuity. Heterogeneity may be associated with different parameter regimes between areas, as in the housing market case, or at different distances from a central point or node, as in the population density case.

The essential point is, then, that regional differentiation is only one possible component of spatial variation. It is not an a priori property of space and should not constitute the exclusive spatial model for spatial data analysis.

2.2.1 (c) Spatial processes

Spatial structure arises from the operation of processes in which spatial relationships enter explicitly into the way the process behaves. Suppose $\mathbf{y}_t = (y_{1,t}, \ldots, y_{n,t})$ defines the state of a system consisting of n areas at time t; $y_{i,t}$ is the value of attribute y at location i at time t. A spatial process is a process where changes of state are due to spatial properties of the attribute. For example we might specify

$$y_{i,t+1} = f(\{y_{j,t}\}_{j \in N(i)}, \{y_{j,t-1}\}_{j \in N(i)})$$

where $N(i)$ specifies the areas adjacent to i. Spatial and temporal dependencies might extend over many lags and assume many functional forms (f). For a formal treatment of these types of models, Bennett (1979) provides a detailed examination in the context of social and environmental systems. Here we are interested in the types of processes which might generate spatial structure in an attribute. There appear to be four important types.

1. In the social sciences, *diffusion process* is a general term classifying processes in which some attribute (a piece of information or rumour, a newly produced commodity or technical development) is taken up by a fixed population. At any time we can specify the distribution of individuals who have adopted and those who have not. The spatial distribution of the population may have important implications for the development of the process (rate of adoption, take up level in the population, pattern of adoption) particularly if the process of adoption depends on imitative behaviour through inter-personal contact or first-hand experience. The same considerations may underlie the spread of certain diseases (Cliff *et al.*, 1981; Cliff *et al.*, 1985), the development of incipient political power (Doreian and Hummon, 1976), or the spread of political information (Johnston, 1986).

2. A second type of process is one involving *exchange and transfer*. Urban and regional economies are bound together by processes of mutual commodity exchange and income transfer. Income earned in the production and sale of a commodity at one place may be spent on goods and services produced elsewhere. Such processes bind the economic fortunes of different towns, regions and countries together. Not only is the process reinforcing at an intra-area scale through the economic multiplier, it is spatially reinforcing through the inter-area multiplier (Paelink and Klaassen, 1979; Haining, 1987b). At the inter-urban scale the processes of wage expenditure on different orders of goods may necessitate visits to different centres. The binding together of the local spatial economy through the process of wage expenditure and other 'spillover' effects may be reflected in

24

the spatial structuring of social and economic variables such as aggregate levels of per capita income. The dividing up of an economic surface into areas of high and low income levels may reflect the extent to which processes of exchange and transfer become focussed and regionalised.

3. A third type of process is one involving *interaction*, in which events at one location influence and are influenced by events at other locations. For example, the determination of prices at a set of retail outlets or markets may in certain circumstances reflect a condition of action and reaction amongst the set of retailers (Haining, 1983a, b). Whether retailer B responds to a price change by another retailer and if so to what extent, depends in turn on the extent to which such price changes may affect B's custom. That assessment may be weighted towards the actions of retailers that are nearby, these being the competitors most likely to have an effect upon B's market share, profitability and ability to survive. A pattern of prices may develop across the retail sites that reflect these underlying competitive interactions.

4. The final group of processes involve *dispersal* or spread. This mechanism is distinguished from processes of diffusion, where some attribute is dispersed through a fixed population, since it is the population itself which disperses and the resultant spatial structure depends on the nature of the dispersal. Such processes may be applied to human populations (Curry, 1977) or the dispersal of seeds from a plant or of physical properties such as atmospheric or maritime pollutants or nutrients in a soil.

We are not, here, concerned with trying to specify formal connections between these processes and particular mathematical structures of spatial dependence in attributes. The possibility of deducing form from process in ways that correspond to realistic social processes seems to be limited; whilst inferring process from form is rendered dangerous by problems of equifinality and the way in which historical attributes may influence the behaviour of the system. Dow *et al.* (1982) comment on the complex network of interdependencies that provide the context within which social phenomena are embedded. Giving a specific example they note: 'how much of voting behaviour is affected by attributes of the voting unit (whether individuals or aggregates) and how much is the result of interactions between them: of the communication process, bandwagon effects, reference group behaviour or other forms of "symbolic interactionism" ' (p. 170). It seems likely that spatial parameters in spatial models (see (2.3) or (2.4)) will be data-dependent quantities, particularly in the case of area attributes which are aggregates of different numbers of primary units. However important they may be in a program of data analysis and building better models for describing a specific spatial data set, they should be interpreted

25

with care for they may have no close and direct association with any process mechanism.

2.2.2 Types of analytical problems

The analytical problems of this book can be broadly classified under three headings: problems in spatial data sampling; problems in providing numerical summaries and characterising the spatial properties of map data; and problems in the analysis of multivariate data sets, in particular the identification of bi- and multivariate relationships.

Spatial data collection, where values are recorded at sample sites on a continuous, or nearly continuous spatial surface, poses the problem of devising an optimal sampling scheme in the sense of achieving a desired level of precision in the estimate of a quantity with the least cost or effort. The term cost may refer either to collection costs, in the case of traditional 'on site' sampling methods, or processing costs, in the case of new optical sampling methods based on photographic or map data for the construction of contour maps or terrain models. Constructing optimal spatial sampling schemes requires that consideration be given to the number of sample sites required and their spatial siting.

In remote sensing, one of the primary goals is accurate scene inference in which characteristics of a true scene are inferred from image data. This problem is usually approached through the use of 'training pixels' – sites where for certain image values the true scene has been determined by ground observation. By having sufficient numbers of such data pairs {(image value, ground truth)} it is then possible to identify the image signature that corresponds to any type of ground cover. The image signature might be characterised, for example, by a mean value together with a measure of variance on each of several bands. Using this information, areas for which the ground truth is not known can then be classified from their image data. A problem arises as to how these training sites should be chosen. When checked, the use of contiguous training pixels, forming say a block, tend to give good local classifications (in the immediate neighbourhood of the training area) but their accuracy declines rather seriously with increasing distance. This effect is largely due to the information content of individual pixel values in the image data. Spatial correlation both in the ground data and arising from the measurement process means that neighbouring pixel values are often strongly correlated and pixel values do not provide an uncontaminated measure of the immediately underlying ground surface. The problem has been recognised for some time (see for example Tubbs and Cobberly, 1978). One solution is

26

to sample pixel values and their associated ground truth using sites rather than blocks. Campbell (1981) for example found that the use of random sampling in the selection of training pixels greatly increased classification accuracies. Random sampling still allows some training pixels to be close together and higher accuracies could probably be attained with systematic sampling schemes. A different solution to this problem is to actually exploit the continuity property of the surface in the classification process. Switzer (1980) for example, has suggested using an augmented data vector in linear discriminant analysis (a widely used technique for generating a classification rule) in which, instead of just matching area ground truth to the single pixel that covers it, the central pixel is used together with its eight neighbours. Since it is also difficult to match up pixel value and ground truth exactly this approach may have an added advantage. In a later paper Switzer (1983) discussed the use of smoothing modifications to standard classification algorithms. In both cases the spatial arrangement properties of the data are now exploited in order to improve classification procedures. These alternative procedures seek a representation for the arrangement properties of maps – so taking us to the second of the three problem areas.

Numerical descriptions of map data encompass both spatial and aspatial properties of the data set. One form of summary disregards the spatial arrangement of the data. Measurements are analysed as a distribution of values (a histogram, frequency diagram, box plot, etc.) and summarised in terms of measures of location (centrality), dispersion, spread, and skewness, and compared with standard models (normal, chi-squared, uniform, poisson, etc.). The next, and in the context of this book the more interesting, step is to examine the spatial arrangement of the values and devise summary measures that characterise the pattern. Two maps might possess an identical distribution of values (same location and skewness properties, etc.), but the layout of these values on the map could be quite different (Figure 2.2). Spatial summary measures should at least distinguish the situation where, say, all large values and all small values cluster in separate parts of the map area from the situation where large and small values alternate or where they are randomly dispersed across the map. Pattern properties of map data may be represented by single summary measures or data plots. More detailed descriptions may be devised in which pattern properties are described and summarised at different scales and over different map segments. The characterisation of distributional and pattern properties of map data provide a basis for map making (Woodcock and Strahler, 1983); smoothing and interpolating map data (Webster, 1985); for providing inputs to characterise the spatial attributes of a model such as spatial variations in precipitation in a hydrological model (Mejia and

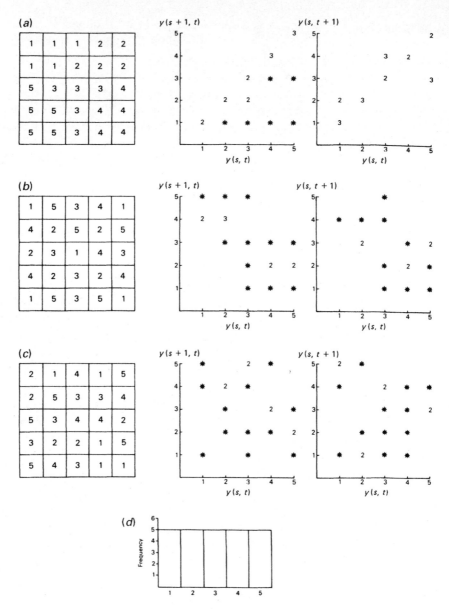

Figure 2.2. Three maps with identical distributional properties but with different spatial arrangement properties. $y(s,t)$ denotes the value at position (s,t) on the map. (*a*) Clustered arrangement of values (similar values together). (*b*) Alternating arrangement of values (different values together). (*c*) Random arrangement of values. (*d*) Histogram of data values for cases *a*, *b* and *c*.

Rodriguez-Iturbe, 1974); for making comparisons between different maps (Haining, 1987c); and for comparing map surfaces at different points in time (Haining, 1978a).

Map description may be extended and a model for the map data constructed. If:

Data = model + error

then in the case of a continuous surface the model may represent different scales of spatial variation. For example:

Model = Large scale trends or gradients + patches or mosaics or 'hot spots'

The error component includes measurement error. If the data refer to areas, the error component may also measure within-area variability which may be a function of the size of the area. If the data is point data the error component is called, in geostatistics, 'nuggett' variation, which includes both measurement error and 'inherent' site variability. The model component describes the systematic (inter-point or inter-area) characteristics in the data, ranging from large scale trends or gradients to local patterns of dependency in which neighbouring areas or sites are correlated. Such patterns may be associated with different processes operating at different scales within the region with some processes responsible for broad scale gradients and others for smaller scale patches.

Data may be collected on two or more variables across the same set of sites or areal partition. Correlation coefficients, like Spearman's rank, Kendall's tau or Pearson's product moment, measure association between two variables (Y_1, Y_2) by examining the set of pairs $\{(y_{1,i}, y_{2,i})\}_i$. These correlation statistics disregard the arrangement properties of the data, indexing similarity between the variables over identical spatial positions. However, association between two spatial variables has another aspect: the degree to which similar (or dissimilar) values of the two variables are spatially close to one another. (An historical perspective on this problem in geography is given in Hubert *et al.*, 1985.) A measure for this indexes similarity between the spatial positions of identically ranked observations. Take for example a study of the areal levels of atmospheric pollution in relation to the areal distribution of health characteristics in a population. Pollution levels disperse and spread, people move over the area; neither are static features and the exposure of any subset of the population to pollution need not necessarily be a product of pollution levels in the particular sub-area where they live. A correlation coefficient might indicate no association and any significant level of association might in part be an artefact of the patterns of spatial dependency in the data. What is needed is an index based

29

on the relative spatial positioning of identical observations in order to show whether similar levels of the two variables are close to one another in geographical space. An index together with an inference framework was devised by Tjøstheim (1978) and a number of generalisations have been developed since then.

Regression models specify a functional relationship between a response variable (Y) and k explanatory variables $(X_1, \ldots, X_k)$. Like the earlier correlation measures, the standard regression model specifies a functional relationship between y_i, (the response variable measured at site i) and $x_{1,i}, \ldots, x_{k,i}$ (the explanatory variables measured at site i), so that

$$y_i = \beta_0 + \beta_1 x_{1,i} + \ldots + \beta_k x_{k,i} + e_i \qquad i = 1, \ldots, n \qquad (2.1)$$

where $\beta_0, \ldots, \beta_k$ are regression coefficients and e refers to the error or disturbance term. In matrix notation (2.1) can be written:

$$\mathbf{y} = \mathbf{X\beta} + \mathbf{e} \qquad (2.2)$$

where as before $\mathbf{X}$ is an $(n \times (k+1))$ matrix of observations on the explanatory variables, with the first column consisting of 1s; $\mathbf{y}$ and $\mathbf{e}$ are $n \times 1$ vectors and $\mathbf{\beta}$ is a $(k+1)$ vector. If $E(\mathbf{ee}^T) = \sigma^2 \mathbf{I}$, where $\mathbf{I}$ is the $n \times n$ identity matrix then the arrangement properties of the attributes are irrelevant to the specification of the model.

The effect of an explanatory variable at any site may, however, not be limited to the specified site. If y_i measures the proportion of people suffering from respiratory diseases in area i and if $x_{k,i}$ measures levels of atmospheric pollution in area i, then y_i may not merely be a function of $x_{k,i}$ but also levels of x_k in areas neighbouring i. Such a definition reflects both the reinforcing effects of extensive tracts of high (or low) levels of pollution but also the fact that people move across areas over the course of time and their exposure to this risk factor is not solely a function of their local environment. Then (2.1) may be augmented with:

$$y_i = \beta_0 + \beta_1 x_{1,i} + \ldots + \beta_k x_{k,i} + \tau \Sigma_{j \in N(i)} x_{r,j} + e_i \qquad i = 1, \ldots, n \qquad (2.3)$$

where $N(i)$ denotes the set of sites or areas that adjoin i (excluding i itself) and X_r is one of the variables $X_1, \ldots, X_k$ and τ is a parameter. We shall call this type of model a regression model with spatially lagged explanatory variables.

Another modification to (2.2) is to note that the effect of the response variable at site i may operate as an explanatory variable at another site (see for example Whittle, 1954 and Mead, 1967 concerning inter-plant competition). Suppose y_i measures the price level for a good at a retail outlet in a city. Local competitors may consider the price level charged at site i as a factor when setting their own price level, an attitude that might be reciprocated by the retailer at *i*. In this case (2.1) is respecified so that:

$$y_i = \beta_0 + \beta_1 x_{1,i} + \ldots + \beta_k x_{k,i} + \rho \Sigma_{j \in N(i)} y_j + e_i \qquad i = 1, \ldots, n \qquad (2.4)$$

where $N(i)$ is as before and ρ is a parameter. We shall call this a regression model with a spatially lagged response variable.

In both (2.3) and (2.4) the arrangement properties of the map enter into the specification of the model. In the case of (2.3) systematic spatial variation in **y** arises from the effects of one (or more) explanatory variables since two neighbouring sites will have at least some neighbours in common. In the case of (2.4) systematic spatial variation in **y** arises for the same reason as in (2.3) except it is now self induced. Draper and Guttman (1980) have referred to (2.3) as the deterministic case for variation in **y** and (2.4) as the stochastic case. Model (2.1) implies no systematic spatial variation in **y** unless it is present in one or more of the explanatory variables, in which case it may be transferred to **y**.

When the analyst wishes to fit a regression model but suspects the presence of numerous, difficult to specify, spatially correlated effects which would invalidate fitting (2.1), then these effects may be included by assuming

$$E(\mathbf{e}\mathbf{e}^T) = \sigma^2 \mathbf{V} \qquad (2.5)$$

where **V** is a non-diagonal matrix describing spatial dependence in the errors. (This model also generates similar forms to (2.3) and (2.4) since (2.2) can now be re-expressed as

$$\mathbf{L}^{-1}\mathbf{y} = \mathbf{L}^{-1}\mathbf{X}\boldsymbol{\beta} + \mathbf{L}^{-1}\mathbf{e} \qquad (2.6)$$

where **L** is such that $\mathbf{L}\mathbf{L}^T = \mathbf{V}$. Note that $E[\mathbf{L}^{-1}\mathbf{e}\mathbf{e}^T(\mathbf{L}^{-1})^T] = \sigma^2 \mathbf{I}$. Forms for **V** will be described in Chapter 3.) The model ((2.2) with the error structure given by (2.5)) is called a regression model with spatially correlated errors.

If heterogeneity of response is anticipated the analyst might wish to allow either (or both) the constant or one or more of the other regression parameters to vary spatially. Suppose an area can be partitioned into just two sub-areas (1 and 2) and it is anticipated that the constant coefficient differs between the areas. Then specify:

$$y_i = \alpha + \beta x_i + \gamma D_i + e_i$$

where D_i is a dummy (0/1) variable with $D_i = 1$ if site *i* is in area 1 and is 0 otherwise. Then for area 1 the constant coefficient is $(\alpha + \gamma)$ and for area 2 it is α. If the slope parameter varies then specify:

$$y_i = \alpha + (\beta + \gamma D_i)x_i + e_i$$
$$= \alpha + \beta x_i + \gamma(D_i x_i) + e_i$$

31

where the slope coefficient is $(\beta + \gamma)$ for area 1 and β for area 2. These models can be extended to allow for more than two areas.

If parameter variation reflects distance from the centre of a city then the dummy variable D_i may be replaced by location co-ordinates for each observation. A further extension is to allow these parameters $(\beta + \gamma D_i)$ to vary stochastically $(\beta + \gamma D_i + v_i)$ where v_i is a random error term. Kau, Lee and Chen (1983) discuss the use of models of this type for describing urban population density gradients where the slope of the gradient is not smoothly non-linear but contains numerous discontinuities. High order trend surfaces or other explanatory variables can be substituted (Casetti and Jones, 1987). Anselin (1988b) considers these models in more detail.

As with analogous forms of time series regression modelling, an analyst may wish to compare not only different regression models in terms of included and excluded variables but also different forms of spatial regression. Part of the motivation for this may be substantive but, in addition, data diagnostics are needed that will help to choose between these different spatial regressions.

2.3 Problems in spatial data analysis

We will now consider ways in which properties of spatial data raise important issues for the statistical analysis of such data. We will start by considering conceptual models of the surface and their implications for inferential analysis. Then we will consider how known or expected properties of spatial surfaces might influence data analysis. Finally we will discuss the nature of spatial data and the special problems that arise in the analysis of point and area data.

2.3.1 *Conceptual models and inference frameworks for spatial data*

The classical inference model assumes that data are the outcome of some well defined experiment. This experiment can be replicated as many times as necessary. An analogue of this classical situation in the case of spatial data is when the surface is considered a single realisation (experiment) of a random process. In addition there may be other random components in the data resulting from the way in which the 'experiment' was observed: only a sample of data values may have been recorded from the surface; there may have been random error in the measurement process.

Inference with spatial data, in this classical context, is concerned with making statements and assessing the evidence regarding properties of the underlying process responsible for the data. When, as is usually the case, only one realisation or 'experiment' is observed some degree of location invariance or stationarity has to be assumed. The essential idea is that although only one realisation of the surface is observed, that realisation, if enough data can be collected, provides sufficient information to identify properties of the underlying process. One form of the stationarity assumption requires that both the mean and the variance covariance properties of the surface are stationary, possibly even isotropic (direction invariant). A weaker form of the stationarity requirement allows the mean to vary as a function of location although it usually depends on only a small number of parameters. In regionalized variable theory only surface increments need be stationary. The form of the spatial covariance function represents the contribution of different scales to overall spatial variation.

The assumption of a 'hypothetical universe' of realisations or 'super-population' is often viewed with considerable skepticism as a model for real spatial data analysis. How might this universe be made susceptible to random sampling and can we be sure the observed map is representative of this universe? What exactly do 5% significance levels or 95% confidence intervals refer to? What if the surface does not satisfy stationarity assumptions? It is sometimes argued that the classical perspective is justifiable when the processes responsible for the observations appear to contain inherent random components and where other instances of the same process can be indicated. These 'replications' may arise at other times across the same area (temporal replication) or in other areas (spatial or spatial-temporal replication). This perspective is explicit in the analysis of geographic mortality data where the observed number of deaths in an area is treated as a binomial or poisson random variable (Cook, Pocock and Shaper, 1982, Pocock, Cook and Beresford, 1981). Galtung (1967) critically discusses the concept of a hypothetical universe. A rationale for the superpopulation perspective is proposed by Barnard, in the discussion in Godambe and Thompson (1971): 'My limited experience in the use of surveys in the social sciences suggests, indeed, that the superpopulation is usually the appropriate one. One is rarely, for example, concerned with the finite *de facto* population of the U.K. at a given instant of time: one is more concerned with a conceptual population of people like those at present living in the U.K.' In those cases where it is difficult to construct other realisations it may still be unsatisfactory to treat the observed surface as representing the population. The process generating the surface appears to have inherent random components in it or, as implied by Bernard, the current population is considered typical of a larger (albeit ill defined) population. This should be recognised in data analysis.

Another model for spatial data which has some features in common with the classical inference model is where the data represent sample observations from a given surface. Henley (1981) describes the following model. The spatial surface is continuously varying but fixed (i.e. not a realisation in the earlier sense). All variation is associated with the actual form of the *real* surface. All differences between places are due to variation in this surface plus a point random error (a 'nuggett effect') which is usually due to observational error. Descriptions of the surface are a reflection of the local form of the surface plus the random errors (there is no theoretical semivariogram). The only random components in this model are due to observational error and the uncertainty that arises from dealing with a sample. The model, which Henley uses as the model for non parametric approaches to geostatistical site interpolation and mapping, rejects assumptions of replication and stationarity as inappropriate for examining earth surface features. This contrasts with earlier models for kriging where the realised surface is but one of many from a 'superpopulation' of surfaces (Matheron, 1971).

The difference between these two analogues of the classical inference model lies in the existence or otherwise of a universe of possible surfaces. Difficulties for classical inference arise if we adopt a deterministic perspective and the data represent an exhaustive survey (all areas or all point sites in a region). The classical inference model still seems tenable if 1. there is random error associated with measurement of surface properties; or 2. the surface is sampled; or 3. the values at the sites are themselves the result of a sample.[1] This might occur, for example, if the sites are towns and data for each town are based on a sample of households or individuals.

Both ways of relating spatial data analysis to the framework provided by classical statistical inference assume that data arise as if from a *controlled* experimental situation. While this does not need to be, a priori, unreasonable, this assumption does carry with it some implicit assumptions, in particular the *axiom of correct specification.* Leamer (1978, p. 4) specifies this axiom as follows:

(a)   The set of explanatory variables that are thought to determine the response variable must be 1. unique, 2. complete, 3. small in number, and 4. observable.

(b)   Other determinants of the response variable must have a probability distribution with at most a few unknown parameters.

(c)   All unknown parameters must be constant.

The analyst should therefore consider the extent to which the situation under which the data were collected meets these criteria.

If there is no experiment defining or specifying the model that should be fit to the data, probabilities are not well defined. (Probabilities are dependent on the model being fit and if we have no reason to prefer one model rather than another, then the probabilities cannot be well defined.) As a consequence, classical inference procedures that depend on evaluating probabilities (in order to construct confidence intervals, significance tests, etc.) are not on firm ground. Leamer defines three groups of data analysts in terms of their response to this situation: believers (those who report results as if they were the outcome of a controlled experiment), agnostics (those who do not deal with standard errors and seek only descriptive summaries of data and discount all results until tried out on another set of data) and pragmatists (who think that the agnostics go too far and that standard errors should be reported, properly enlarged to reflect the type of statistical analysis that has been undertaken).

Leamer (1978) emphasises two other areas of difference with classical inference procedures based on Fisher and Neyman–Pearson inference. First, the purpose of statistical inference in the context of controlled experimentation is to estimate unknown parameters and test hypotheses: 'classical inference apparently allows judgments that are either completely certain or "completely uncertain". We are asked to be certain about the parameter spaces but peculiarly uncertain about the choice of parameters within those spaces' (p. 14). Second, classical judgments are made solely with respect to the current set of data. Now Leamer argues that both the *purpose* of data analysis and the *judgments* made in the process of data analysis are different with non-experimental data and he disputes whether statistical inference, in the sense defined above, is what is required.[2] Data analysts, start with a 'well specified set of certain and uncertain judgments and enlist the data to encourage or discourage subsets of those judgments' (p. 15). These judgments embrace both subjective and theoretical knowledge of the circumstances which may overwhelm or be overwhelmed by the sample evidence. Classical statistical inference, hypothesis testing in an absolute sense, is not a part of real (non-experimental) data analysis, because we cannot be sure exactly what it is we are observing. Leamer develops a Bayesian approach which allows an explicit mixing of prior information with the current information provided by the current set of data. In addition, non-experimental data analysis involves more complex issues of model specification. Since there is no experiment defining the model, the analyst must engage considerable resources in this activity. Using regression modelling as an example he identifies six types of model specification searches. The brackets after each description indicate the type of failure of the axiom of correct specification (see list above) the particular search is designed to deal with.

35

(i)   *Hypothesis testing.* There are several possible regression models justified by theoretical considerations and the analyst tests to see which one is best supported by the evidence. (a1)

(ii)   *Interpretive searching.* One regression model is chosen to describe the data and the analyst tries to make the model fit the data better, perhaps by imposing constraints on the parameters. (a3)

(iii)   *Simplification searches.* A regression model chosen under (i) or (ii) may be very complicated and the analyst tries to reduce the complexity of the model whilst retaining adequacy of fit. (a3)

(iv)   *Proxy searches.* A regression model is chosen but it is known that variables can be measured in many different ways. Different variable definitions are compared to provide the best fit. (a4)

(v)   *Data selection searches.* Fitting a model with different subsets of data or different transformations of the data and selecting the result that appears best. (b,c)

(vi)   *Post-data model construction.* A purely inductive search to try and account as well as possible for observed variation in a response variable; often involves searching for additional variables to improve the level of explanation. (a2)

With such different uses being made of regression modelling it is perhaps not surprising that the interpretation of results cannot be the same in all cases. Although a general objective might be to obtain a good predictive equation for the response variable, or identify explanatory variables that have a significant association with the response variable, or estimate its sensitivity to a given explanatory variable (the sign and size of the associated β parameter), conclusions and inferences should differ between a regression model fitted as a result of say a post-data model construction as opposed to an interpretive search.

Leamer's theory is based on the observation of what real data analysts, in fact economists do. This has involved 'discarding the formal constraints of classical inference [whilst adding] the essential bits of uncertain information through *ad hoc* specification searches' (p. 15). The final result has been a mixture of sample and non-sample information while retaining the verbal commitment to classical inference. Leamer (Chapter 10) points out the inference traps that arise from errors of judgment and warns about 'excessive theoretical development before seeing the facts' (p. 16). His principal message is to insist that the analyst clarifies the purpose of analysis and makes clear the judgmental basis.

Diaconis (1985) reviews alternatives to the usual classical theories of inference in the context of exploratory data analysis, where the purpose of data analysis on non-experimental data is to detect structure and pattern.

Diaconis reviews probability-free and subjective theories of inference (based on Bayesian theories), but does not suggest that these yet provide a basis for non-experimental explanatory inference. Two exceptions to this are bootstrapping (drawing random samples from the observations for constructing confidence intervals) and tests based on data randomisation. The second has found its way into spatial statistics for detecting the presence of structure or pattern in the spatial arrangement of values on a map. If $y = (y_1, \ldots, y_n)$ denotes the set of values on a map then let $\delta(y)$ denote a descriptor of the organisational properties of the values on the map. (Statistics that measure organisational properties are discussed in section 6.2.) By considering the set of all spatial permutations of the vector y and computing $\{\delta(y)\}$ this provides a basis for classifying maps, identifying maps with extreme structures (lying on the tails of the distribution) and providing a measure of map similarity or distance between two maps (y_i, y_j) by computing $(\delta(y_i) - \delta(y_j))$, for example (Costanzo, 1985). The classification of maps with extreme spatial arrangements could be based on conventional cut off levels (1%, 5%, 10%) but it might be better to choose classification levels based on the shape of the frequency distribution. If this frequency distribution is highly peaked with little spread it might be decided that very few map patterns should be classified as extreme structures, whereas we might use a more liberal cut off level if the frequency distribution had a larger variance. This is one way in which non-experimental inference criteria might lead to the adoption of different decision rules from those currently emphasised in the classical inference model.

A limitation of the randomisation framework is that by considering all possible (spatial) permutations of the data across the sites, all of which are considered equally likely, the spatial structure observed in the data is ignored. This is not a problem in pattern classification, but in other contexts where dependency or organisational properties are an integral part of the system, it is often required that inferences allow for the spatial structure in the data. In this case we need models to represent that structure.

The choice of inference framework will arise again later and it is at least arguable that the devising of appropriate inference frameworks is one of the most fundamental issues in spatial analysis. Apart from its practical importance this issue goes to the very heart of how researchers should analyse their data.

2.3.2 *Modelling spatial variation*

Spatial data analysis requires models that describe spatial variation. It is interesting to start by contrasting the nature of space and time from the point of view of descriptive modelling. First the points of similarity: events defined across space or through time are *ordered* and cannot be assumed to be *independent*; continuous spatial and temporal variation both possess different components or *scales* of variation; finally, just as temporal variation may be represented as *stationary or non-stationary* (see Chapter 3) so spatial variation may display similar characteristics with second order variation imposed on a non-constant mean for example.

There are however important differences. Temporal dependency structures and non-stationarities are one-dimensional whereas with the exception of linear spatial systems most spatial dependency structures are two-dimensional. Although temporal and spatial systems are ordered, in temporal systems that ordering is largely constrained to the present being dependent on the past; unknown future states cannot affect the present (except in the sense that events may be influenced by *expectations* of the future). The ordering possibilities for spatial systems are not directional (except in some important but special situations – see Chapter 3), are ambiguous because of the multiplicity of paths and especially in social systems may be related to patterns of interaction which may be highly complex.

Boundary conditions (conditions at the edge or edges of the study area) raise more serious questions for spatial modelling. First, the boundary encircles the area and conditions at the boundary affect a much larger proportion of sites than in the time series case where after some initial state the process evolves without further boundary influences. Second, it may have a complex geometric form. Different types of study area boundary now assume greater significance: for example, whether the boundary is arbitrary (with the same process continuing beyond the edge of the study area) or natural (the process is halted at the boundary as may be the case with a land/sea boundary). Some boundaries are of an intermediate nature across which any process may remain more or less the same but is perhaps subject to some modification (for example, the urban/rural boundary in respect of population migration). There may be internal boundaries within the study area such as bodies of water within a land area.

Discontinuities may arise in a time series: an outbreak of war or an oil price rise or a policy shift in modelling econometric variables. In spatial modelling not only may discontinuities such as geological fault lines, administrative boundaries, social and economic regions, have to be dealt with, they may have complex geometric structures. These discontinuities may take the form of a displacement either side of the discontinuity, a termination of effects as at an administrative boundary, for example, or a shift in parameter regimes. Henley (1981) discusses the problem of kriging

Table 2.2. *Considerations in defining the spatial attributes of a region and regional data prior to data analysis*

1. Observations	Do they come from a continuous or a quasi-continuous surface; a surface with isolated 'patches' which are internally continuous or quasi-continuous; or a set of discrete point sites? Do the data refer to points or areas? Is the data set a sample or an exhaustive coverage? Do the areas partition the region or are they isolated patches? What size are the areas (or distance between sites) in relation both to the types of processes responsible for spatial variation and the reliability of data values? Is the distribution of areas or sites regular or irregular? Are each of the areas homogeneous?
2. Nature of the regional space	What is the boundary type? Is it an isolated region (natural boundary so no external neighbours) or a subregion of a larger region ('artificial' boundary with external neighbours)? What is the boundary shape? Is it smooth, convex (straight line distances defined for all interior sites) or irregular, concave (straight line distances not defined for all interior sites)? Is the area oriented (elongated)? What is the internal geography? Is it homogeneous or heterogeneous. If heterogeneous what is the form of heterogeneity? Are there distinct sub-regions comprising aggregates of points or areas? Are there interior discontinuities (administrative, social and economic divides, linear features such as lines of communication, physical barriers and divides)? What is the scale of the study area in relation to the range of processes operating?
3. Area and site relationships	
(i) External	What is the relationship of each area/site to other regions? Might there be effects originating in other regions influencing spatial variation within the region?
(ii) Internal	What are the relationships between areas/sites within the region? Are there order relations between areas/sites? If there are order relations what are their form?
(iii) Intra-site	Might there be area/site specific effects or common effects imposed on subsets of areas/sites? Might any spatial heterogeneity be due to different responses between sub-areas to given influences?

in the presence of fault lines. In remote sensing, Woodcock and Strahler (1983) distinguish between those elements of scene due to continuous background variation (soil, rock, snow) and those elements due to discrete entities (buildings, lines of communication, trees, hedgerows).

Table 2.2 summarises geographical attributes of a region and regional data which will influence how data analysis proceeds. It combines some of the statistical characteristics discussed above with areal characteristics, the importance of which are dependent on the nature of the problem. In the analysis of environmental problems describing spatial variation in pollution levels, a model might need to take particular care with the nature of the space and the relationship of the area to major (internal and external) sources of pollution. In the analysis of intra-urban spatial price variations, specification of inter-site relationships might be of considerable importance since they define local patterns of competition. The effect of external corporate influences might need to be specified if different groups of retailers were owned by different parent companies. Certain intra-urban character-istics might also be important such as the alignment of the principal roads, social area characteristics and whether the boundaries of the urban area are sharp with few competitive retailers just outside the immediate study area.

2.3.3 *Statistical modelling of spatial data*

The analysis of spatially referenced point and area data often raises distinctive problems. Examples are introduced here that are in addition to, but often interact with, the wider problems of statistical modelling.

2.3.3 *(a) Dependency in spatial data*

Suppose data have been collected at *n* sites across a region with the intention of estimating the population mean. The pattern of sites is rather irregular (Figure 2.3). Suppose that the underlying correlation structure is as shown below the map. The distance scales for the graph and the map are the same. Intuitively we would expect that the information provided about the population mean by the cluster of points at B suffered from duplication. That is, a certain amount of the information carried by each observation is duplicated by other observations in the cluster. The same information duplication would be evident in the case of the two points in the cluster marked A and D although the amount of duplication would be perhaps less serious. An analogy might be drawn between this type of information duplication and that which would occur if a data file of observations was augmented by merely duplicating some of the observations already in the data file. Clearly in the latter case the 'new observations' contain no information that is not currently in the file. Although less serious than this analogy, the duplication of spatial information arising from the underlying correlation pattern poses broadly the same problem when we try to estimate the mean on the basis of the n observations. The information content is less than would have been obtained from n *independent* observations. Put the other way round, there is an implicit loss of information relative to the case of n independent observations on the same process. Intuitively, the amount of information provided by the two sites at cluster A is not twice as much as that provided at C, nor is the information content of cluster B six times greater than the lone site at C. The same problem arises with aggregate areal data where again inference procedures based on the assumption that

Figure 2.3. Information provided by observations on a spatial surface.

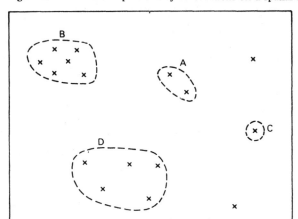

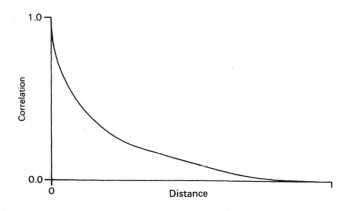

41

the n observations carry equal and independent amounts of information (like separate and independent experiments) are likely to mislead us.

To make these ideas a little more definite consider the problem of estimating a constant spatial mean (μ) where the observations ($y_1 \ldots y_n$) are independent drawings from a normal distribution with mean μ and variance σ^2. The maximum likelihood estimator of μ ($\hat{\mu}$) is

$$\hat{\mu} = (1/n)\Sigma_{i=1,\ldots,n}\, y_i = (\mathbf{1}^T\mathbf{1})^{-1}(\mathbf{1}^T\mathbf{y}) \tag{2.7}$$

where $\mathbf{1}$ is a vector of 1s. Suppose now the vector $\mathbf{y}$ is a sample from a multivariate normal distribution with constant mean μ and dispersion matrix $\sigma^2\mathbf{V}$. The subscript on $\mathbf{y}$ refers to the location of the observation, the off diagonal elements of $\sigma^2\mathbf{V}$ represent the spatial covariances between locations and each diagonal element represents the variance at that location. The maximum likelihood estimator of μ ($\tilde{\mu}$) is now

$$\tilde{\mu} = (\mathbf{1}^T\mathbf{V}^{-1}\mathbf{1})^{-1}(\mathbf{1}^T\mathbf{V}^{-1}\mathbf{y}) \tag{2.8}$$

Let $\sigma^2\mathbf{V} = \sigma^2(\mathbf{I} - \tau\mathbf{W})^{-1}$ where $\mathbf{I}$ is the ($n \times n$) identity matrix, σ^2 is the scale parameter which for the present discussion is set to one, τ is a parameter and $\mathbf{W} = \{w_{i,j}\}$ where w_{ij} lies between 0 and 1 and depends on whether sites i and j are neighbours and the strength of the relationship. (This model is discussed in Chapter 3.) In Figure 2.3 large values for w_{ij} would probably be specified between sites within cluster A, but small or zero values between sites in cluster A and the site at C. (Note however this does not imply sites in A and the site at C are independent.) Now

$$(\mathbf{1}^T\mathbf{V}^{-1}\mathbf{1})^{-1} = (n - \tau\Sigma\Sigma_{i,j}w_{i,j})^{-1} \tag{2.9}$$

If the observations are independent, $\tau = 0$, and the denominator of (2.8) is n, which coincides with (2.7). In the case $\tau > 0$, (2.9) measures the information loss in the estimation of μ that the analyst must work with as a result of spatial dependence. Further:

$$(\mathbf{1}^T\mathbf{V}^{-1}\mathbf{y}) = \frac{[1,\ldots,1]}{} \begin{bmatrix} y_1 - \tau\Sigma_j w_{1,j} y_j \\ \vdots \\ y_n - \tau\Sigma_j w_{n,j} y_j \end{bmatrix} \tag{2.10}$$

and ($\tau\Sigma_j w_{i,j} y_j$) is a measure of the information on y_i contained in the other observations. This duplication will tend to be large when τ is large and when w_{ij} is large.

For general dispersion matrix $\sigma^2\mathbf{V}$

$$\text{Var}(\tilde{\mu}) = \sigma^2(\mathbf{1}^T\mathbf{V}^{-1}\mathbf{1})^{-1} \tag{2.11}$$

42

where Var($\bar{\mu}$) denotes the variance of $\bar{\mu}$; so data dependency also affects confidence interval and significance testing.

Consider the regression model with correlated errors given by (2.5) where V is known, then the equivalent results are:

$$\hat{\beta}_V = (X_*^T X_*)^{-1}(X_*^T y_*)\tag{2.12}$$

where $X_* = L^{-1}X$, $y_* = L^{-1}y$ and $LL^T = V$ and hence

$$\hat{\beta}_V = (X^T V^{-1} X)^{-1}(X^T V^{-1} y)\tag{2.13}$$

with

$$\text{Var}(\hat{\beta}_V) = \sigma^2 (X^T V^{-1} X)^{-1}$$

Again V plays the role of adjusting for the information content in the sample; in (2.8) and (2.13) it downweights the influence of sets of highly correlated observations on the estimate. Of course V is usually not known and then there is the further (substantial) problem of finding a good model for V.

2.3.3 (b) Spatial heterogeneity: regional subdivisions and parameter variation

In regression modelling, parameters may differ between subregions or segments of the study area, reflecting perhaps the influence of processes operating at larger scales. Data for m areal units are available and the problem is to consider different aggregations and estimate parameters that may vary across these aggregations. If not specified in advance, the number of aggregated units or segments (k) will need to be selected and a decision made as to which units are to be aggregated to form the segments. Even if it is largely unnecessary (as well as impractical) to consider all possible aggregations (of which there will be many even for moderately large k and m) nonetheless the analyst is usually faced with the problem of choosing one segmentation from a set of alternatives. (Unlike studies involving race or sex where sample segmentation is unambiguous there is rarely one clear regional segmentation.)

Cliff *et al.* (1975) suggest three criteria for constructing k regional aggregations from m ($m > k$) units: (i) there should be as few segments as possible (criterion of simplicity); (ii) there should be intra-segment homogeneity (criterion of similarity); (iii) contiguous units should be aggregated (criterion of compactness). Their method, which is based on maximising a function composed of two indices that measure (i) and (ii), is used by

43

Goodman (1981) in an analysis of house price variation. This method of regionalisation is only one of a number that could be used. If the process of region building is viewed as a special case of classification (with the added constraint of locational contiguity) then the collection of techniques for classification can be used. Classification methods for regionalisation have been of much interest to geographers and the reader is referred to Semple and Green (1984) and Balling (1984) for reviews of available methods. Other methods have been suggested for selecting segmentations. Brueckner (1981) applies Quandt's switching regression technique to a one-dimensional system. The selection of one segmentation from a set of possible alternatives could be treated as a non-nested hypothesis testing problem (Anselin, 1986).

Some segmentations may be inappropriate for regression analysis. There may be insufficient variation on one or more of the variables in one or more of the regions; there may be large differences in the number of observations in the different segments; in some regions there may be fewer observations than parameters to be estimated (Johnston, 1984, p. 207). These problems may make it difficult to compare estimates between different segments and may turn out to be an important criterion in selecting segmentations to study. If areas are needlessly segmented the price is larger parameter estimator variance. Further, if regression residuals are spatially correlated, then the usual likelihood ratio tests and F tests which assume independence between observations (see for example Johnston, 1984, Chapter 6; Lin 1985; Schulze, 1987) are no longer valid for hypothesis testing (Casetti and Jones, 1987; Anselin, 1988b). If parameters vary stochastically error variances are not constant (Kau, Lee and Chen, 1983).

2.3.3 (c) Spatial distribution of data points and boundary effects

Consider the distribution of data points in Figure 2.3. A trend surface model is to be fit. Because the spatial coverage is uneven, while model fitting will be heavily influenced by those parts of the map where there are most sample points, in the sparsely sampled parts of the map the form of the surface may depend excessively on a small number of points which have high leverage.

Let y and $\hat{y}$ denote the vectors of observed and predicted values of the response. If a standard trend surface model is fitted then $\hat{y} = A\beta$ where A is the matrix of location co-ordinates (see page 75). Since $\beta = (A^TA)^{-1}(A^Ty)$

$$\hat{y} = A(A^TA)^{-1}A^Ty$$

so that each diagonal element of $H = A(A^TA)^{-1}A^T$ measures the influence of that data point on the fitted surface at that point. The ith diagonal element is the leverage of the *i*th data point and the larger the value the greater the influence of that observation on the fit. This is undesirable if there is any error at one or more of these points either in recording the location of the site or in measuring the response or if an inappropriate model is fit.

The points with high leverage are isolated points such as the observation at *C*. (The same occurs in standard regression analysis where there are a small number of observations on the explanatory variables that are detached from the main cluster of values.) If an estimator such as (2.13) is used to counter the effects of spatial dependence, it will help to compensate for the effects of data clusters in model fitting, since it downweights data clusters relative to uncorrelated observations. However by downweighting such clusters, leverage effects at isolated sites are likely to be made worse.

Boundaries may also raise problems for model fitting. If boundaries are artificial, observations close to the boundary will usually lose some of their neighbours. Spatial models that are fitted using sets of adjacent observations may need to allow in the fitting procedures for the fact that sets near the boundary are incomplete. Different assumptions about the unobserved boundary values will lead to different model fits. In regression modelling extra-regional influences may have more effect on observations close to the boundary than on those in the interior so boundary observations might have larger variances. The analyst might try to counter by collecting data beyond the bounds of the study region and fitting a (spatially) larger model; model specification and estimation procedures should examine sensitivity of results to different boundary assumptions and error assumptions; model evaluation should distinguish between boundary and interior sites in the presentation and examination of diagnostics. Where isolated sites are close to the boundary, surface fitting may need to deal simultaneously with both large leverage and large residual values at the boundary. In standard regression modelling the DFITS measure (Belsley, Kuh and Welsch, 1980) summarises this dual property.

2.3.3 (d) Assessing model fit

Not only are estimation and significance testing stages of data analysis affected by spatial attributes. The assessment of the goodness of fit of a model involves the comparison of model and data properties and the analysis of model residuals. In the case of fitting a spatial model, part of that evaluation must consider the *spatial* properties of the residuals. The presence of spatial structure in the distribution of residuals is usually an indication of failure to account for important elements of the problem. Analysis of the spatially

45

referenced set of residuals and the identification of structure in these may
help in the specification of an improved model.

2.3.3 (e) Distributions

Non-normal empirical distributions arise in many practical situations of
spatial data analysis. Count, ratio and presence/absence data frequently
arise in census data (Evans, 1981) and mortality data (Lovett, Bentham and
Flowerdew, 1986). Theoretical arguments indicate that synthetic aperture
radar data may follow a Rayleigh distribution (Besag, 1986). Whilst the
multivariate normal distribution might be expected to provide a statistical
model for dependent normal data with the variance covariance matrix
providing a means of specifying the structure of inter-site correlation,
equivalent models for Poisson, negative binomial, binomial (etc.) distributed variables are not so readily specified, nor are their properties
identifiable from standard texts. Dacey (1968) encountered this problem in
trying to model spatially correlated count data describing settlement
distributions in Puerto Rico. He showed that whilst the simple frequency
counts follow closely a negative binomial distribution, neighbouring count
values were correlated and there was no apparent way of constructing a
correlated negative binomial distribution. (Since that time such models have
been developed, although such are the constraints on the appropriate
negative binomial model that it still does not solve this particular problem.)
Data transformations may be required in order to enable an attack to be
made on certain kinds of problems where spatial correlation is suspected in
the distribution of count data (Cressie and Read, 1989).

2.3.3 (f) Extreme data values

Extreme values arise in some geographical problems not because of any
inherent errors in the data but because of the top heavy or 'primate'
structure of the underlying system. Cox and Jones (1981) make the point
that 'Dublin often appears as an outlier in plots of socio-economic data for
the Irish republic because it is genuinely different from the rest of the
country' (p. 138). Such observations usually have high leverages in the fit
of regression models. Their relationship to the rest of the data may be
impossible to assess. This could lead to the fitting of an inappropriate model
which then provides a bad description of the bulk of the data.

The problem of assessing the sensitivity of parameter estimates to
extreme values in regression will be discussed in Chapter 8. In standard
regression modelling with independent errors (2.1) diagnostics have been

developed to assess the influence of individual observations by deleting cases in turn (Belsley, Kuh and Welsch, 1980; Wrigley, 1983). Where errors are correlated (2.5) such data deletion methods cannot be used without modification. Spatial dependence between observations also raises problems for the implementation of data sampling methods like bootstrapping.

2.3.3   (g) Model sensitivity to the areal system

Observations on areal units create a distinctive set of problems for data analysis. Consider first a continuous spatial surface where observations refer to areal aggregates that form a regular grid. The results of any statistical analysis will be conditional on the scale, orientation and origin of the grid as well as the scale of the study area. Properties of the surface at scales smaller than the sampling grid will not be detectable since they will have been filtered out while processes operating at scales larger than the study area will display insufficient variation within the study area; a re-orientation or new origin for the grid will involve aggregating (or integrating) different parts of the map to form the set of observations. Scale and origin problems are also encountered in time series analysis but although most time series do not have a natural origin they do have a natural grid (days, weeks, months, years . . .) which provides a framework for analysis. Spatial surfaces neither have a natural origin nor a natural grid so there is more freedom of choice in how to lay down any grid, and grid orientation is an additional variable that does not arise at all in time series analysis. How serious the orientation and origin effects are for spatial analysis are not known and will presumably depend on the variability of the surface including its correlation structure. One might suspect that where spatial correlation is strong in all directions over distances up to the size of the grid or where grid size is small, analysis will probably not be greatly affected by choice of origin or orientation.

Social data are often available only in respect of irregular areal units formed by aggregating over different numbers of primary units (e.g. households). All too often aggregates are made available in which primary units have been grouped arbitrarily (whilst preserving contiguity) or to fulfil objectives that are not relevant to the study. A better method is to group primary units in a way that preserves intra-area homogeneity with respect to those socio-economic variables that are to be examined in terms of their association with the response variable (see section 2.3.3(b) for methods) so that values are representative of areas. (This property may be difficult to satisfy for large areas.) Areal units should not differ greatly in size for statistical reasons (see section 2.3.3(h)) but also because different processes may be operating at different scales. Even so, many different spatial aggregations, both in terms of the number of areal units and the pattern of aggregation, could be constructed. The results of any statistical analysis based on such data must be conditional on the scale of the study area and the areal partitioning, including its scale (an example involving correlation is given in Cliff and Ord, 1981, pp. 131–4). Even the term 'number of observations' (n) has an artificial quality and is quite different from the term 'number of experiments or trials' in classical inference, for any area can be partitioned in many different ways whilst preserving the value of n; alternatively n can be increased or decreased altering the original set of observations only marginally or quite radically. In the case of density data the correlation between two variables tends to increase with the level of aggregation although the increase is moderated if the primary units are positively correlated in space and is more severe if the units are negatively correlated (Arbia, 1986). Data variability declines with increasing aggregation and the variance of the two variables appears in the denominator of the correlation coefficient. The sampling variances of regression parameter estimators also increase with aggregation.

To what extent are the results of statistical analyses affected by keeping the number of areal units fixed but aggregating different primary units? Although it seems intuitively obvious that there will be some affect, again, exactly how serious this problem is, is unclear. Work attempting to show the volatility of statistical results to such variation, controls for scale by holding the number of zones constant and permitting different internal boundary configurations. This does not seem to control for intra-area variability and area size which ought to be important conditions for any zoning system. The problem of zonal effects termed by geographers the 'modifiable areal units problem' is vigorously reviewed and solutions discussed in Openshaw and Taylor (1981). The variability of statistical results arising from the nature of the areal system is a worrying one since it undermines regional comparative research (where two areas have different zoning systems) and makes it difficult to replicate studies and findings in one area by carrying out parallel studies elsewhere (Johnston, 1984). It also casts doubt on the stability of results arising from analyses of aggregated data. It emphasises the importance of designing a zoning system that reflects the purpose of the study and also suggests the advisibility (where possible) of performing several analyses at any given scale with alternative plausible zoning systems.

Boundary and 'frame' effects (the geometric structure of the study area) may affect statistical analysis and the interpretation of results.[3] Where the study area is not closed or isolated, what happens within the area may be affected by outside influences (for example in the spread of a contagious disease). Statistical descriptions of patterns, inasmuch as they require inter-site measures, will require assumptions about conditions beyond the edge of the study area. Boundary effects are of interest in the definition of the models to be described in Chapter 3 but are also important in trend surface and nearest neighbour analyses (Haggett, 1980; Ripley, 1984).

2.3.3 (h) Size–variance relationships in homogeneous aggregates

When forming areal aggregates from different numbers of primary units or integrating over different sized areas, to what extent should variances be expected to be identical for all sized areas (whether size is measured in terms of square kilometres or population size or numbers of households, for example)? To what extent should observations based on a small number of households be considered as reliable and given as much weight in analysis as observations based on a large number? Data which represent an areal average may show an inverse relation between size and variance since any value for a large area is an average over more primary units. Size–variance relationships can be investigated by plotting data values against area size.

Suppose Y_i is a density variable then we might expect

$$\text{Var}(Y_i) = \sigma^2/n_i \qquad n_i > 0$$

where n_i is a measure of the size of area i. It is possible to allow for these considerations in a fixed effects regression model (2.1) where we may now write that:

$$\text{Var}(e_i) = \sigma^2/w_i \qquad w_i = n_i$$

If the errors are independent

$$E(\mathbf{e}\mathbf{e}^{\mathsf{T}}) = \sigma^2(\mathbf{W}^+)^{-1}$$

where $\mathbf{W}^+$ is a diagonal matrix with elements $[w_1, \ldots, w_n]$. The (weighted) least squares estimator for $\boldsymbol{\beta}$ is:

$$\hat{\boldsymbol{\beta}}_{\mathbf{w}} = (\mathbf{X}_+{}^{\mathsf{T}}\mathbf{X}_+)^{-1}(\mathbf{X}_+\mathbf{y}_+) \tag{2.14}$$

where $\mathbf{X}_+ = \mathbf{C}\mathbf{X}$, $\mathbf{y}_+ = \mathbf{C}\mathbf{y}$ and $\mathbf{C}^{\mathsf{T}}\mathbf{C} = \mathbf{W}^+$, and $\mathbf{C}$ is a diagonal matrix with $\mathbf{C} = \text{diag}[w_1^{\frac{1}{2}}, \ldots, w_n^{\frac{1}{2}}]$. The matrix $\mathbf{W}^+$ plays the role of downweighting the observations with large error variances. The larger w_i the smaller the variance of observation i, the more reliable the observation and so the more emphasis that is placed on it in the regression problem (Weisberg, 1985, pp. 81–3. This argument and approach was first used on spatial data, to increase the influence of observations associated with large areas, by Robinson, 1956.) For (2.14), σ^2 estimated from the data:

$$\text{Var}(\hat{\boldsymbol{\beta}}_{\mathbf{w}}) = \tilde{\sigma}^2(\mathbf{X}^\mathsf{T}\mathbf{W}^+\mathbf{X})^{-1}$$

where $\tilde{\sigma}^2 = (n-k)^{-1}(\mathbf{y} - \mathbf{X}\hat{\boldsymbol{\beta}}_{\mathbf{w}})^\mathsf{T}\mathbf{W}^+(\mathbf{y} - \mathbf{X}\hat{\boldsymbol{\beta}}_{\mathbf{w}})$.

If $E(\mathbf{e}\mathbf{e}^\mathsf{T})$ is not diagonal as in (2.5), $\mathbf{V}$ known, then

$$\begin{aligned}\tilde{\boldsymbol{\beta}}_{\mathbf{vw}} &= [(\mathbf{CL}^{-1}\mathbf{X})^\mathsf{T}(\mathbf{CL}^{-1}\mathbf{X})]^{-1}[(\mathbf{CL}^{-1}\mathbf{X})^\mathsf{T}(\mathbf{CL}^{-1}\mathbf{y})]\\ &= [(\mathbf{X}^\mathsf{T}(\mathbf{L}^{-1})^\mathsf{T}\mathbf{W}^+\mathbf{L}^{-1}\mathbf{X})]^{-1}[(\mathbf{X}^\mathsf{T}(\mathbf{L}^{-1})^\mathsf{T}\mathbf{W}^+\mathbf{L}^{-1}\mathbf{y})]\end{aligned} \quad (2.15)$$

This (generalised) least squares estimator for $\boldsymbol{\beta}$ downweights the contribution of highly correlated clusters of observations and observations with large variances. In some cases the model for spatial correlation also makes an implicit adjustment for variance differences between sites, an issue to be discussed in Chapters 3 and 7. For (2.15)

$$\text{Var}(\tilde{\boldsymbol{\beta}}_{\mathbf{vw}}) = \tilde{\sigma}^2(\mathbf{X}^\mathsf{T}(\mathbf{L}^{-1})^\mathsf{T}\mathbf{W}^+\mathbf{L}^{-1}\mathbf{X})^{-1}$$

where $\tilde{\sigma}^2 = (n-k)^{-1}[(\mathbf{L}^{-1}(\mathbf{y} - \mathbf{X}\tilde{\boldsymbol{\beta}}_{\mathbf{vw}}))^\mathsf{T}\mathbf{W}^+(\mathbf{L}^{-1}(\mathbf{y} - \mathbf{X}\tilde{\boldsymbol{\beta}}_{\mathbf{vw}}))]$.

Other cases can be included. For example, if Y_i is a total of n_i observations then we might expect:

$$\text{Var}(Y_i) = \sigma^2 n_i \qquad \text{so that } w_i = n_i^{-1}$$

Alternatively, the variance of Y_i might be proportional to some explanatory variable, then:

$$\text{Var}(Y_i) = \sigma^2 X_i \qquad \text{so that } w_i = x_i^{-1}$$

An empirical check to decide whether these modifications are needed in regression modelling is to plot the residuals, $\mathbf{L}^{-1}(\mathbf{y} - \mathbf{X}\tilde{\boldsymbol{\beta}}_{\mathbf{vw}})$, against n_i or x_i.

2.4 A statistical framework for spatial data analysis

2.4.1 Data adaptive modelling

Data analysis can be described as consisting of two broad phases: exploratory and confirmatory data analysis. Hoaglin *et al.* (1983) characterise exploratory data analysis (EDA) as a phase in which patterns and structures in the data are uncovered and hypotheses proposed. It is the first encounter with the data and is concerned with broad features and 'soft' models and the detection of data attributes that might create difficulties for naive methods of statistical analysis based on strict distributional assumptions. 'An important element of the exploratory approach is flexibility, both in tailoring the analysis to the structure of the data and in responding to patterns that successive steps of analysis uncover' (p. 1). On the other hand, confirmatory data analysis (CDA) is closer to traditional statistical inference (providing confidence intervals and testing hypotheses) but also includes sensitivity or influence analyses (assessing robustness of model fit and statistical conclusions to the data and particular subsets of the data), residual analysis, evidence from other comparable studies and collecting new data in order to validate the results of CDA on one set of data. [4] 'In brief, EDA emphasises flexible searching for clues and evidence, whereas CDA stresses evaluating the available evidence' (p. 2). These phases of data analysis are not used sequentially, indeed their alternating use may be highly desirable.

Perhaps the most striking feature of EDA is its use of resistant methods of data analysis in order to make preliminary identification of data properties. The first stages of data analysis must work with the unrefined data. Resistant methods of analysis are insensitive to a small number of 'atypical' values. The purpose of EDA is to reveal the presence of such values and to provide an identification of data structures that are not strongly influenced by their presence. In the case of single data sets such measures of structure might include the properties of location, scale, skewness and distribution. In a multivariate context this might extend to the identification of variable relationships (scatterplots of y against x) or the detection of scale–level (variance–mean) relationships.

EDA may be extended to exploratory modelling of the data in the sense of finding some simple model (or fit) to summarise data structure. In the terminology of EDA:

$$\text{Data} = \quad \text{fit} \quad + \quad \text{residuals}$$
$$\text{(smooth)} \quad \text{(rough)}$$

and a further EDA, this time on the residuals, may indicate ways in which the model can be improved. This use of EDA may be particularly important in areas where formal CDA test procedures are difficult to implement or of unknown reliability. Certain test procedures, in regression for example, are developed to test for the failure of one model requirement and are invalid or of unknown reliability when two or more assumptions fail. EDA, and in particular graphical plots, may provide the analyst with helpful insights into the presence of structure or pattern in the data and improve model specification (Good, 1983).

EDA also provides important evidence on whether conventional, parametric methods of data analysis will retain their desirable properties or whether robust or resistant methods of analysis would provide a safer foundation for estimation and inference. The methods of EDA are used to detect data outliers, symmetric non-Gaussian and skew (heavy tailed) distributions; for methods that depend on strict distributional assumptions being met can be disproportionately influenced by relatively small departures from the stated assumptions. The advocates of this approach to data analysis lay less stress on the novelty of methods (indeed many of the methods are far from new) as on the framework and integrated nature of the approach. Critics, on the other hand, point to the range and diversity of methods used and warn of reporting results that are mere artefacts on the data. Mulaik (1985) provides a good overview of approaches to EDA which might help to minimise the risk of reporting meaningless patterns. Two practical suggestions with large data sets that involve a form of experimentation are the use of cross validation (using a random sample of the data and then comparing with the rest of the data) and bootstrapping (using samples drawn from the data to assess the variability of results).

Because EDA encourages hypothesis formulation it implies the need for a methodology for hypothesis testing. CDA is concerned with significance and hypothesis testing, estimation and prediction. The model is fitted and diagnostic checks performed; that is, underlying model assumptions are checked and model sensitivity to the data assessed. The properties of the residuals play an important role in evaluating the model, identifying problems and identifying how the model might be improved. An important contribution to CDA has emerged in the development of robust and resistant methods of parameter estimation. These will be described in more detail later in this chapter but the importance of robust methods is that, rather than making strong distributional assumptions or treating all distributions equally (as in nonparametric statistics), they provide a best-compromise method over a *class* of possible distributions. Resistant methods are used to assess the sensitivity of model fit to data attributes.

Figure 2.4 shows in flow chart form a framework for data adaptive modelling into which EDA and CDA fit. Only step 1 is evidently the domain of EDA and subsequent steps may be either within an EDA or CDA framework. The issues of 'purpose' and 'judgment' stressed by Leamer (1978) are particularly important at stages 2 and 7 (see section 2.3(a)). As the analyst moves away from a 'soft' model towards a 'firmer' model supported by additional evidence, then the emphasis moves towards significance and hypothesis testing, which lies within the domain of CDA.[5]

The modelling framework emphasises the role of data and data properties in modelling. But data analysis does not take place in a substantive vacuum. What is the role of the subject field within this framework other than specifying the problem and the types of data to collect? Four aspects seem particularly important.

1. First, it may suggest the types of data problems that are likely to occur. Certain variables might be known to generate highly skewed data distributions or extreme values. Unreliable values might be anticipated in certain types of data because of inherent measurement problems, for example, in situations where measuring devices have to be left unattended or in census or social surveys where certain questions are known to produce unreliable responses. When rates are computed from different sized base populations (e.g. standardised mortality rates), an untypically large (or small) value can arise due to the small size of the base population. Census data sets usually suffer from problems of multicollinearity when used in regression modelling.

2. Second, the subject field can help in the identification of a 'soft' model for the data which is robust to known data problems and also help to

Figure 2.4. Data adaptive modelling. After Martin (1987).

Data Adaptive Modelling

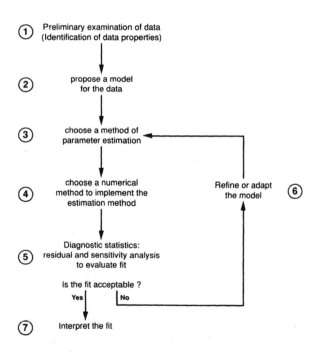

distinguish between plausible alternatives where data evidence is ambiguous. It may suggest, for specific models, which statistical assumptions may be the most controversial thus helping to identify what assumptions may need the most careful checking. These are important steps towards the development of 'firm' models. The role of subject theory can be much stronger than this, particularly in areas of well developed theory. In a regression context, for example, well developed theory may specify variables, the functional expression, whether parameters are likely to be constant or variable over the surface, variable transformations and parameter constraints. If not able to specify a unique model, theory may be useful in specifying a small range of alternatives for the analyst to consider. In Bayesian estimation, subject matter theory may specify a prior distribution for the parameters. It is useful then, to distinguish between statistical and modelling assumptions in data analysis. Statistical assumptions are properties required of the data in order to ensure the validity of the statistical procedures. Such assumptions can be checked in the current data set, and if found violated, steps must be taken to ensure that the assumptions are met. Modelling assumptions do not relate to data properties and cannot be checked out against the data. Rather they relate either to the scientific background or 'to a gut feeling that "given the information in the data we are going to need to stick to some reasonable looking assumptions to provide the hope of sensible conclusions"' (Leonard, 1983, p. 22).

3. Third, the subject field may specify what population attributes we require estimates for. This provides the substantive motivation for the analysis but, equally important, may help to guide the analyst in making statistical decisions in the light of what is being sought from the data.

4. Finally, the subject field is important in evaluating the statistical models and interpreting output. If no sensible interpretation can be given to the final model, it is unlikely to be a satisfactory model, neither of practical nor theoretical use to the analyst.

We have previously mentioned the importance of robust and resistant methods of estimation and in the next section we describe some basic approaches and indicate how they link up with the issues discussed here.

2.4.2 Robust and resistant parameter estimation

The distinction between robust and resistant methods of parameter estimation does not seem to be clear cut, not least perhaps because the terms are sometimes used interchangeably. An estimator is called resistant if its behaviour is insensitive to a small number of extreme values. On the other hand, a robust estimator is used when underlying distributional assumptions (usually Gaussian) have been violated. For example the distribution of measurements may be symmetric but not Gaussian or they may be skewed (heavy tailed). Robustness is the more specific term and is a property which only holds for an estimator with respect to a specified class of distributions. At their simplest, resistant estimation procedures may simply involve data editing (deleting 'unreliable' values) and then applying standard methods. However, robust estimators are designed to have high efficiency with respect to a specified class of distributions; that is, have small estimator variance or, in the case of hypothesis testing, have high power for a fixed level of significance.[6]

Robust estimation is discussed in detail in Hoaglin *et al.* (1983, 1985) and Hampel *et al.* (1986). An important feature of most robust estimators, in particular the more refined examples, is that they do not have closed form expressions but are obtained after iteration using model residuals to specify an appropriate weighting. Robust estimators of location (which will be of interest in Chapters 6 and 7) and robust regression estimators are now discussed. Robust estimators of scale are discussed in Iglewicz (1983).

2.4.2 (a) Robust estimation of the centre of a symmetric distribution

The location problem is only well defined in the case of symmetric distributions. If the distribution of measurements is mildly non-symmetric then it is usual to still use the estimators that are discussed here. In the case of severe forms of asymmetry a data transformation may make the data symmetric (Emerson and Stoto, 1983) but this still leaves the question of what attribute of the original data set is being estimated, a problem that cannot be separated from the subject matter content of the problem.

The simplest estimators of the centre of a symmetric distribution are L-estimators based on linear combinations of the order statistics. This class of estimators, which includes familiar non-parametric techniques: trimmed mean, median, trimean estimators, are simple to use and can be quite effective but often have low efficiency. We concentrate here on M-estimators, which are weighted mean estimators where the weights depend on the data.

M-estimators are based on relative deviations of the set of observations $(y_1, \ldots, y_n)$ from a current estimate of the centre, denoted T_0. The measure of relative deviation is taken as:

$$u_i = (y_i - T_0)/cS$$

where c is a tuning constant (values in the range 6 to 9 are often recommended) and S is a measure of scale (often estimated by the median absolute deviation from the median – MAD).

In the case of an M-estimator the new estimate is T^* and

$$T^* = \Sigma_i w(u_i) y_i / \Sigma w(u_i)$$

where $w(u)$ is a symmetric weighting function conventionally chosen so that $w(0) = 1$ and decreasing as $|u|$ increases.

M-estimators iterate to convergence as follows. If T is the estimate at step n and T^* is the estimate at step $n+1$, at convergence:

$$
\begin{aligned}
0 = T^* - T &= (1/\Sigma w(u_i))[\Sigma w(u_i) y_i - \Sigma w(u_i) T] \\
&= (cS/\Sigma w(u_i))[\Sigma w(u_i)\{(y_i - T)/cS\}] \\
&= (cS/\Sigma w(u_i))[\Sigma w(u_i) u_i]
\end{aligned}
$$

so that at convergence $\Sigma w(u_i) u_i = 0$. Thus we seek an estimate of T such that:

$$\Sigma_i[w((y_i - T)/cS)][(y_i - T)/cS] = 0 \qquad (2.16)$$

Different types of M-estimators are specified by assuming different forms for $w(.)$. They are chosen in order to balance robustness and efficiency (Li, 1985, p. 293) and will be discussed below in the context of regression modelling. The use of the term M-estimators comes from maximum likelihood, for these estimators 'generalize the idea of the maximum likelihood estimator of the location parameter in a specified distribution' (Goodall, 1983, p. 340).

What are the implications of this branch of robust estimation for spatial data analysis? The semi-variogram of geostatistics (see Chapter 3) defined by

$$\gamma(h) = E[(Y(s) - Y(s+h))^2] \qquad h \geqslant 0$$

and the spatial covariance function

$$C(h) = E[(Y(s) - E[Y(s)])(Y(s+h) - E[Y(s+h)])]$$

where E denotes mathematical expectation of the random variables (Y), can both be interpreted as measures of centrality of functions of the variables and might be estimated robustly by the methods described here. This will be discussed further in Chapter 6.

2.4.2 (b) Robust estimation of regression parameters

Consider the standard (fixed effects) multiple regression model with uncorrelated errors, given by (2.2). The data consist of $\{(y_i, \mathbf{x}_i)\}$ where $\mathbf{x}_i$ is the ith row of X. The M-estimator for the parameter vector β is that value, $\tilde{\beta}_m$ such that

$$\Sigma_i(y_i - \mathbf{x}_i\beta)w(u_i)\mathbf{x}_i^\mathsf{T} = 0 \qquad (2.17)$$

where $w(u_i) = w(y_i - \mathbf{x}_i\beta)$.

A computational solution to (2.17) is W-estimation (where W means weighted). Since (2.17) can be viewed as the normal equations for a weighted least squares regression problem we can write

$$\mathbf{W}^* = \mathrm{diag}[w(u_1), \ldots, w(u_n)]$$

then

$$\tilde{\beta}_m = (\mathbf{X}^\mathsf{T}\mathbf{W}^*\mathbf{X})^{-1}(\mathbf{X}^\mathsf{T}\mathbf{W}^*\mathbf{y}) \qquad (2.18)$$

where $\mathbf{W}^*$ depends on the residuals $\{u_i\}$ through the chosen weighting function. Implementation of (2.17) may be as follows. Obtain the OLS residuals and use these to obtain $\mathbf{W}^*$. A new estimate of β is obtained ($\tilde{\beta}_m$) and hence new residuals, and the procedure repeated until convergence occurs. This procedure is sometimes referred to as iterated re-weighted least squares, in contrast to (ordinary) weighted least squares where $\mathbf{W}^+$ is specified once only. Note that the weighting function in (2.18) uses the residuals from the previous iteration. Convergence is not guaranteed. There are other estimation procedures which although theoretically the same as (2.17) give different results because they are solved by different computing methods (Li, 1985).

Table 2.3 lists the main robust estimators for regression and Figure 2.5 graphs the weight functions. The same weighting functions are used in (2.16) for estimating the location parameter in which case the ordinary least squares (OLS) estimator gives the mean and the least absolute residual (LAR) estimator gives the median. Selecting a weighting function involves criteria in addition to the balance between robustness and efficiency since regression analysis is carried out for many different reasons. For example, the aim might be to get a good estimate of β, or predict $\mathbf{y}$ or to produce good residuals. Li (1985) who reviews robust regression in detail suggests using the OLS estimator to obtain an initial estimate, iterating to a moderate degree using the Huber M-estimator and then finishing with Tukey's bi-weight. Confidence intervals in the case or robust estimation of (2.2) can be obtained using:

$$\mathrm{Var}(\tilde{\beta}_m) = \tilde{\sigma}^2{}_\mathrm{R}(\mathbf{X}^\mathsf{T}\mathbf{X})^{-1}$$

where $\tilde{\sigma}^2{}_\mathrm{R}$ is a robust estimator of σ^2 such as:

$$\tilde{\sigma}^2{}_\mathrm{R} = [(0.6745)^{-1}S]^2$$

57

Table 2.3. *Robust estimators: weighting functions*

Estimator	$w(t)$	Range of t	Comment
OLS	1	$\lvert t \rvert < \infty$	Sensitive to outliers.
LAR	$\text{sign}(t)/t$	$\lvert t \rvert < \infty$	Insensitive to outliers but sensitive to high leverage values. May not give unique solutions.
Huber	1	$t \leqslant k$	A class of estimators.
	k/t	$t > k$	Estimates can be unstable or heavily influenced by an anomalous data value at a high leverage point.
Andrews	$(A/t)\sin(t/A)$	$t \leqslant \pi A$	Multiple solutions and
	0	$t > \pi A$	sensitive to choice of starting
Tukey's	$(1-(t/B)^2)^2$	$t \leqslant B$	point.
bi-weight	0	$t > B$	

OLS = ordinary least squares, LAR = least absolute residual.

where S is the MAD of the residuals (Li, 1985). (Note that this contrasts with weighted least squares since the underlying model is the standard regression model with constant variance and $\mathbf{W}^*$ simply downweights extreme values.)

It is of interest to consider briefly how issues of robustness to outliers and distribution properties affect estimation in the case of spatial regression models. A robust form of the estimator (2.12) is

$$\tilde{\boldsymbol{\beta}}_{\mathbf{V}m} = (\mathbf{X}_*^\mathsf{T}\mathbf{W}^*\mathbf{X}_*)^{-1}(\mathbf{X}_*^\mathsf{T}\mathbf{W}^*\mathbf{y}_*)$$
$$= [\mathbf{X}^\mathsf{T}(\mathbf{L}^{-1})^\mathsf{T}\mathbf{W}^*(\mathbf{L}^{-1})\mathbf{X}]^{-1}[\mathbf{X}^\mathsf{T}(\mathbf{L}^{-1})^\mathsf{T}\mathbf{W}^*(\mathbf{L}^{-1})\mathbf{y}] \qquad (2.19)$$

Initially $\mathbf{L}^{-1}(\mathbf{y}-\mathbf{X}\tilde{\boldsymbol{\beta}}_{\mathbf{V}})$ are the residuals used to determine $\mathbf{W}^*$. Clusters of correlated observations and observations with extreme values are downweighted in the estimation procedure.

Of course $\mathbf{V} = \mathbf{L}\mathbf{L}^\mathsf{T}$ will usually not be known and we might wish to estimate $\mathbf{V}$ robustly as well. This could involve several stages of robust estimation: stage 1 robust estimation of the covariances (or semi-variogram); stage 2 robust estimation of a model to describe the empirical covariances. How important it is to estimate $\mathbf{V}$ by robust methods depends partly on how important it is to get a good estimate of $\mathbf{V}$ and on the sensitivity of $\tilde{\boldsymbol{\beta}}_{\mathbf{V}m}$ to the estimate of $\mathbf{V}$. In geostatistical interpolation (kriging) there is currently much interest in robust estimation of the semi-variogram, of a model for the semi-variogram and finally of the interpolated values. The

58

decision on whether to estimate **V** robustly is based on inspecting the *regression* residuals $(\mathbf{y} - \mathbf{X}\boldsymbol{\beta})$ where $\boldsymbol{\beta} = (\mathbf{X}^T\mathbf{X})^{-1}(\mathbf{X}^T\mathbf{y})$.

The methods discussed above assume that the extreme values are associated with unusual responses which are then downweighted. In experimental work where levels of the explanatory variables are controlled this may be the most appropriate assessment. In non-experimental work

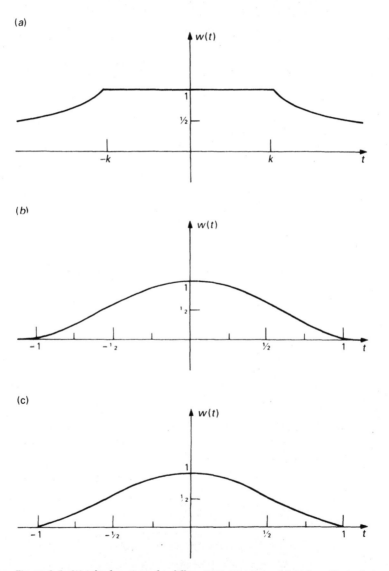

Figure 2.5. Weight functions for different W-estimators. (*a*) Huber, (*b*) Andrews $(\pi A = 1)$, (*c*) Tukey's bi-weight $(B = 1)$.

59

the analyst may have equal uncertainty about measurements associated with both response and explanatory variables. If the result of mismeasurement of an explanatory variable at one or more sites results in high leverages at those sites then this is clearly undesirable, particularly if the analyst is uncertain about the choice of model, and robust estimation (as outlined here) may not deal adequately with the problem. What is needed is an estimation procedure that is resistant to high leverage effects. Li (1985) reviews developments in what is termed 'bounded influence regression' which seeks to counter the effects of outliers in the space of explanatory variables. However, developments here still seem at an early stage.

The incorporation of robust procedures into fitting a spatial regression model such as (2.1) with errors given by (2.5) (where V is unknown and non-diagonal) raises theoretical issues (how should robust procedures be adapted and how should V itself be estimated) and computational issues – it could be time consuming to implement a fully interactive procedure with no guarantee of satisfactory convergence properties. We consider these issues in Chapter 6.

Robust and resistant methods could be helpful in assessing the sensitivity of results to boundary effects. Earlier it was suggested that larger residual values might be found at the boundary in which case different forms of downweighting for boundary observations, including those based on residual size, could be worth exploring. This would seem to be a suitable line of attack in those situations where the same model can be used to describe boundary and non-boundary observations but where an estimation procedure is needed that is responsive to the different problems that might be associated with these two sets of observations. Where the analyst believes that different models are needed to describe boundary and non-boundary observations then such data adaptive methods may not deal adequately with the problem. (This remark could equally apply to all the issues raised here.) Chapter 3 will give examples of spatial models that have different properties for boundary sites as opposed to interior sites.

As a last comment on robust and resistant methods, it is not the intention here to suggest that extreme observations must be statistically pummelled into line with the rest of the data. Areas with extreme values may be highly informative and interesting and in no sense 'wrong'! The aim is rather to develop methods that act as a check on standard least squares or maximum likelihood estimation procedures and that ensure such areas do not unduly influence the fit of a model to the rest of the data thereby distorting the general picture. In this sense they are considered part of sensitivity analysis and moreover may provide information useful in developing an improved model (Besag, 1981b).

NOTES

1   A number of other suggestions have been made, and Upton and Fingleton (1985, p. 325) note that 'great imagination has gone into turning what appears to be a population into a sample, thereby making statistical theory relevant'. Different sources of randomness may be invoked but with a single sample it is impossible to disentangle the contribution of sampling variation from measurement error.

2.   With classical testing procedures so weighted against the null hypothesis the analyst should, Leamer (1978) suggests, 'trouble himself not with the results of classical hypothesis testing but rather with the question of why he bothered to test an obviously false hypothesis in the first place' (p. 89). Box (1983) makes the point more forcefully: 'there is evidence in the social sciences that excessive reliance upon this theory [the Neyman–Pearson theory for testing statistical hypotheses] alone . . . has led to harmful distortion of the conduct of scientific investigation in these fields' (p. 53).

3.   The influence of frame effects (area shape and size) on the results of principal components analyses of univariate space–time data series are discussed in Richman (1986) and Jolliffe (1987).

4.   EDA, according to Good (1983) is 'more an art, even a bag of tricks, than a science', whereas CDA is concerned with formulae and the evaluation of these formulae within the context of rules which determine the conclusions that can be drawn. CDA can therefore be 'trivialised' in the sense of being reduced to standard computer algorithms. EDA is not as yet so susceptible to trivialisation for it involves a greater degree of subjectivity and judgment on the part of the user. EDA encourages the creative playing with data prior to applying routine CDA-type procedures.

5.   In passing we should note step 4. Particularly in the area of spatial analysis this is not always a straightforward issue. Fitting many spatial models (see Chapters 3 and 4) often requires the use of numerical methods to maximise the likelihood function. Many algorithms are available for function maximisation, as even a cursory glance at the NAG (Numerical Algorithms) software manual will reveal. Should the algorithm use only function values to find the maximum, or first derivatives as well? Results, in the author's experience, are not always the same. Regression parameter estimates may require two inversions: V^{-1} and $(X^T V^{-1} X)^{-1}$. The problems of evaluating $(X^T X)^{-1}$ in standard regression modelling are discussed in Wetherill *et al.* (1986, Chapter 3) but spatial analysis may require the careful choice of the inversion routine for V also.

The presence of outliers in a data set may be indicative of an underlying heavy tailed distribution or unequal variances. On the other hand the analyst may be satisfied that the distribution is symmetric, variances are equal, and that the outliers are simply the result of a data error or sampling variation, for example. In the first case the analyst may wish to propose a new error model; in the second the initial model is retained and resistant methods applied to the estimation phase of analysis. Where the data evidence is ambiguous the analyst will need to decide which route to follow and subject matter theory may be involved.

PART B

Parametric models for spatial variation

3

Statistical models for spatial populations

This chapter introduces parametric statistical models for describing spatial variation. The approach followed is rather different from the presentation usually given of models for describing temporal variation. First, time is one-dimensional and the flow of time from past to present to future is usually assumed to impose a natural ordering or direction on patterns of interaction. Space on the other hand is two-dimensional and usually possesses no such equivalent ordering so that spatial models must allow for a wider range and more complex structures of interaction.[1] Second, time series models such as autoregressive models can be specified in terms of covariances or joint probabilities or conditional probabilities. The three formulations are equivalent, and each gives rise to simple functional expressions. But spatial models with simple autoregressive representations in terms of variate relationships generally have covariance and correlation functions that are far from elementary; conditional and joint probability specifications of nearest neighbour spatial models are not equivalent and give rise to different orders of autoregressive model. Third, if the border value of a first order temporal autoregressive series is its initial value, the border or edge effects are of order $1/n$ where n is the length of the observed series. In the case of spatial data the boundary encircles the region. For an $n = N \times M$ rectangular lattice there are usually at least $2N + 2M - 4$ border sites, so specifying what happens at the edges is likely to be of more importance in the case of spatial models. Fourth, surface discontinuities and other forms of heterogeneity associated with spatial partitioning are likely to be more pervasive and geometrically more complex than those found in time series analysis.

In the first section different forms of spatial stationarity are discussed as well as methods for imposing order relations between sites. Section 3.2 examines models for continuous random variables emphasising the multivariate normal distribution. Section 3.3 describes models for discrete random variables. Models for boundary states and the effects of edge weighting schemes are discussed in sections 3.4 and 3.5.

3.1   Models for spatial populations: preliminary considerations

3.1.1   *Spatial stationarity and isotropy*

A spatial stochastic process is a collection of random variables (Y) indexed by a set T. This index set may consist of pairs of real numbers (specifying points on a continuous surface) or pairs of integers (for example specifying the locations of pixels on an image) or just labels (identifying regularly or irregularly scattered point sites or areas). A stochastic process is defined by specifying the joint distribution function for any finite collection of random variables (Ripley, 1981, p. 9). This distribution may possess certain attributes and we start by defining stationarity and isotropy. In practice, spatial analysis deals with specific areas which are physically bounded or a sub-area of a larger region which is itself bounded. For such situations we should strictly refer to 'quasi' or 'local' stationarity and isotropy, where the required properties are restricted to a specific area and are not assumed to hold for points outside this domain or for distances that exceed the diameter of the area (Journel and Huijbregts, 1978, p. 33).

3.1.1   *(a) Second order (weak) stationarity and isotropy*

Let T denote a regular (lattice) arrangement of point sites. Suppose the area is bounded and let $s(= (s_1, s_2))$ and t denote any two points in T. Let $C(s,t)$ and $R(s,t)$ denote the covariance and correlation respectively between the random variables at the two points s and t. Define

$$C(s,t) = E\{[Y(s) - E(Y(s))][Y(t) - E(Y(t))]\}$$
$$R(s,t) = C(s,t)/(C(s,s)C(t,t))^{\frac{1}{2}}$$

where E denotes expectation. $E(Y(s))$ is the mean of Y and is independent of s. $C(s,s)$ is called the variance of $Y(\mathrm{Var}(Y(s))$ and is also independent of s. If, in addition, C and R are independent of location the process is said to be weakly spatially stationary if $C(s,t)$ depends only on the relative locations of s and t, that is the length and orientation of the vector $h = (h_1, h_2)$ between s and t. So $C(s,t) = C(h)$ and $R(s,t) = R(h)$. If C and R do not depend on the orientation of h but only on its length then the process is weakly isotropic and we can write $C(r)$ where $r = (h_1^2 + h_2^2)^{\frac{1}{2}}$. Note that these definitions assume that the variance is positive and finite (Ripley, 1981, pp. 9–10).[2]

Strict spatial stationarity and strict isotropy are more stringent versions of the weak definitions and require the equivalence of distribution functions under translation and rotation with respect to the lattice. When the random variables are normally distributed weak spatial stationarity implies strict spatial stationarity and weak isotropy implies strict isotropy since Gaussian processes are completely specified by their first two order moments.

In the case of a spatially continuous process (defined at all points in the plane) then the covariance function of a weakly stationary process depends only on the length and orientation of the vector between any two points in the plane. The covariance function of a weakly isotropic process depends only on the length of the vector (Matérn, 1960, pp. 11–13).

3.1.1 *(b) Second order (weak) stationarity and isotropy of differences from the mean*

Models satisfying the first definition are of limited usefulness because they require a location invariant mean. A weaker assumption is to define stationarity of the differences from the mean. This will allow many types of surfaces that are non-stationary in the first sense to be modelled using results for stationary processes. So it may be reasonable to model certain non-stationary features (such as the presence of growth poles in an economic surface, discharge points in a polluted environment, and changes in the balance of controlling factors underlying the outbreak of an epidemic) through the mean, leaving deviations *around* the mean to be represented by a weakly stationary model. But some caution is needed before taking this step. For example, Cliff and Ord (1981, p. 143) suggest that in the case of an active spatial diffusion process spreading outwards from a source and frozen in time the spatial covariance structure is unlikely to be location invariant, differing with distance (and possibly direction) from the source point.

3.1.1 *(c) Second order (weak) stationarity and isotropy of increments*

A still weaker assumption is to define stationarity with respect to surface increments using the intrinsic random functions (IRF's) of Matheron (1973). Consider the following case of an IRF of order 0. Suppose Y represents a spatially continuous process with a constant mean $(E(Y)=m)$. Y need not be stationary, indeed variances (and covariances) need not be finite. Now suppose for all points $\mathbf{u}=(u_1,u_2)$ on the surface and for any separation distance r, the increments $(Y(\mathbf{u}+r)-Y(\mathbf{u}))$ form a weakly stationary process. Since the mean of the process is constant then the mean of the increments is zero. To satisfy the stationarity requirement, the variance of the increments must be finite and independent of **u**. The variance of the increments is called the variogram and is written

$$2\gamma(r) = \text{Var}(Y(\mathbf{u}+r) - Y(\mathbf{u}))$$
$$= E[(Y(\mathbf{u}+r) - Y(\mathbf{u}))^2] \tag{3.1}$$

where $\gamma(r)$ is called the semi-variogram. Note that if the Y process is weakly stationary, then $\gamma(r) = (C(0) - C(r))$. For many such processes, as $r \to \infty$, $C(r) \to 0$ and $\gamma(r) \to C(0)$, so $C(r) = \gamma(\infty) - \gamma(r)$.

Now following Matheron (1973) we introduce general IRF's of order k. These are processes which include polynomials of order k (the case $k=0$ denoting a constant) which must be filtered out in forming the increments. As before let $Y(\mathbf{u})$ be a spatially continuous process. A finite linear combination written $\Sigma_{i=1,n}\lambda_i Y(\mathbf{u}_i)$ is called an increment of order k if it filters out polynomials of order k. For this the $\{\lambda_i\}$ must satisfy the condition:

$$\Sigma_{i=1,n}\lambda_i u_{i_1}{}^{P_1} u_{i_2}{}^{P_2} = 0 \text{ for all } P_1, P_2 \geqslant 0 \text{ and } P_1 + P_2 \leqslant k \tag{3.2}$$

This implies:

$k=0$ (constant) $\Sigma_{i=1,n}\lambda_i = 0$
$k=1$ (linear trend) $\Sigma_{i=1,n}\lambda_i = 0$; $\Sigma_{i=1,n}\lambda_i u_{i_1} = 0$;
$\qquad \Sigma_{i=1,n}\lambda_i u_{i_2} = 0.$
$k=2$ (quadratic trend) The conditions for $k=1$ and in addition
$\qquad \Sigma_{i=1,n}\lambda_i u_{i_1}{}^2 = 0$; $\Sigma_{i=1,n}\lambda_i u_{i_2}{}^2 = 0$;
$\qquad \Sigma_{i=1,n}\lambda_i u_{i_1} u_{i_2} = 0.$

and other cases follow in a similar fashion. Now $Y(\mathbf{u})$ is said to be an intrinsic random function of order k (IRF$-k$) if the increments $\Sigma\lambda_i Y(\mathbf{u}_i)$ form a weakly stationary process (call this process Z) and the $\{\lambda_i\}$ filter out polynomials of order k. For example, in two dimensions where a site is labelled (i,j),

$$Z = \Sigma_i\Sigma_j\lambda_{ij}Y(i,j) = Y(i-1,j) + Y(i+1,j) + Y(i,j-1) + Y(i,j+1) - 4Y(i,j)$$

is an increment of order $k=1$, but since $\Sigma\lambda_{ij}i^2$ and $\Sigma\lambda_{ij}j^2$ are non-zero it is not an increment of order 2. (The analogy is sometimes drawn between the increment process and differencing in time series analysis.)

The properties of the increment process, Z, depend on the properties of Y and the coefficients $\{\lambda_i\}$. The principal idea is that even though the observed process is non-stationary the increments (the realisations of the Z process) are stationary. This final set of models allows for non-stationarity both in the mean and in the deviations from the mean, stationarity being required only of the increments of the deviations.

The concept of stationarity is of questionable value for processes operating on irregularly scattered point sites or across continuous space and observed in terms of an irregular areal partition. Even if there exists an underlying continuous space process which is stationary, if the observations arise from different areal integrations then variances and covariances will not be the same for all areas. In practice the assumption of stationarity could be made but it could not be checked from the data. We now consider how to define inter-area relationships for such irregular systems. These definitions have implications for how spatial dependency is represented.

3.1.2   Order relationships in one and two dimensions

We consider the construction of a graph for a study region (D) within which there are n fixed and labelled sites. The purpose of constructing the graph is to specify relationships between the sites. (Where the 'sites' refer to areas it may be necessary to specify points, such as the centroid to represent each area). The first problem is to choose an edge structure which indicates for each site its set of neighbours. Next, numerical values (weights) are assigned to the edges.

In some applications, a hierarchy of edge systems are specified. These are associated with a hierarchy of models which are then fitted to the data. No relationship is implied between the system of edges associated with the best fitting model and possible underlying mechanisms responsible for the site values. The modelling objective is purely descriptive, that is, describing the dependency properties of the data. However, particularly in certain non-experimental sciences, edge systems are sometimes chosen and statistical analysis used to test hypotheses about the role of different mechanisms. The internal geography of the region (the presence of barriers, internal boundaries, crenulations of the regional boundary, etc.) may also influence the specification of the graph. Interesting epidemiological examples are given in Haggett (1976) and Cliff *et al.* (1985, pp. 182–5) where tests for spatial correlation are carried out with different graphs each reflecting different assumptions about the routes by which infectious diseases might be spread through a spatially distributed population.

Figure 3.1(a) shows a one-dimensional system of sites in D. The edge system (Figure 3.1(b)) joins sites in terms of proximity so that t is connected to $t+1$ and $t-1$. Note that an edge must join two different sites (no loops are allowed) and that the two boundary sites have only one edge each. An alternative edge system shown in Figure 3.1(c) connects t to $t \pm 1$ and $t \pm 2$ with fewer edges for the two sites on each end. Time series modelling assumes graphs of the type shown here except that because of the natural ordering of time the edges of the graph can also be given a direction as for example in Figure 3.1(d). Figure 3.2(a) shows a lattice of sites in D. The edge system in Figure 3.2(b) joins neighbours in terms of proximity so that (s_1,s_2) is joined to $(s_1 \pm 1,s_2)$ and $(s_1,s_2 \pm 1)$, again with modifications at the edges. In Figure 3.2(c) sites are joined as in Figure 3.2(b) but with additional diagonal connections.

Figures 3.3(a) and 3.3(b) show a system of irregularly arranged sites and an edge structure largely based on proximity. Many different criteria can be used in selecting an edge structure:

(a)   *Proximity:*

(i) Distance: each site is linked to all other sites within a specified distance.

(ii) Nearest neighbours: each site is linked to its k ($k = 1,2,3, \ldots$ etc.) nearest neighbours.

(iii) Gabriel graphs: 'any two (sites), A and B are said to be contiguous if and only if all other localities are outside the A–B circle, that is, the circle on whose circumference A and B are at opposite points' (Matula and Sokal, 1980). See Figure 3.4(a).

(iv) Delaunay triangulation: all sites which share a common border in a Dirichlet partitioning of the area (see Figure 3.4(b)) are joined by an edge (Besag, 1975).[3] Where the n sites reference sub-areas that already partition D then the edge system may be based on whether the areas have a border in common.

(b)   *Interaction:* all sites between which there is a flow (measured for example by traffic movements, patterns of person to person contact, telephone calls, income transfers, etc.) are linked by an edge.

The use of proximity criteria seems most appropriate where inter-site connections in terms of the flow of information and material are not limited

Figure 3.1. A system of sites and edges in one dimension.

(a)   (b)

(c)   (d)

70

to special transport networks. The Gabriel graph has been frequently used in biology (typically in studies of gene flow) and its properties, and in particular its intermediate status with respect to nearest neighbour linkage systems and Delaunay triangulations together with computational issues, are described in Matula and Sokal (1980). Griffith (1982) constructs a Dirichlet partitioning of grain elevator sites in Manitoba to reflect empirical evidence that farmers minimise hauling distances between their farms and the grain elevator collection points.

The use of interaction data, on the other hand, often reflects the presence of distinct transport networks by which information and material flow between sites. Particularly in social and economic systems, the resulting linkage patterns may often appear spatially quite irregular, even chaotic, when they reflect the spatial manifestation of an underlying set of relationships that are not subject to simple distance constraints. For example, linkages derived from assumed inter-urban relationships in a central place system fall into this class; if individual sites refer to industrial units a linkage system may be chosen to reflect input–output relationships between the units. Tinkler (1973) constructs graphs for collections of villages to reflect the sequencing of periodic markets.

Figure 3.2. A system of sites and edges in two dimensions: lattice case.

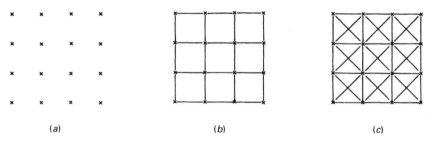

(a)　　　　　　　　　(b)　　　　　　　　　(c)

Figure 3.3. A system of sites and edges in two dimensions: non lattice case.

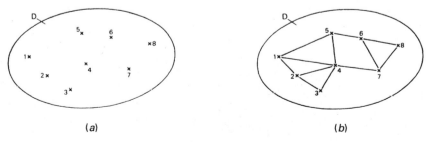

(a)　　　　　　　　　(b)

71

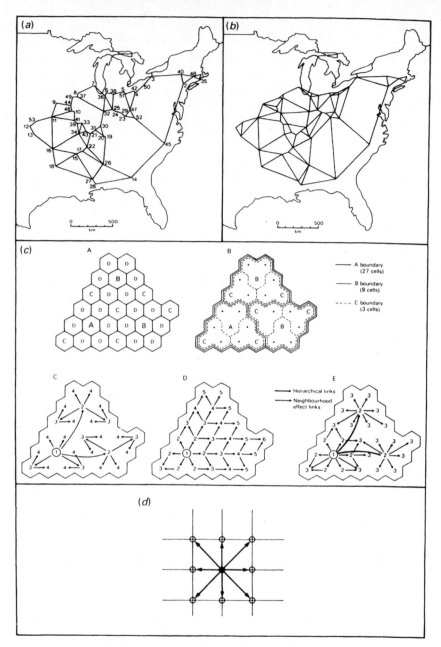

Figure 3.4. Graph systems: (a) Gabriel graph, (b) Delaunay triangulation, (c) Central place system of sites, (d) Directed neighbourhood system. (a) After Matula and Sokal (1980). (c) After Cliff *et al.* (1981).

In social and economic systems some interaction may take place between all sites so that it is frequently important to simplify the system of edges by introducing a cut off level with flows below a certain level not generating an edge. Alternatively, the matrix of flows may be analysed to identify the key sites and linkages within the system. (Methods for achieving this are described in Nystuen and Dacey, 1961; Holmes and Haggett, 1977; and Gatrell, 1979.)

In some edge systems defined by the above criteria it may be reasonable to assume a direction to the edges. Income transfers between sites that represent urban places are often size related and function related, with income moving as a result of consumer expenditure from small low order urban places to large higher order urban places (Haining, 1981b, 1987a). Central place theory (Berry, 1967) implies directional relationships between urban places that differ in terms of their size and hence their order (1,2, . . .) and Cliff *et al.* (1981) discuss different spatial diffusion mechanisms which have directional properties (Figure 3.4(*c*), (*d*)). Physical systems such as hillslope and fluvial systems will also show directional characteristics.

For graphs where sites are separated by uniform distances any differential weighting of the edge structure is usually unnecessary. In this case the join structure of the sites can be summarised by an $n \times n$ matrix $\Delta = \{\delta_{ij}\}$ where row and column i reference site i and where

$$\delta_{ij} = \begin{cases} 1 \text{ if site } j \text{ is linked to site } i \text{ by an edge} \\ 0 \text{ if site } j \text{ is not linked to site } i \text{ by an edge} \\ 0 \text{ if } i = j \end{cases}$$

An interesting property of this matrix is that $\Delta \times \Delta = \Delta^2 = \{\delta_{ij}^{(2)}\}$ identifies the total number of different ways of moving in two steps (where a step is defined as one move along an edge) between any pair of sites i and j. In general $\Delta^k = \{\delta_{ij}^{(k)}\}$ identifies the total number of ways of moving in k steps between any pair of sites (including circular and redundant routes). Note also that $\delta_{ij} = \delta_{ji}$ if the edges have no direction but if there is directionality then if $\delta_{ij} = 1$ then $\delta_{ji} = 0$. Some of the earlier ways of defining edges produce directional edges (e.g. the proximity criterion based on nearest neighbours when used on an irregular distribution of sites).

In the case of an irregular arrangement of sites, edge weighting can be used to reflect, for example, different degrees of proximity, different lengths of common boundary, different levels of interaction and so forth. The choice of weighting, like the edge structure itself, should be chosen to reflect the problem. As before the (weighted) edge structure can be summarised by an $n \times n$ matrix $W = \{w_{ij}\}$ where

Statistical models for spatial populations

$$w_{ij} \begin{cases} \neq 0 \text{ if site } j \text{ is linked to site } i \text{ by an edge} \\ = 0 \text{ if site } j \text{ is not linked to site } i \text{ by an edge} \\ = 0 \text{ if } i = j \end{cases}$$

It may not be necessary for $w_{ij} = w_{ji}$ but in general we do arrange that as edge length increases or interaction falls then $w_{ij} \downarrow 0$. In some cases row or column sums of W are scaled to one in order to standardise across regions with different numbers of neighbours or other attributes. Examples:

(i) $w_{ij} = d_{ij}^{-\gamma}$ $(\gamma \geqslant 0)$

where d_{ij} is the distance (euclidean, block metric, time, etc.) between i and j and γ is a constant.

(ii) $w_{ij} = \exp(-d_{ij}^{\gamma})$

where exp denotes the exponential function.

(iii) $w_{ij} = (l_{ij}/l_i)^{\tau}$ $(\tau \geqslant 0)$

where l_{ij} is the length of the common border between areas i and j and l_i is the perimeter of the border of area i (excluding the boundary of D), and τ is a constant.

(iv) $w_{ij} = (l_{ij}/l_i)^{\tau}/d_{ij}^{-\gamma}$

The effect of edge and weighting structures on the statistical properties of spatial models will be examined in section 3.5.

Imposing a non-directional edge system on the group of sites in D creates subsets of sites with the property that each member of the set is linked to every other member of the set. Such sets of sites are called 'cliques' (Besag, 1974). Consider Figure 3.3(b) and define each individual site as a clique of size 1. So there are 8 cliques of size 1. Further there are 12 cliques of size 2 ({1,2}, {1,4}, {1,5}, {2,3}, {2,4}, {3,4}, {4,5}, {4,7}, {5,6}, {6,7}, {6,8}, {7,8}) and four cliques of size 3 ({1,2,4}, {1,4,5}, {2,3,4}, {6,7,8}). There are no cliques of size four or larger. In Figure 3.2(b) in addition to the cliques of size 1 there are cliques of size 2 of the form:

$$\{(s_1,s_2), (s_1 \pm 1, s_2)\} \text{ and } \{(s_1,s_2), (s_1,s_2 \pm 1)\}$$

There are no cliques of size three or larger. Figure 3.2(c) shows cliques up to and including size 4.

Finally we define the set of acquaintances of a site. Site k is said to be an acquaintance of site j $(j \neq k)$ if for some i there is an edge from j to i and an edge from i to k. The set of neighbours is included in the set of acquaintances. In Figure 3.3(b) the set of acquaintances of site 8 include the neighbours (sites 6 and 7) and also sites 4 and 5.

3.2 Population models for continuous random variables

This section examines permissible models for describing spatial populations. A continuous valued random variable (Y) has been measured at n sites. The observations are denoted $y^T = (y_1, \ldots, y_n)$. We shall define models that can be used to describe spatial variation in Y. One component of this variation is first order or mean variation which is represented by the n dimensional vector μ. The other component is second order variation about μ. We emphasise the case where Y satisfies a multivariate normal (MVN) distribution and consider models for the dispersion matrix V. Such models can be obtained by specifying a permissible form for V directly or by specifying a variate interaction model which expresses relationships between the random variables and from which V can be derived. Models specified in terms of covariances assume stationarity and differ in terms of such characteristics as their rates of decrease with distance and whether they are isotropic or not. Models specified in terms of variate interactions need not assume stationarity and differ in terms of such characteristics as order, whether they have unilateral or bilateral forms, are symmetric or asymmetric, isotropic or not. It is possible that dependence is present only in one direction so that an interaction model reduces to a series of independent and identical one-dimensional bilateral schemes, one per row (or column) of the spatial data set.

The final part of this section looks at semi-variogram models and generalised covariance functions for IRF-k models. Specification of models for finite regions requires consideration of boundary conditions. However, we defer treatment of representations for the boundary until section 3.4.

3.2.1 *Models for the mean of a spatial population*

3.2.1 *(a) Trend surface models*

The mean is represented as a polynomial function of a specified order, that is $\mu = A\theta$ where A is a matrix of location co-ordinates for the n sites and θ is a vector of trend surface parameters so that:

$$
A = \begin{bmatrix}
1 & a_{11} & a_{12} & a_{11}^2 & a_{12}^2 & a_{11}a_{12}\cdots\cdots\cdots\cdots a_{11}{}^p a_{12}{}^q \\
1 & a_{21} & a_{22} & a_{21}^2 & a_{22}^2 & a_{21}a_{22}\cdots\cdots\cdots\cdots a_{21}{}^p a_{22}{}^q \\
\cdot & \cdot & \cdot & \cdot & \cdot & \quad \cdot \cdots\cdots\cdots\cdots \quad \cdot \\
\cdot & \cdot & \cdot & \cdot & \cdot & \quad \cdot \cdots\cdots\cdots\cdots \quad \cdot \\
\cdot & \cdot & \cdot & \cdot & \cdot & \quad \cdot \cdots\cdots\cdots\cdots \quad \cdot \\
1 & a_{n1} & a_{n2} & a_{n1}^2 & a_{n2}^2 & a_{n1}a_{n2}\cdots\cdots\cdots\cdots a_{n1}{}^p a_{n2}{}^q
\end{bmatrix}
$$

where (a_{i1}, a_{i2}) defines the location of site i and

$$\boldsymbol{\theta}^\mathsf{T} = (\theta_{00} \quad \theta_{10} \quad \theta_{01} \quad \theta_{20} \quad \theta_{02} \quad \theta_{11} \ldots \ldots \theta_{pq})$$

The sum $(p+q)$ represents the order (k) of the trend surface: Zero order $(p+q=0)$ which is the same as the constant mean case $(\boldsymbol{\mu} = \mu \mathbf{1})$; linear or first order $(p+q=1)$ which generates a sloping plane surface; quadratic or second order $(p+q=2)$; cubic or third order $(p+q=3)$ and so on. Successively higher orders generate surfaces of increasing complexity (Figure 3.5). Since the co-ordinate scheme used is purely a convenience the fitted surface should not depend on it. The only invariant trend surfaces are full surfaces (zero, linear, quadratic, cubic, . . ., etc.).

Ripley (1981, p. 35) notes that other families of surfaces for representing mean variation are rarely used. For example, Fourier functions are only useful for periodic surfaces and require the estimation of large numbers of coefficients.

3.2.1 (b) Regression model

In this model the mean is represented as a function of a set of k explanatory variables so $\boldsymbol{\mu} = \mathbf{X}\boldsymbol{\beta}$ where $\mathbf{X}$ is an $n \times (k+1)$ matrix and $\boldsymbol{\beta}$ is a $(k+1)$ dimensional vector of regression parameters where

$$\mathbf{X} = \begin{bmatrix} 1 & x_{11} & x_{12} & \cdots \cdots & x_{1k} \\ 1 & x_{21} & x_{22} & \cdots \cdots & x_{2k} \\ \cdot & \cdot & \cdot & \cdots \cdots & \cdot \\ \cdot & \cdot & \cdot & \cdots \cdots & \cdot \\ \cdot & \cdot & \cdot & \cdots \cdots & \cdot \\ 1 & x_{n1} & x_{n2} & \cdots \cdots & x_{nk} \end{bmatrix}$$

and $\boldsymbol{\beta}^\mathsf{T} = (\beta_0 \beta_1 \beta_2 \ldots \beta_k)$. The term x_{ij} is the observation on explanatory variable j $(j=1, \ldots, k)$ at site or area i. Regional effects can be introduced through dummy explanatory variables. Suppose the n areas can be partitioned into t $(t<n)$ subsets corresponding to t regional groupings of the areas. Regional effects can be estimated by using $t-1$ dummy variables D_j $j=1, \ldots, t-1$ as additional explanatory variables where

$$D_{ij} \begin{cases} = 1 \text{ if area } i \text{ is in region } j \\ = 0 \text{ if area } i \text{ is in region } l \ (l \neq j) \end{cases}$$

The fitting of such regression models is described in Johnston (1984, Chapter 6); Lin (1985); and Schulze (1987).

The trend surface model represents continuous first order spatial variation. First order spatial variation in the standard regression model is

76

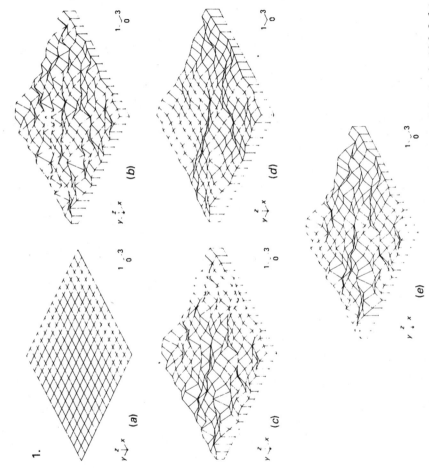

Figure 3.5. Orders of trends surface model with different error structures. 1.0 order $\theta = 10.0$. 2. 1st order $\theta^T = (20.0, -0.2, -0.2, -0.3)$. 3. 2nd order $\theta^T = (0.0, 4.0, 4.0, -0.2, -0.2, -0.1)$. (*a*) Trend Surface, (*b*) Trend Surface plus N(0,1) independent errors. (*c*) Trend Surface plus SAR ($\rho = 0.15$) errors. (*d*) Trend Surface plus SAR ($\rho = 0.24$) errors. (*e*) Trend surface plus moving average ($\nu = 0.15$) errors.

77

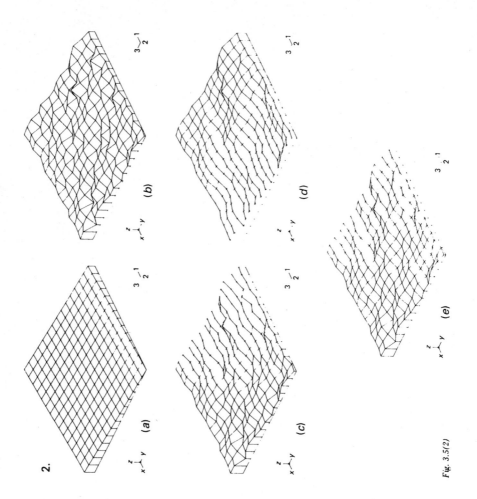

2.

(a)   (b)

(c)   (d)

(e)

Fig. 3.5/2)

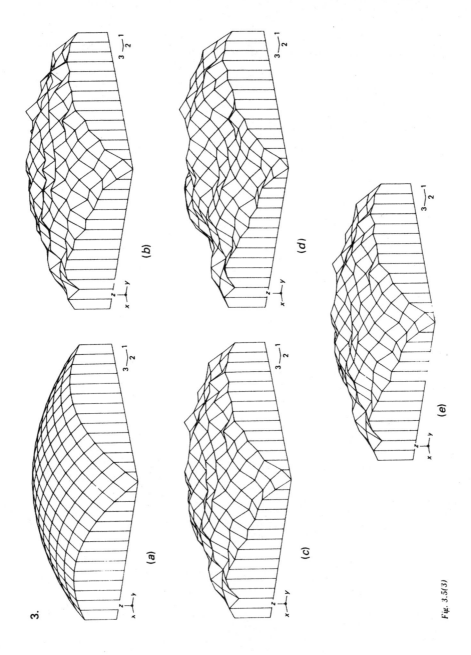

3.

(a)

(b)

(c)

(d)

(e)

Fig. 3.5(3)

generally discontinuous between areas but the nature of the discontinuity depends on the spatial variation in the explanatory variables. The use of regional dummy variables introduces some continuity within the region with discontinuities at the regional boundaries.

3.2.2   *Models for second order or stochastic variation of a spatial population*

3.2.2   (a) *Interaction models for* **V** *of a MVN distribution*

Since $\mathbf{V} = \{v_{ij}\}$ is a dispersion matrix v_{ij} denotes the covariance between Y_i and Y_j and v_{ii} denotes the variance of Y_i. Since $\mathbf{V}$ must be symmetric and positive definite any interaction scheme must ensure these conditions. We start by considering the joint or simultaneous approach to specifying spatial interaction schemes.

Let $\mathbf{e}$ be an $n \times 1$ vector of (unobservable) random variables with mean zero and $n \times n$ dispersion matrix $\sigma^2 \mathbf{V}_e$. This vector is often referred to as the vector of errors. We consider schemes of the form

$$\mathbf{B}(\mathbf{Y} - \boldsymbol{\mu}) = \mathbf{C}\mathbf{e} \qquad\qquad (3.3)$$

where $\mathbf{B}$ and $\mathbf{C}$ are $n \times n$ non-singular matrices with $\mathbf{B} = \{b_{ij}\}$ and $b_{ii} = 1$ for all i. The dispersion matrix for $\mathbf{Y}$ in (3.3) is

$$\begin{aligned}\mathbf{V} = E[(\mathbf{Y} - \boldsymbol{\mu})(\mathbf{Y} - \boldsymbol{\mu})^{\mathsf{T}}] &= E[\mathbf{B}^{-1}\mathbf{C}\mathbf{e}\mathbf{e}^{\mathsf{T}}\mathbf{C}^{\mathsf{T}}(\mathbf{B}^{-1})^{\mathsf{T}}] \\ &= \sigma^2[\mathbf{B}^{-1}\mathbf{C}\mathbf{V}_e\mathbf{C}^{\mathsf{T}}(\mathbf{B}^{-1})^{\mathsf{T}}]\end{aligned}$$

Defining spatial models in this way, with $\mathbf{V}_e$ a diagonal matrix, and where spatial attributes are introduced through particular parametrisations of $\mathbf{B}$ and $\mathbf{C}$ is usually referred to as the 'joint' specification where the joint density of $\mathbf{Y}$ can be expressed as a product of terms where each factor specifies a site and the set of neighbours with which it interacts. So if $\phi(y_1, \ldots, y_n)$ denotes the joint density of $\mathbf{Y}$ we require

$$\phi(y_1, \ldots, y_n) = \Pi_{j=1,\ldots,n} Q_j(y_j; \{y_{N(j)}\})$$

where $\{y_{N(j)}\}$ denotes the set of neighbours of j with which j is assumed to interact. The neighbours of j are those sites joined to j by an edge (section 3.1.2).

To see this factorisation let $f(e_1, \ldots, e_n)$ denote the joint density of $(e_1, \ldots, e_n)$ then since $\mathbf{V}_e$ is diagonal

$$f(e_1, \ldots, e_n) = \Pi_{j=1,\ldots,n} f(e_j)$$

80

Suppose $C = I$ in (3.3) so that the model is specified solely in terms of non zero b_{ij}'s. It then follows (see for example Alexander, 1961, Theorem 39.14)

$$\phi(y_1, \ldots, y_n) = |B| f(\Sigma_{i=1,\ldots,n} b_{1i}(y_i - \mu_i), \ldots, \Sigma_{i=1,\ldots,n} b_{ni}(y_i - \mu_i))$$
$$= |B| \Pi_{j=1,\ldots,n} f(\Sigma_{i=1,\ldots,n} b_{ji}(y_i - \mu_i))$$
$$= |B| \Pi_{j=1,\ldots,n} f(b_j(y - \mu)) \tag{3.4}$$

where b_j is the jth row of B and non-zero entries in b_j are associated with those sites that are joined to j by an edge, and

$$|B| = \begin{vmatrix} (\partial e_1/\partial y_1) \cdots \cdots \cdots (\partial e_1/\partial y_n) \\ \cdot \qquad\qquad\qquad \cdot \\ \cdot \qquad\qquad\qquad \cdot \\ \cdot \qquad\qquad\qquad \cdot \\ \cdot \qquad\qquad\qquad \cdot \\ (\partial e_n/\partial y_1) \cdots \cdots \cdots (\partial e_n/\partial y_n) \end{vmatrix}$$

If the e process is normal then

$$\phi(y_1, \ldots, y_n) = |B| \Pi_{j=1,\ldots,n} (2\pi\sigma^2)^{-\frac{1}{2}} \exp[-(1/2\sigma^2)(b_j(y - \mu))^2]$$
$$= |B| (2\pi\sigma^2)^{-(n/2)} \exp[-(1/2\sigma^2)\{(y - \mu)^\top B^\top B(y - \mu)\}] \tag{3.5}$$

We now examine some important special cases. Throughout we shall set $\sigma^2 V_e = \sigma^2 I$. The adjustment for the case $\sigma^2 V_e = \Sigma$ where Σ is a diagonal matrix and $\text{diag}[\Sigma] = [\sigma_1^2, \ldots, \sigma_n^2]$, so error variances are not necessarily equal for all i, can be introduced using (3.3).

Example 3.1. Simultaneous autoregressive (SAR) models. (Whittle, 1954)

$$C = I \qquad B = (I - S)$$

where $(I - S)$ is invertible and the diagonal elements of S are zero.

$$V = \sigma^2[(I - S)^\top (I - S)]^{-1}$$
$$\text{Cov}(e, Y) = E[e(Y - \mu)^\top] = \sigma^2(I - S^\top)^{-1}$$

and

$$Y_i = \mu_i + \Sigma_j s_{ij}(Y_j - \mu_j) + e_i$$

The parametrisation of S depends on the edge structure imposed on the set of sites. Directional and non-directional schemes, with and without weighting, can be allowed, the only restriction being that $(I - S)$ must be non-singular.

Example 3.1a. Single parameter SAR schemes

Let $\mathbf{S} = \rho\mathbf{W}$ where ρ is a constant and $\mathbf{W} = \{w_{ij}\}$. For a rectangular lattice we could let $\mathbf{W} = \boldsymbol{\Delta}$ so for interior sites

$$\delta_{ij} = \begin{cases} 1 & \text{if } j \text{ is immediately to the north, south, east or west of } i \\ 0 & \text{otherwise} \end{cases}$$

The graph structure is shown in Figure 3.2(b). For an interior site and assuming a constant mean the model can be expressed

$$\begin{aligned} Y(s_1, s_2) = \mu + \rho(Y(s_1 + 1, s_2) + Y(s_1 - 1, s_2) + Y(s_1, s_2 + 1) \\ + Y(s_1, s_2 - 1) - 4\mu) + e(s_1, s_2) \end{aligned}$$

Given any general graph structure and matrix $\mathbf{W}$ we can write

$$Y_i = \mu_i + \rho \Sigma_{j \in N(i)} w_{ij}(Y_j - \mu_j) + e_i$$

where $N(i)$ are the neighbours of i, those sites joined to i by an edge.

To ensure invertibility of $(\mathbf{I} - \rho\mathbf{W})$ there are restrictions on the value of ρ. If $\{\omega_1, \ldots, \omega_n\}$ denote the n eigenvalues of $\mathbf{W}$ then providing $\omega_{\min} < 0$ and $\omega_{\max} > 0$ then $1/\omega_{\min} < \rho < 1/\omega_{\max}$ where $\omega_{\max}$ and $\omega_{\min}$ are the largest and smallest eigenvalues. Since $\omega_{\max}$ and $\omega_{\min}$ depend on n as well as the edge structure and its weighting so does the permissable range of values of ρ. For square lattices, as $n \to \infty$ then $\omega_{\max} \uparrow 4$ and $\omega_{\min} \downarrow -4$ so $|\rho| < 0.25$. For the case where all the row sums are standardised to a constant c, $\omega_{\max} = c$ and $\omega_{\min} \leqslant -1/c$ and so $\rho < c$. Note that these conditions do not ensure local stationarity of the model even for square lattices. For example, the diagonal elements of $\mathbf{V}$ are not identical (Figure 3.6(a)) and covariances are also a function of location. Figure 3.5 shows various single parameter SAR models superimposed on different orders of trend surface.

Example 3.1b. Asymmetric schemes for regular lattices

If spatial interaction differs between the north–south and east–west directions the SAR model can be rewritten with different parameters in the two directions. If the two parameters are written ρ_{s_1} and ρ_{s_2} then $|\rho_{s_1}| + |\rho_{s_2}| < 0.5$ for large $n \times n$ lattices. Other schemes of this type are described in Whittle (1954).

Example 3.1c. Higher order schemes. *(Whittle, 1954; Brandsma and Ketellapper, 1979a)*

Higher order models can be defined where Y is a function of more then one interaction parameter but where different parameters measure different distance effects. For example, in the case of the graph in Figure 3.2(c) we could write $\mathbf{S} = \rho_1 \mathbf{W}_1 + \rho_2 \mathbf{W}_2$ where $\mathbf{W}_1$ defines vertical and horizontal edges and W_2 defines diagonal edges, with ρ_1 and ρ_2 the associated parameters. We would usually expect $|\rho_1| > |\rho_2|$.

A hierarchy of such models can be developed, for example

$$S = \rho_1 W_1 + \rho_2 W_2 + \ldots + \rho_k W_k$$

with $\{W_i\}$ specifying neighbours at different distances.

Example 3.2 Moving average (MA) models. (*Haining, 1978d*)

$$C = (I + M) \qquad B = I$$

where the diagonal elements of M are **zero**.

$$V = \sigma^2[(I + M)(I + M)^\top]$$
$$\text{Cov}(e, Y) = E[e(Y - \mu)^\top] = \sigma^2(I + M)^\top$$

and

$$Y_i = \mu_i + \Sigma_j m_{ij} e_j + e_i$$

Figure 3.6. Variances for individual sites (areas) on a 7×7 lattice for (*a*) the single parameter SAR ($\rho = 0.15$) and (*b*) moving average (MA) ($v = 0.15$) models with $\sigma^2 = 1$. After Haining (1988).

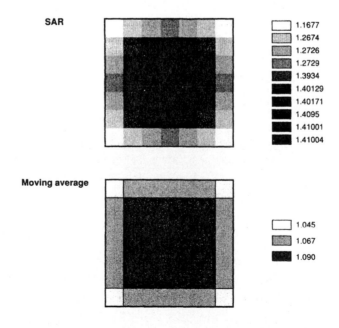

A moving average model can be represented as a simultaneous autoregressive scheme (and is said to be invertible) if $(I+M)$ is non-singular, although this representation of the process is usually parametrically very complicated.

Example 3.2a Single parameter MA schemes
Let $M = vW$ where v is a constant and W is as defined above. For the case of a rectangular lattice with the edge structure shown in Figure 3.2(*b*) then with a constant mean, the equation for an interior site is:

$$Y(s_1,s_2) = \mu + v(e(s_1+1,s_2) + e(s_1-1,s_2) + e(s_1,s_2+1) + e(s_1,s_2-1)) + e(s_1,s_2)$$

Except for sites affected by the boundary (see Figure 3.6(*b*)):

$$Var(Y(s_1,s_2)) = 1 + 4v^2$$
$$Cov(Y(s_1,s_2),Y(s_1\pm1,s_2)) = Cov(Y(s_1,s_2),Y(s_1,s_2\pm1)) = 2v$$
$$Cov(Y(s_1,s_2),Y(s_1\pm2,s_2)) = Cov(Y(s_1,s_2),Y(s_1,s_2\pm2)) = v^2$$
$$Cov(Y(s_1,s_2),Y(s_1\pm1,s_2\pm1)) = 2v^2$$
$$Cov(Y(s_1,s_2),Y(s_1\pm k,s_2\pm l)) = 0 \text{ otherwise}$$

Unlike the SAR models, correlation declines rapidly and becomes zero after a certain distance. For this model for large lattices an autoregressive representation exists if $|v| < 0.25$; so the largest first order autocorrelation in this model is < 0.40. Cliff and Ord (1981, p. 150) define a moving average process in which correlations are zero beyond the first adjacency. (See also the nearest neighbour model of Kiefer and Wynn, 1981, where correlation is only non-zero at lag 1.) As with the SAR model higher order and non-symmetric versions can be defined (Figure 3.7).

Figure 3.5 shows a MA model ($v = 0.20$) superimposed on different orders of trend surface.

Example 3.3 Errors in variables schemes. *(Whittle, 1954)*
If Y represents for example a low order SAR model with mean μ and dispersion matrix V_y and if E represents, independent (of Y), superimposed normal errors with mean zero and dispersion matrix kI, where k is a constant then $Y^+ = Y + E$ has mean μ and dispersion matrix

$$V_{y+} = V_y + kI$$

Errors in variables schemes are appropriate where there is intrinsic site variability (such as in the case quoted by Whittle) or measurement error (or both). It is often termed 'nugget' variance in the geostatistical literature.

The previous set of spatial models were constructed by expressing the

84

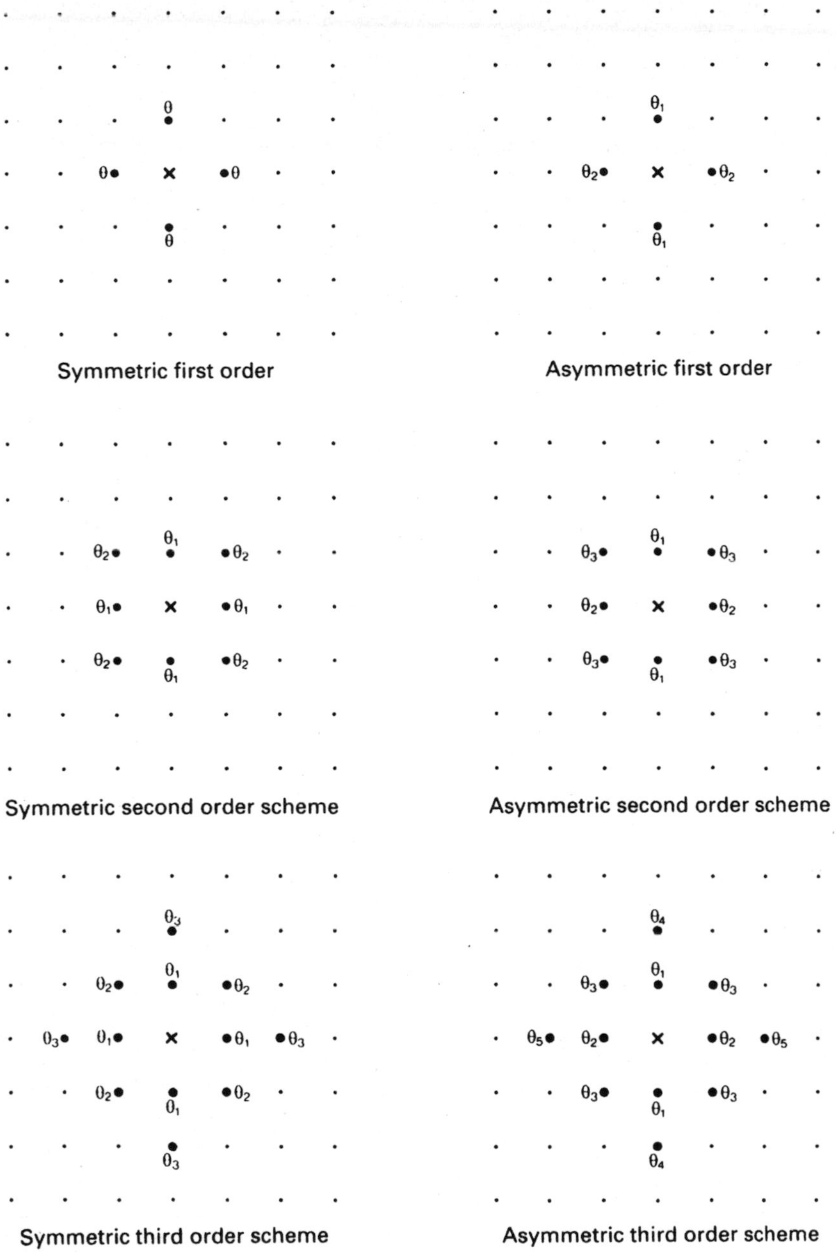

Symmetric first order

Asymmetric first order

Symmetric second order scheme

Asymmetric second order scheme

Symmetric third order scheme

Asymmetric third order scheme

Figure 3.7. A hierarchy of bilateral models: (*a*) lattice case.

First order model

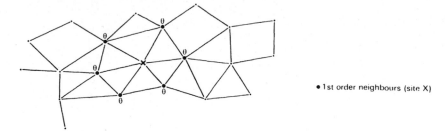

● 1st order neighbours (site X)

Second order model

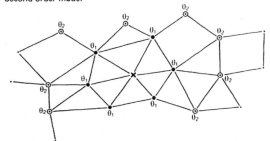

⊙ 2nd order neighbours (site X)

Third order model

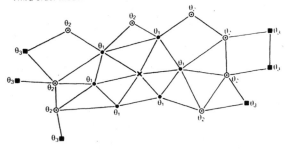

■ 3rd order neighbours (site X)

(*b*) non-lattice case.

joint probability density function of the random variables as a product of *joint* density functions each consisting of one of the random variables together with its set of neighbours. We now consider the conditional approach to specifying interaction schemes in which spatial models are constructed by first specifying the *conditional* probability of each Y_i given its set of neighbouring sites. If

$$\Pr\{Y_i = y_i \mid \{y_j\}_{j \in N(i)}\} \tag{3.6}$$

86

denotes the conditional probability of y_i given the values at neighbouring sites then when (3.6) defines a unique joint probability structure given by

$$\Pr\{y_1, \ldots, y_n\} \tag{3.7}$$

the joint probability distribution is called a Markov Random Field. This is considered a natural generalisation of the Markov property from time series to spatial arrays (Kindermann and Snell, 1980). However, in order to ensure a unique joint probability structure it is necessary to impose restrictions on the functional form of the conditional probabilities. These restrictions are identified by the theorem of Hammersley and Clifford and discussed in Besag (1974). We sketch the idea. Define

$$Q(\mathbf{y}) = \ln\{p(\mathbf{y})/p(0)\}$$

where $p(\mathbf{y})$ is the joint density for any realisation $(\mathbf{y})$ and $p(0)$ is the joint density for the realisation consisting entirely of zeros. It is assumed that $p(0) > 0$, this is called the positivity condition. The Hammersley–Clifford theorem identifies the most general form of $Q(\mathbf{y})$ that ensures a valid probability structure for the system. The required form of $Q(\mathbf{y})$ is:

$$Q(\mathbf{y}) = \Sigma_i y_i G_i(y_i) + \Sigma\Sigma_{i<j} y_i y_j G_{i,j}(y_i,y_j) + \\ \Sigma\Sigma\Sigma_{i<j<k} y_i y_j y_k G_{i,j,k}(y_i,y_j,y_k) + \ldots + y_1 \ldots y_n G_{1,\ldots,n}(y_1, \ldots,y_n)$$

where summations are from $1, \ldots, n$ and the $G_{i,j,\ldots,s}$ may be non-zero if and only if the sites $i,j, \ldots,s$ form a clique (see section 3.1.2). Now valid conditional probabilities can be written down since

$$p(y_i \mid \text{neighbours of } i)/p(y_i = 0 \mid \text{neighbours of } i) = \exp(Q(\mathbf{y}) - Q(\mathbf{y}_i)) \tag{3.8}$$

where $\mathbf{y}_i = (y_1, \ldots,y_{i-1},0,y_{i+1}, \ldots,y_n)$.

Here we shall only be interested in the class of auto-models, that is, models where

$$Q(\mathbf{y}) = \Sigma_i y_i G_i(y_i) + \Sigma\Sigma_{i<j} \beta_{ij} y_i y_j \tag{3.9}$$

and $\beta_{ij} = 0$ unless sites i and j are neighbours and $\beta_{ij} = \beta_{ji}$. This restriction implies that if the edge structure generates cliques of size three or larger, the interactions that are generated by three-way links are modelled in terms of their pairwise components. From (3.9):

$$\exp[Q(\mathbf{y}) - Q(\mathbf{y}_i)] = \exp[y_i(G_i(y_i) + \Sigma_i \beta_{ij} y_j)] \tag{3.10}$$

Automodels are then further classified according to the conditional density function assumed. For example, let $(Y_i, \ldots,Y_n)$ be a vector of normal random variables at n sites with $E[\mathbf{Y}] = \boldsymbol{\mu}$. For each i define the conditional density:

$p(y_i \mid \text{all other site values}) =$
$(2\pi\sigma_i^2)^{\frac{1}{2}}\exp[-(1/2\sigma_i^2)\{(y_i - \mu_i - \Sigma_{j \neq i}k_{ij}(y_j - \mu_j)\}^2]$

This implies (see for example Morrison, 1967, p. 88):

$E[Y_i \mid \text{all other site values}] = \mu_i + \Sigma_{j \neq i}k_{ij}(y_j - \mu_j)$
$\text{Var}[Y_i \mid \text{all other site values}] = \sigma_i^2$

The $\{k_{ij}\}$ are non-zero only if i and j are neighbours. The joint density of **Y** is MVN($\mathbf{\mu}$,**V**) where

$$\mathbf{V} = (\mathbf{I} - \mathbf{K})^{-1}\mathbf{M}; \quad \mathbf{M} = \text{Diag}[\sigma_1^2, \ldots, \sigma_n^2]$$

However, since **V** must be symmetric a further condition is that

$$\sigma_j^2 k_{ij} = \sigma_i^2 k_{ji} \tag{3.11}$$

Suppose we now set $\mathbf{M} = \sigma^2\mathbf{I}$ and $(\mathbf{I} - \mathbf{K}) = \mathbf{B}$ then the joint density of **Y** is

$$|\mathbf{B}|^{\frac{1}{2}}(2\pi\sigma^2)^{-(n/2)}\exp[-(1/2\sigma^2)(\mathbf{y} - \mathbf{\mu})^{\mathsf{T}}\mathbf{B}(\mathbf{y} - \mathbf{\mu})] \tag{3.12}$$

and this should be contrasted with (3.5). These properties are described in Morrison (1967, pp. 86–8) for example. Model (3.12) has been termed the conditional autoregressive (CAR) model or autonormal scheme.

Example 3.4a. Single parameter CAR schemes
Let $\mathbf{M} = \sigma^2\mathbf{I}$ and $\mathbf{K} = \tau\mathbf{W}$. For a rectangular lattice we may let $\mathbf{W} = \mathbf{\Delta}$ and write

$$E[Y(s_1,s_2) \mid \text{all other sites}] = \mu + \tau(y(s_1 - 1,s_2) + y(s_1 + 1,s_2)$$
$$+ y(s_1,s_2 - 1) + y(s_1,s_2 + 1) - 4\mu)$$

For more general lattices note that since $(\mathbf{I} - \mathbf{K})$ must be symmetric as well as positive definite then $w_{ij} = w_{ji}$ unless the diagonal elements of **M** vary so that (3.11) holds. Figure 3.8 shows the (unconditional) variance for this model when $\sigma^2 = 1$, $\tau = 0.15$ on a 7×7 lattice.

To ensure invertibility of $(\mathbf{I} - \mathbf{K})$, the domain of τ is constrained by the eigenvalues of **W** as described in example 3.1a.

Example 3.4b. Asymmetric schemes for regular lattices. (Besag, 1974; 1978b)
Following the same approach as for SAR models we can write

$$E[Y(s_1,s_2) \mid \text{all other sites}] = \mu + \tau_{s_1}(y(s_1 - 1,s_2) + y(s_1 + 1,s_2) - 2\mu)$$
$$+ \tau_{s_2}(y(s_1,s_2 - 1) + y(s_1,s_2 + 1) - 2\mu)$$

with the condition $|\tau_{s_1}| + |\tau_{s_2}| < 0.5$ for square ($n \times n$) lattices as $n \to \infty$.

Example 3.4c. Higher order schemes (Besag, 1974)

As with SAR schemes, higher order CAR models can be defined by introducing additional parameters for different distances. For example, let $K = \tau_1 W_1 + \tau_2 W_2$ with W_1 and W_2 defined as previously.

Finally we comment on the relation between CAR and SAR models: 1. any SAR model is a CAR model by setting $K = S + S^T - S^T S$. 2. Any CAR model is an SAR model but in a rather unnatural way. Obtain L^T where $LL^T = (I - K)$ which is the Cholesky decomposition of $(I - K)$. Now set $S = (I - L)^T$. The CAR model can be derived from (3.3) by setting $B = (I - K)$, $C = I$ and $V_e = (I - K)$ but this again seems an unnatural derivation (Martin, 1987). 3. The matrix V^{-1} identifies the conditional mean and variance of any Y_i given all the other site values since if $V^{-1} = \{v^{ij}\}$ then

$$E[Y_i \mid \text{all other site values}] = \mu_i - (v^{ii})^{-1} \Sigma_{j \neq i} v^{ij} (y_j - \mu_j) \qquad (3.13)$$

and

$$\text{Var}[Y_i \mid \text{all other site values}] = (v^{ii})^{-1}$$

We now contrast the conditional expectation of the first order CAR model given in example 3.4a with the conditional expectation of the first order SAR model defined in example 3.1a. For the SAR model

$$\begin{aligned}
V^{-1} &= \sigma^{-2}[(I - \rho W)^T (I - \rho W)] \\
&= \sigma^{-2}[I - 2\rho W + \rho^2 W^2]
\end{aligned}$$

Consideration of W and W^2 shows that for an interior site (i) $v^{ii} = \sigma^{-2}(1 + 4\rho^2)$ and (set $\mu = 0$)

Figure 3.8. Variances for individual sites (areas) on a 7×7 lattice for the single parameter CAR ($\tau = 0.15$) model with $\sigma^2 = 1$. After Haining (1988).

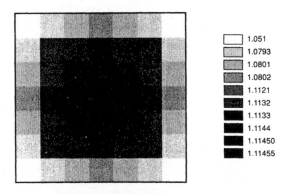

☐	1.051
▨	1.0793
▨	1.0801
▨	1.0802
▨	1.1121
▨	1.1132
■	1.1133
■	1.1144
■	1.11450
■	1.11455

$$E[Y(s_1,s_2)\mid \text{all other sites}]=(1+4\rho^2)^{-1}\{2\rho[y(s_1+1,s_2)+y(s_1-1,s_2)$$
$$+y(s_1,s_2+1)+y(s_1,s_2-1)]-2\rho^2[y(s_1-1,s_2-1)+y(s_1+1,s_2-1)$$
$$+y(s_1-1,s_2+1)+y(s_1+1,s_2+1)]-\rho^2[y(s_1-2,s_2)+y(s_1+2,s_2)$$
$$+y(s_1,s_2-2)+y(s_1,s_2+2)]\}$$

These twelve lattice sites are the set of acquaintances of site (s_1,s_2) and it is clear that the SAR model is not a first order conditional scheme.

Errors in variables schemes can be defined for conditional schemes as described in example 3.3 (Besag, 1977b).

3.2.2 (b) Interaction models for other multivariate distributions

Models for other continuous distributions can be defined. However, in the case of conditionally specified models the consistency condition imposed by the Hammersley–Clifford theorem means that whilst it may be possible to generate other types of continuous automodels, in practice many of them may not be of much practical interest. Besag (1974, p. 202) comments on the negative exponential scheme and notes problems that apply to all gamma type distributions.

3.2.2 (c) Direct specification of V

Since V consists of variances down the diagonal and covariances in the off-diagonals, V can be specified directly through functions that preserve the symmetry and positive definite property of V. We shall describe some important examples of such functions which generate isotropic covariance functions. For detailed discussion of the mathematical conditions for permissable forms of covariance models see Ripley (1981, pp. 10–13) and Christakos (1984).

Since the covariance function $C(\ )$ must be positive definite and symmetric then:
(i) $C(0)\geqslant 0$ (non-negative variance)
(ii) $C(r)=C(-r)$ (the covariance is an even function)
(iii) $|C(r)|\leqslant C(0)$
where r is the distance between two points. It is often the case that social and environmental data have covariance functions with the following common features:
(i) $C(r)\to 0$ as r increases
(ii) $|C(r_1)|<|C(r_2)|$ if $r_1>r_2$
(iii) $C(r)\geqslant 0$ for all r

Covariance functions are usually monotonically decreasing and periodicity is not often of serious importance. We classify the shapes of covariance functions particularly in terms of their behaviour near the origin, which is important in interpolation.

(i) Covariance functions linear near the origin

Perhaps the simplest model is the triangular correlation function where

$$C(r) = \begin{cases} \sigma^2(1-\gamma r) & \text{for } r \leqslant 1/\gamma \\ 0 & \text{for } r > 1/\gamma \end{cases}$$

This covariance function is positive definite in 1 but not 2 or 3 dimensions. Ripley (1981, p. 55) describes a process due to Zubrzycki which in two dimensions yields the following covariance function:

$$C(r) = \begin{cases} \sigma^2\{1-(2/\pi)[(r/R)(1-(r^2/R^2))^{\frac{1}{2}}+\sin^{-1}(r/R)]\} & r \leqslant R \\ 0 & r > R \end{cases}$$

The resulting covariance function for the same process defined in three dimensions is simpler. In this case

$$C(r) = \begin{cases} \sigma^2\{1-(3|r|/2\alpha)+(|r|^3/2\alpha^3)\} & |r| \leqslant \alpha \\ 0 & |r| > \alpha \end{cases} \tag{3.14}$$

This latter function is the 'spherical model' frequently used in geostatistics even for two-dimensional problems.[4] In both cases σ^2 is a scalar constant. The correlation function $R(r)$ of the spherical model is illustrated in Figure 3.9.

(ii) Covariance functions non-linear near the origin

The exponential function is a valid covariance function in two dimensions:

$$C(r) = \sigma^2 \exp[-ar^\beta] \qquad a>0; \ \beta>0$$

with

$$R(r) = \exp[-ar^\beta]$$

When $\beta = 1$ the function decays steeply at the origin (downward convex) but when $\beta = 2$ the function is downward concave or cusped with a zero derivative at the origin.

Whittle (1954) introduced the model:

$$C(r) = \sigma^2 \eta r K_1(\eta r) \qquad \eta > 0 \tag{3.15}$$

with

$$R(r) = \eta r K_1(\eta r)$$

where $K_1(x)$ is a modified Bessel function of the second kind of order 1 and can be approximated by $(\frac{1}{2}\pi/x)^{-\frac{1}{2}} \exp[-x]$. The covariance function arises

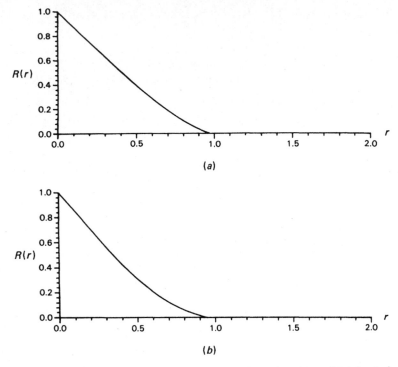

Figure 3.9. Correlation function of the spherical model. Model defined in two (*a*) and three (*b*) dimensions.

from a continuous space analogue of a first order SAR model defined on an infinite square lattice (see example 3.1a). This covariance function also has a zero derivative at the origin. These correlation functions are illustrated in Figure 3.10.

(iii) Covariance functions discontinuous near the origin
It is sometimes the case that empirical correlation functions display a sharp discontinuity at the origin. This was observed, for example, by Whittle (1954) in an analysis of data from a uniformity trial of 1000 orange trees. The discontinuity was thought to be due to intrinsic variability in the plants themselves superimposed on locally smooth variations in soil fertility. The correlation function for such a process can be written as:

$$R^+(r) = \begin{cases} 1 & r=0 \\ \alpha R(r) & r>0 \end{cases}$$

where $R(r)$ is any one of the previous smoothly varying correlation functions and $0 \leqslant \alpha \leqslant 1$. Figure 3.11 gives an illustration.

92

(iv) Covariance function behaviour over longer distances

Another potentially important consideration in choosing a covariance function for spatial data is the rate of decay of the function over longer distances. The main examples covered so far either have covariance functions which become zero at a fixed distance or decay rapidly at an exponential rate. They would not cover the situation where covariance

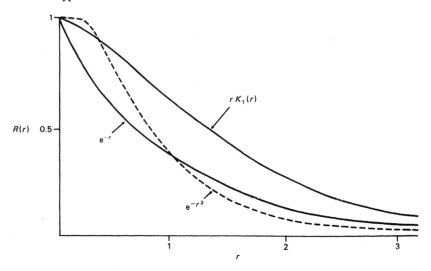

Figure 3.10. Correlation functions: Negative exponential (e^{-r^2}) and Bessel ($rK_1(r)$) type.

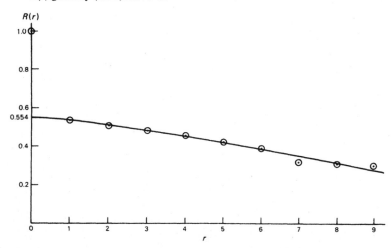

Figure 3.11. Correlation function which is discontinuous at the origin and with $R(r)$ given by (3.15) for $r > 0$.

decreases more slowly, say as a negative power of distance (r), that is, $r^{-\lambda}(\lambda > 0)$. A possible model is:

$$C(r) = \sigma^2(r+\theta)^{-\lambda}$$

where θ and λ are positive constants.

(v) Covariance functions with linear structures

A general approach to specifying models for $\mathbf{V}$ is to assume a structure of the form:

$$\mathbf{V} = \lambda_1 \mathbf{B}_1 + \lambda_2 \mathbf{B}_2 + \ldots + \lambda_m \mathbf{B}_m \tag{3.16}$$

where the $(n \times n)$ $\mathbf{B}_i$ matrices (n denoting the number of sites) are known and the $\{\lambda_i\}$ are constants. The $\{\mathbf{B}_i\}$ are the principal idempotents of $\mathbf{V}$ in particular:

$$\mathbf{B}_i^2 = \mathbf{B}_i; \qquad \mathbf{B}_i \times \mathbf{B}_j = 0 \text{ (if } i \neq j); \qquad \Sigma_{i=1,\ldots,m} \mathbf{B}_i = \mathbf{I}$$

Equation (3.16) is called a decomposable structure. The matrix $\mathbf{B}_i$ is obtained from the eigenvector ($\mathbf{x}_i$) associated with the eigenvalue λ_i of $\mathbf{V}$ that is $\mathbf{B}_i = \mathbf{x}_i \mathbf{x}_i^{\mathsf{T}}$. Streitberg (1979) has discussed spatial models of this type. He suggests starting with a symmetric weights matrix $\mathbf{W}$ which is used to construct the eigenvalues and eigenvectors. Since for a system with n sites $\mathbf{W}$ will have n eigenvalues and eigenvectors he suggests grouping eigenvalues with similar values. This will keep the problem manageable and since eigenvalues often appear in pairs or clusters for such matrices, may prove a reasonably objective procedure. Streitberg (1979, p. 157) makes some suggestions.

Linear covariance functions also arise in IRF-k theory to be discussed next. These types of covariance functions are important because parameter estimation and statistical inference are often easier after such a decomposition.

3.2.2 (d) Intrinsic random functions

We now describe models for representing spatial variation when only the increments of the observed process are stationary. A common approach is through the semi-variogram. This is appropriate when the mean is constant, when it is not it is usual to specify the mean separately. Alternatively, polynomial generalised covariance functions (PGCFs) can be specified which are the covariance functions of IRF-k processes. This group of processes includes constant mean processes as the special case $k=0$. On the other hand, there is a rich and varied range of semi-variogram models which are usually easier to interpret than PGCFs. They deserve treatment in their own right.

From the definition of the semi-variogram (3.1) it follows that $\gamma(0) = 0$ and $\gamma(\mathbf{h}) = \gamma(-\mathbf{h}) \geqslant 0$. Usually as $\mathbf{h}$ increases, $\gamma(\mathbf{h})$ increases at least for small $\mathbf{h}$. However, if the process is stationary then $\gamma(\mathbf{h})$ is a bounded function in which $\gamma(\mathbf{h})$ increases as $\mathbf{h}$ increases but reaches a plateau known as a 'sill'. Such variograms are called 'transition' models. Pairs of observations are correlated if they are within a distance, a, of one another called the 'range'. Correlation decreases with increasing distance, and for distances exceeding the range observations are uncorrelated. These points are illustrated in Figure 3.12 for an isotropic transition scheme.

Not just any function can be a model for a semi-variogram. Just as only positive definite functions can be models for the covariance function, only conditional nonpositive definite functions can be models for the semi-variogram. These functions ensure that the variances of the increments are positive subject to satisfying the conditions on the $\{\lambda_i\}$ (see section 3.1). Christakos (1984) discusses conditions and tests for a model to be a semi-variogram, in practice most use is made of a relatively small class of 'safe' models which we now discuss.

If the process is stationary then

$$\gamma(\mathbf{h}) = C(0) - C(\mathbf{h})$$

so that any of the previous covariance functions can be used as models for transition semi-variograms. More generally, semi-variograms are characterised by whether they possess a sill and also by their behaviour near $\mathbf{h} = 0$. Four types are illustrated in Figure 3.13. The parabolic case is characteristic of highly regular spatial variability whilst the nuggett effect denotes a discontinuity at the origin (since $\gamma(0) = 0$) and is analogous to the

Figure 3.12. Properties of a transition semi-variogram.

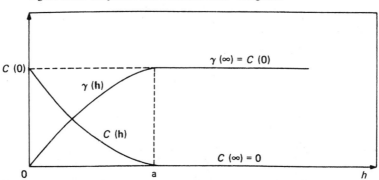

discontinuous correlation function models described earlier. Other semi-variogram models can be constructed and the list further extended by allowing 'hybrid' forms. Figure 3.14 identifies some common semi-variogram models (see also Journel and Huijbregts, 1978, pp. 161–71; McBratney and Webster, 1986). The de Wijsian scheme in which the semi-variogram increases logarithmically with distance reflects a mild departure from a transition scheme. A strong departure from weak stationarity is indicated by a steep increase in $\gamma(\mathbf{h})$ with increasing $\mathbf{h}$. On the other hand there are restrictions on the steepness of the increase and $\gamma(\mathbf{h})$ must necessarily increase more slowly than $|\mathbf{h}|^2$ (Matheron, 1971). The presence of such a steep increase is taken to imply the existence of some order of polynomial trend in the surface which should be removed – or a different class of models fitted. This alternative class is now considered.

If the observed process (Y) is an IRK-k then (see section 3.2) if the increment process (Z) where

$$Z = \Sigma_{i=1,\ldots,n} \lambda_i Y(\mathbf{u}_i)$$

is formed from authorised linear combinations (that is the $\{\lambda_i\}$ satisfy the conditions for filtering out polynomials of order k in the Y process) then surfaces of this type can be modelled by observing that since Z is stationary Var(Z) is finite and independent of location and

$$\mathrm{Var}(Z) = E[(\Sigma_{i=1,\ldots,n} \lambda_i Y(\mathbf{u}_i))^2]$$
$$= \Sigma_{i=1,\ldots,n} \Sigma_{j=1,\ldots,n} \lambda_i \lambda_j K_{ij}(\boldsymbol{\psi})$$

where $K_{ij}(\boldsymbol{\psi})$ is called the PGCF depending on the parameter $\boldsymbol{\psi}$. (It is not a covariance function in the earlier sense, but the covarance function of a process that is only stationary in the increments.) Delfiner (1976) gives an approximate expression for this function for all k when the process is isotropic. The function is linear in the parameters and can be expressed (if there is no nugget effect)

$$K(r, \boldsymbol{\psi}) = \Sigma_{d=0,\ldots,k} \psi_d |r|^{2d+1} \tag{3.17}$$

Figure 3.13. Common semi-variogram forms.

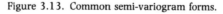

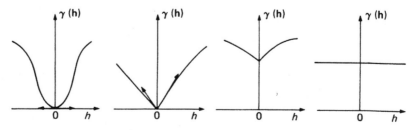

where r denotes the distance between sites and k is the order of the IRF field. For the case $k=0$

$$K(r,\psi) = -\psi_0|r| \qquad \psi_0 \geqslant 0$$

If there is an additional nugget effect representing small scale variability then an additional term $c\delta(|\mathbf{r}|)$ is added to (3.17) where c is a positive constant and δ is Kronecker's delta. For the case $k=1$ (with nuggett variance):

$$K(r,\psi) = c\delta(|r|) + \psi_0|r| + \psi_1|r|^3 \tag{3.18}$$

and to ensure a conditional positive definite function $c \geqslant 0$, $\psi_0 \leqslant 0$, $\psi_1 \geqslant 0$ (Kitanidis, 1983). Further examples are given in Delfiner (1976). If the process is non-isotropic it is assumed this has been accounted for in the $\{\lambda_i\}$ (see Delfiner, 1976). Figure 3.15 shows realisations of IRF-k fields.

Figure 3.14. Semi-variogram models.

Model	Equation (Constants: A, B, a, b, c, C_0)	Comment	Graph
De Wijsian	$\gamma(h) = A \ln(h) + B$	Mildly non-stationary; increasing without an upper bound. B > 0 implies nugget effect.	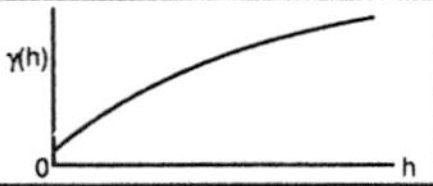
Linear	$\gamma(h) = Ah + B$ $A > 0 \quad B \geq 0$	Strongly non-stationary B > 0 implies nugget effect.	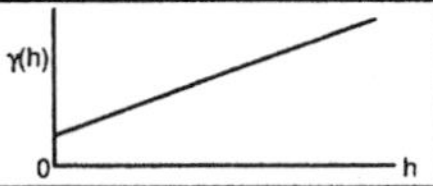
Log - log	$\gamma(h) = B + h^A$	Strongly non-stationary B > 0 implies nugget effect.	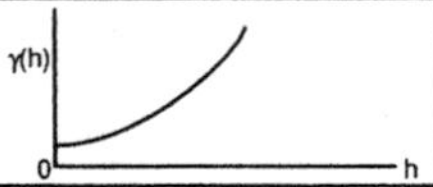
Exponential	$\gamma(h) = C_0 + C[1 - \exp(-|h|/a)]$	Approaches asymptotically an upper limiting value C_0 > 0 implies nugget effect.	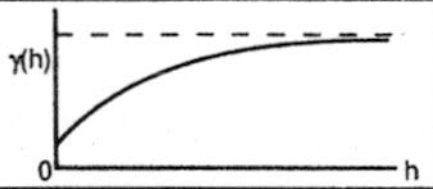
Hole	$\gamma(h) = C[1 - \sin(ah)/ah]$	Increasing to a plateau with oscillation.	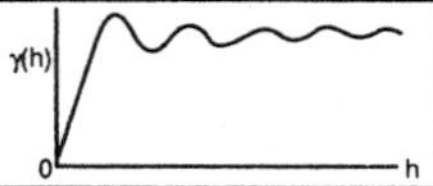
Spherical	$\gamma(0) = 0$ $\gamma(h) = C[3h/2a - (h^3/2a^3)]$ $+ C_0 \quad h \leq a$ $\gamma(h) = C + C_0 \quad h > a$	Stationary (transition) model.	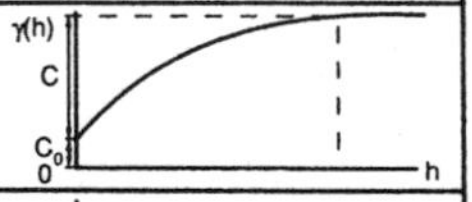
Gaussian	$\gamma(h) = C_0 + C[1 - \exp(-3h^2/a^2)]$	Approaches asymptotically an upper limiting value C_0 > 0 implies nugget effect.	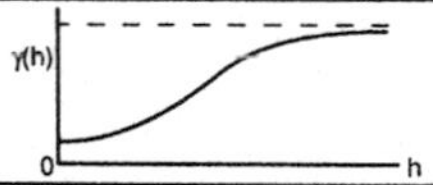

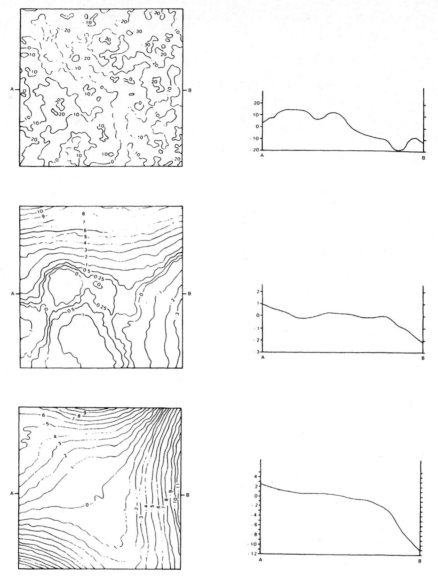

Figure 3.15. Realisations of IRF-k fields ($k = 0$, 1, 2). After Matheron (1973).

3.3   Population models for discrete random variables

We examine models that have been constructed explicitly for discrete random variables. They originate from the conditional specification of variate interaction schemes by assuming different forms for the conditional probability distribution function. Again we consider only auto-models.

Example 3.5 The autologistic model for binary observations. (Besag, 1974)
Define:

$$Q(\mathbf{y}) = \Sigma_{i=1,\ldots,n} \, \alpha_i y_i + \Sigma\Sigma_{1 \leqslant i < j \leqslant n} \, \beta_{ij} y_i y_j$$

and from (3.8) and (3.10)

$$P\{Y_i = y_i \,|\, \{y_j\}_{j \neq i}\}/P\{Y_i = 0 \,|\, \{y_j\}_{j \neq i}\} = \exp[y_i(\alpha_i + \Sigma_{j \neq i}\beta_{ij}y_j)]$$

where $\beta_{ij} = \beta_{ji}$ and $\beta_{ij} = 0$ unless i and j are neighbours. Since $Y_i = 0$ or 1,

$$P\{Y_i = y_i \,|\, \{y_j\}_{j \neq i}\} = \exp[y_i(\alpha_i + \Sigma_{j \neq i}\beta_{ij}y_j)]/[1 + \exp(\alpha_i + \Sigma_{j \neq i}\beta_{ij}y_j)] \quad (3.19)$$

This is called the autologistic model. The $\{\beta_{ij}\}$ are interaction parameters referring to the cliques of size two. The $\{\alpha_i\}$ are site effects and can be chosen to model trend or the influence of other variables. Some realisations are shown in Figure 3.16 where $\alpha_i = \alpha$ and $\beta_{ij} = \beta\delta_{ij}$ for different α and β and β increasing. The model is related to a logistic model and can be considered an extension of that model to spatial data.

Example 3.6 A classification model for c unordered and interchangeable colours (Besag, 1986)
From (3.9) let

$$Q(\mathbf{y}) = (\Sigma_{k=1,\ldots,c} \, \alpha_k n_k - \Sigma_{1 \leqslant k < l \leqslant c} \, \beta_{k,l} n_{k,l})$$

where n_k is the number of sites of colour k, n_{kl} is the number of distinct neighbour pairs coloured k and l and the αs and βs are parameters.

$$P\{Y_i = k \,|\, \{y_j\}_{j \neq i}\} \propto \exp[Q(\mathbf{y}) - Q(\mathbf{y}_i)]$$
$$= \exp[\alpha_k - \underset{l \neq k}{\Sigma_{l=1,\ldots,c}}\beta_{k,l}u_i(l)]$$

where $\mathbf{y}_i$ and $\mathbf{y}$ are two realisations differing only at site i, $u_i(l)$ is the number of neighbours of site i having colour l and $\beta_{k,l} = \beta_{l,k}$.

If the colours are unordered and $\beta_{k,l} = \beta$, and if they are also interchangeable then the α coefficient is redundant. So

$$P\{Y_i = k \,|\, \{y_j\}_{j \neq i}\} = \exp[\beta u_i(k)]/\Sigma_{l=1,\ldots,c} \, \exp[\beta u_i(l)]$$

Statistical models for spatial populations

Example 3.7 The autobinomial model for naturally ordered colours.
(Besag, 1974, 1986; Cross and Jain, 1983)

If the colours are naturally ordered (such as grey levels on an image) the value at any site can be represented as the result of c trials with the number of successes $(0,1,2, \ldots ,c)$. If the conditional distribution at each site is binomial then:

$$P\{Y_i=k\,|\,\{y_j\}_{j\neq i}\} = \binom{c}{k}\theta_i^{\,k}(1-\theta_i)^{c-k} \qquad k=0,1,2, \ldots ,c$$

and

$$\theta_i = \exp[\alpha_i+\Sigma\beta_{ij}y_j]/(1+\exp[\alpha_i+\Sigma\beta_{ij}y_j])$$

For $c=1$ this gives the autologistic model. Realisations of the autobinomial are given in Cross and Jain (1983).

Unfortunately, some of the models that can be constructed in this way seem not to be of practical interest. When the consistency conditions are

Figure 3.16. Realisations of the autologistic model. After Cross and Jain (1983).

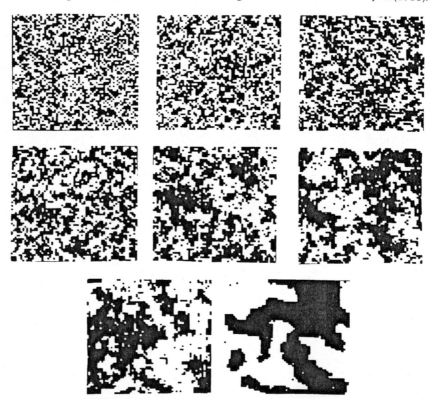

enforced in the case of the auto-Poisson and auto-Pascal distributions for example then the interaction parameters $\{\beta_{ij}\}$ are forced to be negative so that the distributions can only describe 'competitive' and not 'co-operative' interaction (Besag, 1974).

3.4   Boundary models for spatial populations

When developing a spatial model on a finite number of sites, regional boundary effects are likely to be important. Martin (1987) postulates two aspects of the boundary value problem:

(i)   the effect of different boundary formulations on the covariance matrix of the process;

(ii)   the effect of boundary forms on the properties of estimators of model parameters.

There are various ways of handling boundary effects. One is to treat the process as locally stationary, that is as a restriction of a stationary process (defined at least conceptually on a much larger region) to the study region – call this D. This is equivalent to taking the marginal distribution of the infinite lattice process for all sites in D.[5] Another approach is to define the process on D and specify a separate model to take care of the boundary. This is equivalent to taking the conditional distribution on D given the behaviour of the sites outside D.

Which sites are the boundary sites of a region depends on the type of model as well as the configuration of sites in D. Boundary sites can be divided into the observed and unobserved set. Figure 3.17(a) shows a rectangular lattice with a region D marked off to represent the observed sites. Outside this area is an unobserved area we shall denote D^c. Suppose the edge structure for the lattice is that given by Figure 3.2(b). There are those boundary sites inside D which are observed but have neighbours in D^c and those boundary sites in D^c which are unobserved but have neighbours in D. The first group will be called observed boundary sites, the second group unobserved boundary sites. It can also be seen from these figures that the two sets of observed and unobserved boundary sites depend on the nature of the edge structure, for if the edge structure of Figure 3.2(c) were to be extended to include for each site those sites that are two steps removed in the four directions then the two sets of boundary sites would increase accordingly. In the case of an irregular distribution of sites in D (Fig. 3.17(b)) the two sets of boundary sites can only be determined once the arrangement of sites in D^c is known together with the edge structure between sites in D and D^c.

(*a*)

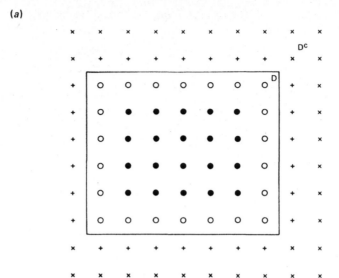

(*b*)

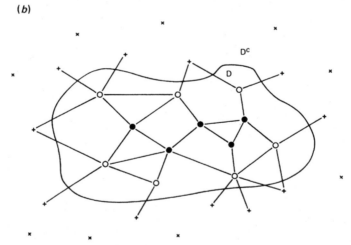

Figure 3.17. Boundary sites for a region: (*a*) lattice case, (*b*) non-lattice case. D –
● = interior site, ○ = (observed) boundary site; D^c – + = (unobserved) boundary
site, × = other (unobserved) site.

The dispersion matrix $V = \{v_{ij}\}$ of a stationary process restricted to D has the form:

$$v_{ij} = \begin{cases} \text{constant} & i=j \\ v(\mathbf{h}_{ij}) & i \neq j \end{cases} \qquad (3.20)$$

where $v(\mathbf{h}_{ij})$ means that the entry depends only on the vector $\mathbf{h}$ between sites i and j. In the case of an isotropic process, only the length of the vector is important. Any of the isotropic covariance functions of section 3.2 can be models for V, however, care must be taken in constructing locally stationary versions of the variate interaction schemes. These models were defined on finite lattices. As already noted their dispersion matrices do not satisfy (3.20) so they are not locally stationary. One strategy here might be to identify the covariance (or correlation) functions for stationary forms of these models and use these as models for V. Doing this will also extend the range of permissible models for V. An alternative strategy is to identify functional forms of the dependence equations, or conditioning sets, which will ensure that the models are locally stationary. Both approaches face difficulties in practice.

Example 3.8 Single parameter CAR scheme
Consider the model defined in example 3.4a on a rectangular lattice and with $\mu = 0$. Let the process be defined on a doubly infinite lattice and let $R(k_1, k_2)$ denote the correlation at lag (k_1, k_2). Then for $|\tau| < 0.25$

$$R(k_1, k_2) = \psi \int\int_{-\pi}^{\pi} [\cos(k_1 x)\cos(k_2 y)][1 - 2\tau(\cos(x) + \cos(y))]^{-1} dx dy \qquad (3.21)$$

where ψ is such that $R(0,0) = 1.0$ (see Moran 1973a,b). The decay of spatial correlation can be approximated by a monotonically decreasing function of distance, the decay being very slow when τ is large (Besag, 1981a). Table 3.1 gives the correlations for this model for different τ. Some results for anisotropic schemes are given in Table 3.2.

The conditioning sets for the observed boundary sites of locally stationary versions of the model are not the same as given in example 3.4a. Recall from (3.13) that the conditioning set for site i includes all those sites (j) for which $v^{ij} \neq 0$. Kunsch (1983, theorem 2) shows that for locally stationary versions of this model

$$V^{-1} = \sigma^{-2}(I - (K + \Gamma)) \qquad (3.22)$$

where $\Gamma = \{\gamma_{ij}\}$ and $\gamma_{ij} = 0$ except when sites i and j are both on the observed boundary of D. The conditioning set for any boundary site will therefore

Table 3.1. *Spatial correlations for the stationary form of the single parameter CAR model (model 3.4a)*

		k_1					
	k_2	0	1	2	3	4	5
(a) $\tau=0.24$							
	0	1.0	0.434	0.219	0.121	0.070	0.042
	1	0.434	0.294	0.179	0.108	0.065	0.040
	2	0.219	0.179	0.126	0.083	0.053	0.034
	3	0.121	0.108	0.083	0.059	0.040	0.027
	4	0.070	0.065	0.053	0.040	0.029	0.020
	5	0.042	0.040	0.034	0.027	0.020	0.015
(b) $\tau=0.23$							
	0	1.0	0.368	0.155	0.071	0.034	0.017
	1	0.368	0.224	0.118	0.060	0.031	0.016
	2	0.155	0.118	0.074	0.042	0.023	0.012
	3	0.071	0.060	0.042	0.027	0.016	0.009
	4	0.034	0.031	0.023	0.016	0.010	0.006
	5	0.017	0.016	0.012	0.009	0.006	0.004
(c) $\tau=0.20$							
	0	1.0	0.265	0.077	0.024	0.008	0.002
	1	0.265	0.126	0.049	0.018	0.006	0.002
	2	0.077	0.049	0.023	0.010	0.004	0.001
	3	0.024	0.018	0.010	0.004	0.002	0.000
	4	0.008	0.006	0.004	0.002	0.001	0.000
	5	0.002	0.002	0.001	0.000	0.000	0.000
(d) $\tau=0.15$							
	0	1.0	0.171	0.030	0.005	0.001	0.000
	1	0.171	0.055	0.014	0.003	0.000	0.000
	2	0.030	0.014	0.004	0.001	0.000	0.000
	3	0.005	0.003	0.001	0.000	0.000	0.000
	4	0.001	0.000	0.000	0.000	0.000	0.000
	5	0.000	0.000	0.000	0.000	0.000	0.000

need to include a large number of the other boundary sites to produce a locally stationary field. Furthermore these terms are not negligible.

Example 3.9 Single parameter SAR schemes
Consider the model defined in example 3.1a on a rectangular lattice and with $\mu=0$. Let the process be defined on a doubly infinite lattice and let $R(k_1,k_2)$ denote the autocorrelation at lag (k_1,k_2). Then for $|\rho|<0.25$

Table 3.2. *Spatial correlations for the stationary form of the two parameter CAR model (model 3.4b)*

	k_2	k_1				
		0	1	2	3	4
(a) $\tau_{s_1}=0.20$; $\tau_{s_2}=0.10$						
	0	1.0	0.220	0.049	0.114	0.002
	1	0.121	0.050	0.164	0.004	0.001
	2	0.016	0.009	0.003	0.001	0.000
	3	0.002	0.001	0.000	0.000	0.000
(b) $\tau_{s_1}=0.40$; $\tau_{s_2}=0.075$						
	0	1.0	0.563	0.328	0.197	0.121
	1	0.252	0.213	0.161	0.115	0.080
	2	0.084	0.079	0.067	0.054	0.041
	3	0.031	0.030	0.027	0.023	0.019
	4	0.012	0.012	0.011	0.010	0.008

Evaluated using NAG routine D01DAF.

$$R(k_1,k_2)=\psi \int\int_{-\pi}^{\pi} [\cos(k_1 x)\cos(k_2 y)][1-2\rho(\cos(x)+\cos(y))]^{-2}dxdy$$

where ψ is such that $R(0,0)=1.0$ (Whittle 1954). Table 3.3 shows the correlations for this model for different values of ρ. Like the previous example the correlations decay slowly (for large ρ) at a rate that is approximately a function of distance. An isotropic continuous space analogue of this model exists (Whittle, 1954, pp. 447–8) and for this model

$$R(r)=\alpha r K_1(\alpha r)$$

where $\alpha=((1/\rho)-4)$, and K_1 is a modified bessel function of the second kind of order 1 (Abramowitz and Stegun, 1984, Chapter 9).

As with the CAR model the interaction equation will be altered for the boundary sites of the locally stationary versions of the model.

Table 3.4 shows selected low order correlations ($v_{ij}/\sqrt{v_{ii}v_{jj}}$, where i and j are north–south neighbours) for the SAR model for small (square) lattices to show the differences that can arise between stationary and non-stationary versions of the same model.

Two processes which do have straightforward correlation structures on an infinite lattice and which therefore could be used as models for V in

Table 3.3. *Spatial correlations for the stationary form of the single parameter SAR model (model 3.1a)*

				k_1			
	k_2	0	1	2	3	4	5
(a) $\rho = 0.24$							
	0	1.0	0.839	0.648	0.483	0.353	0.254
	1	0.839	0.748	0.600	0.457	0.338	0.245
	2	0.648	0.600	0.503	0.396	0.300	0.221
	3	0.483	0.457	0.396	0.322	0.251	0.189
	4	0.353	0.338	0.300	0.251	0.200	0.155
	5	0.254	0.245	0.221	0.189	0.155	0.123
(b) $\rho = 0.23$							
	0	1.0	0.740	0.488	0.308	0.190	0.115
	1	0.740	0.609	0.429	0.281	0.177	0.109
	2	0.488	0.429	0.324	0.224	0.147	0.093
	3	0.308	0.281	0.224	0.163	0.111	0.073
	4	0.190	0.177	0.147	0.111	0.038	0.025
	5	0.115	0.109	0.093	0.073	0.025	0.017
(c) $\rho = 0.20$							
	0	1.0	0.546	0.251	0.109	0.046	0.019
	1	0.546	0.366	0.191	0.090	0.039	0.017
	2	0.251	0.191	0.113	0.058	0.027	0.012
	3	0.109	0.090	0.058	0.032	0.016	0.008
	4	0.046	0.039	0.027	0.016	0.009	0.004
	5	0.019	0.017	0.012	0.008	0.004	0.002
(d) $\rho = 0.15$							
	0	1.0	0.349	0.097	0.025	0.006	0.001
	1	0.349	0.165	0.056	0.016	0.004	0.001
	2	0.097	0.056	0.023	0.007	0.002	0.000
	3	0.025	0.016	0.007	0.002	0.001	0.000
	4	0.006	0.004	0.002	0.001	0.000	0.000
	5	0.001	0.001	0.000	0.000	0.000	0.000

Evaluated using NAG routine D01DAF.

locally stationary processes are the moving average process (Haining, 1978d) and the doubly geometric process (Martin, 1979). Results for the moving average process have been given in section 3.2, the correlation structure decays rapidly to zero. Figure 3.18 shows a graph of the correlation function together with that of the single parameter CAR and SAR schemes.

Table 3.4. *Lag one (R(0,1)) correlations for finite lattice schemes*

(a) SAR model

	ρ					
	0.24	0.23	0.20	0.15	0.10	0.075
Asymp.	0.839	0.740	0.546	0.349	0.213	0.155
11×11	0.762	0.686	0.523	0.343	0.211	0.154
	(.82,.60)	(.73,.56)	(.54,.45)	(.34,.31)	(.21,.20)	(.15,.15)
9×9	0.746	0.675	0.518	0.341	0.211	0.154
	(.81,.60)	(.73,.56)	(.54,.45)	(.34,.31)	(.21,.20)	(.15,.15)
7×7	0.721	0.657	0.510	0.339	0.210	0.154
	(.79,.60)	(.72,.56)	(.54,.45)	(.34,.31)	(.21,.20)	(.15,.15)
5×5	0.680	0.627	0.496	0.335	0.209	0.153
	(.75,.60)	(.69,.56)	(.53,.45)	(.34,.31)	(.21,.20)	(.15,.15)

(b) CAR model

	τ					
	0.24	0.23	0.20	0.15	0.10	0.075
Asymp.	0.434	0.368	0.265	0.171	0.105	0.077
11×11	0.388	0.342	0.256	0.168	0.104	0.077
	(.43,.29)	(.36,.27)	(.26,.22)	(.17,.15)	(.10,.10)	(.07,.07)
9×9	0.378	0.336	0.254	0.168	0.104	0.076
	(.42,.29)	(.36,.27)	(.26,.22)	(.17,.15)	(.10,.10)	(.07,.07)
7×7	0.365	0.328	0.251	0.167	0.104	0.076
	(.41,.29)	(.36,.27)	(.26,.22)	(.17,.15)	(.10,.10)	(.07,.07)
5×5	0.343	0.313	0.245	0.165	0.104	0.076
	(.39,.29)	(.35,.27)	(.26,.22)	(.17,.15)	(.10,.10)	(.07,.07)

Correlations for the site at the centre of the lattice. The figures in parentheses are the maximum and minimum correlations on the lattice. Asymp. = asymptotic (an infinite number of lattice points in *both* directions).

The doubly geometric process is the product of two independent one-dimensional processes:

$$Y(\mathbf{s}) = Y_1(s_1)Y_2(s_2)$$

where

$$Y_1(s_1) = \lambda Y_1(s_1 - 1) + e_1(s_1)$$
$$Y_2(s_2) = v Y_1(s_2 - 1) + e_2(s_2)$$

The model can be written:

$$Y(\mathbf{s}) = \lambda Y(s_1 - 1, s_2) + v Y(s_1, s_2 - 1) - \lambda v Y(s_1 - 1, s_2 - 1) + e(s_1, s_2)$$

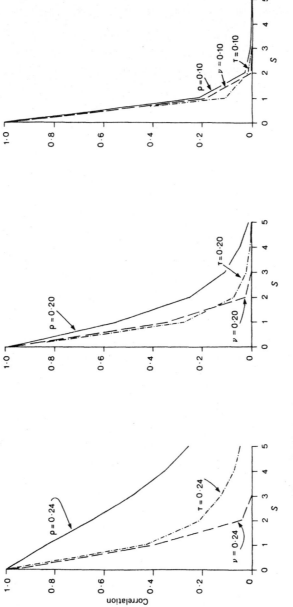

Figure 3.18. Correlation function for stationary forms of the SAR (ρ), CAR (τ) and MA (ν) models along the axis $R(s,0)$; $s = 0,1,2,\ldots,5$. After Haining (1978d).

and its correlation for lag (s_1, s_2) is:

$$R(s_1, s_2) = \lambda^{|s_1|} \gamma^{|s_2|}$$

so that the correlation structure is anisotropic and decays rapidly at a geometric rate.[6]

A second approach to boundary specification is to define the process on D with a separate model for the boundary. The following description follows Martin (1987). Let $\mathbf{Y} = (Y_1, \ldots, Y_n)^\top$ denote the n random variables associated with the sites in D and let $\mathbf{Z} = (Y_{n+1}, \ldots, Y_{n+h})^\top$ denote the sites in D^c that border D. Let $\mathbf{Y}^* = (\mathbf{Y} \mid \mathbf{Z})$. The sites in D^c have not been observed and assume:

$$\mathbf{B}^*(\mathbf{Y}^* - \boldsymbol{\mu}^*) = \mathbf{C}^* \mathbf{e}^* \tag{3.23}$$

where $\mathbf{B}^*$ and $\mathbf{C}^*$ are $((n+h) \times (n+h))$ and $\mathbf{e}^*$ is $((n+h) \times 1)$. This describes a linear dependence equation for $\mathbf{Y}^*$. The dispersion matrix for $\mathbf{Y}^*$ is:

$$\mathbf{V}_{Y*} = (\mathbf{B}^*)^{-1} \mathbf{C}^* \mathbf{V}_{e*} (\mathbf{C}^*)^\top (\mathbf{B}^{*-1})^\top$$

The dispersion matrix for $\mathbf{Y}$ ($\mathbf{V}_Y$) is obtained from $\mathbf{V}_{Y*}$ by deleting the rows and columns associated with $\mathbf{Z}$. Note however that if:

$$\mathbf{V}_{Y*}^{-1} = \begin{bmatrix} \mathbf{V}^{YY} & \mathbf{V}^{YZ} \\ \mathbf{V}^{ZY} & \mathbf{V}^{ZZ} \end{bmatrix}$$

then

$$\mathbf{V}_Y^{-1} = \mathbf{V}^{YY} - \mathbf{V}^{YZ}(\mathbf{V}^{ZZ})^{-1}\mathbf{V}^{ZY} \equiv \mathbf{V}^{YY} - \boldsymbol{\Gamma}$$

from the theory for partitioned matrices.

Example 3.10
For the finite lattice autonormal process $\mathbf{V}^{YY} = (\mathbf{I}_D - \mathbf{K}_D)$; $\mathbf{V}^{YZ} \equiv \mathbf{R}$; $\mathbf{V}^{ZY} \equiv \mathbf{R}^\top$; $\mathbf{V}^{ZZ} = (\mathbf{I}_{D^c} - \mathbf{K}_{D^c})$. So

$$\boldsymbol{\Gamma} = \mathbf{R}(\mathbf{I}_{D^c} - \mathbf{K}_{D^c})^{-1} \mathbf{R}^\top \text{ and } \mathbf{V}_Y^{-1} = (\mathbf{I}_D - (\mathbf{K}_D + \boldsymbol{\Gamma}))$$

In the case of a first order (single parameter) autonormal model

$$\mathbf{V}_Y^{-1} = (\mathbf{I}_D - (\mathbf{K}_D + \mathbf{V}^{YZ}(\mathbf{V}^{ZZ})^{-1}\mathbf{V}^{ZY}))$$

The autonormal model of section 3.2 (defined on a finite D) comes about by setting $\mathbf{V}^{YZ}(\mathbf{V}^{ZZ})^{-1}\mathbf{V}^{ZY} = 0$. This happens for example if $\mathbf{V}^{YZ} = \mathbf{V}^{ZY} = 0$, that is the process in D is 'disconnected' from the process in D^c. It also comes about by setting the (unobserved) border values to the mean of the process since $(v^{ij})^{-1}(y_j - \mu_j) = 0$, for j an unobserved border value, if $y_j = \mu_j$. Thus the earlier models defined on D contained implicit boundary assumptions which have now been made explicit. Other forms can be examined by introducing different assumptions on $\mathbf{V}^{YZ}$, $\mathbf{V}^{ZZ}$ and $\mathbf{V}^{ZY}$.

Following time series terminology we could refer to these two approaches to specifying boundary conditions as the 'stochastic' and 'fixed' border cases. But with so many boundary sites in the spatial case, choosing the mean for *all* sites in the fixed border case may seem restricting. In principle there seems no reason why, for example, the unobserved boundary sites should not be chosen so that the gradient at the boundary is zero or the gradient between observed and unobserved boundary sites is equal to the gradient between the observed boundary sites and the immediate interior sites. If the region D is bounded by water (on one or more sides) again unobserved 'sites' could be set to zero (Tobler, 1979c). A further variant for the case of a bounded region is to postulate two different models, one for the interior sites of D and one for observed boundary sites. The latter sites might be expected to interact more intensely with their edge neighbours because they have fewer to interact with. Thus both interior and boundary sites of D may follow the same model but with different interaction parameters. In all these cases, however, it is necessary to determine the effect of these boundary assumptions on the exact form of $\mathbf{V}_Y^{-1}$.

3.5   Edge structures, weighting schemes and the dispersion matrix

We examine how the choice of edge structure and weighting scheme affects the dispersion matrix $\mathbf{V}$ of interaction models. Specification of edges and weights affects properties of $\mathbf{W}$. Consider the SAR model

$$\mathbf{V} = \sigma^2[(\mathbf{I} - \rho\mathbf{W})^{\mathsf{T}}(\mathbf{I} - \rho\mathbf{W})]^{-1} \tag{3.24}$$

From example 3.4c the conditional variance at site i is given by $\sigma^2/(1 + \rho\Sigma_j w_{ji}{}^2)$ so that in general it will be smaller for strongly connected, particularly interior, sites. In the case of unconditional variances the situation is different. In the subsequent discussion $\sigma^2 = 1$. Figure 3.19(a) shows an arrangement of sites in a region. The pattern is quite regular but there is a distinction between sites 1 to 4 on the periphery and the cluster of four interior sites 5 to 8.

Case a: Figure 3.19(b) shows a possible edge system and $\mathbf{W}$ is specified so that w_{ij} is one when two sites are joined by an edge and zero otherwise. All the sites have three neighbours. If (3.24) is evaluated for any ρ all the variances will be found to be equal and the covariances equal between sites separated by an equal number of steps along the edge structure. Moreover, since each site has three neighbours, scaling **W** (dividing each entry in **W** by its row sum) does not alter this.

Case b: Suppose now we weight the edges to reflect relative distances between sites. These distances are given in Figure 3.19(*c*) and now:

$$w_{ij} = \begin{cases} d_{ij}^{-1} & \text{if } i \text{ and } j \text{ are neighbours} \\ 0 & \text{if } i = j \text{ and otherwise} \end{cases}$$

The weighting system now reflects relative position, and evaluating (3.24) we find that

$$\mathrm{Var}(Y_1) = \ldots = \mathrm{Var}(Y_4) < \mathrm{Var}(Y_5) = \ldots = \mathrm{Var}(Y_8)$$

The covariances are also altered: for example $\mathrm{Cov}(Y_1,Y_2)$ is now less than $\mathrm{Cov}(Y_1,Y_4)$ and both are less than $\mathrm{Cov}(Y_1,Y_5)$. Scaling **W** reduces differences in the variances and in some cases alters the relative magnitudes of the covariances since scaling in this case effects a further transformation in the relative positioning of the eight sites dampening the effect of the peripheral siting of sites 1 to 4 relative to sites 5 to 8.

Case c: Next, suppose that to reflect the proximity of the four central sites two additional edges are introduced (Figure 3.19(*d*)). There is now a highly connected interior set of sites with four neighbours each and a more weakly connected set of peripheral sites with three neighbours each. Now define **W**

Figure 3.19. A system of sites and edges. (*a*) Arrangement of sites, (*b*) possible edge system for the sites, (*c*) distance weighted edge system, (*d*) edge system with more interior edges.

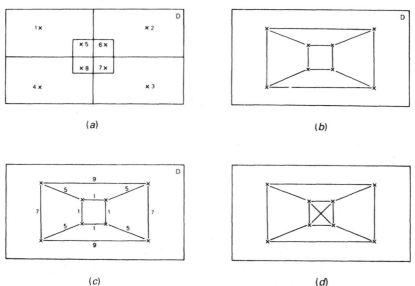

111

with zeros and ones as in Case a and evaluate (3.24). The variances for the four interior sites are equal and greater than the variances for the four peripheral sites (which are also identical). In addition the covariance structure is stronger between the interior sites than between the peripheral sites. Since the numbers of neighbours are unequal between the two groups scaling tends to reduce the differences in the variances and also the relative magnitudes of the covariances.

Case d: If the distance weighting transformation of Case b is applied to the edge structure of Case c the pattern of results shifts relative to c in the same way as Case b does to Case a. If scaling is introduced this will again dampen differences.

In general, strongly connected sites tend to have higher unconditional variances, an effect which is emphasised for the interior sites of highly connected clusters. Peripheral sites and weakly connected sites will tend to have lower variances. If the sites are areas as in Figure 3.19(a) then weakly connected sites will tend to be the larger areas (since distances to neighbours will be longer) and *vice versa.* If the variable is a density variable an assumption of lower variances in larger areas is not unreasonable. If the sites refer to point sites, again it might be reasonable to assume that isolated sites have lower variances than those subject to intense local interaction. However, it is important to realise that these properties are a consequence of the way in which edges and weights are specified.

The autonormal model requires that

$$\sigma_j^2 k_{ij} = \sigma_i^2 k_{ji}$$

(see section 3.2). If $k_{ij} = k_{ji}$ then $\sigma_j^2 = \sigma_i^2$; that is, the conditional variance of the process is the same for all sites. This will arise if k_{ij} is zero or one (depending on whether i and j are neighbours or not) or if k_{ij} is an inverse distance weighting. However if $k_{ij} \neq k_{ji}$ (which will occur if the weighting scheme is scaled, for example), then $\sigma_j^2 \neq \sigma_i^2$. Besag (1975) gives an example

$$k_{ij} = (l_{ij}/l_i)d_{ij}^{-\alpha}\phi$$

where ϕ and α are parameters, l_{ij} is the common boundary length between regions i and j, d_{ij} is distance and l_i is the perimeter of the border of region i. Then $\sigma_i^2 = \sigma^2/l_i$. He suggests that this consequence of the choice of weighting scheme (with conditional variance being required to decline with increasing area) may be desirable if the variable being analysed is a density variable. In other situations we might wish the conditional variance to be a function of the population in each area rather than the size of each area. Cressie and Chan (1989) give the example where

$$k_{ij} = (n_j/n_i)^{\frac{1}{2}} d_{ij}^{-\alpha} \phi$$

and j is a neighbour of i and n_j is the population of area j. It then follows that $\sigma_j^2 = \sigma^2/n_j$.

3.6   Conclusions: issues in representing spatial variation

The models of regionalised variable theory are used to represent variation on continuous stationary and non-stationary spatial surfaces and surfaces that are only stationary in their increments. When surfaces are stationary (possibly after the removal of a trend) then covariance models can be used. Data analysis proceeds by selecting from within the class of permissible models a functional form that accords best in terms of empirical and theoretical model selection criteria.

Variate interaction schemes are widely used in the analysis of data arising from discrete and continuous space processes. Although the origins of interaction models are in the description of spatial variation arising from agricultural uniformity trials (Whittle, 1954; Ripley, 1981), their use in the social and environmental sciences, has often been as part of a larger regression model rather than for describing spatial variation *per se*. Analysis proceeds by specifying a graph structure (edges and weights), boundary conditions (often specified implicitly rather than explicitly) and choosing an appropriate probability distribution. The correlation structure in the data as well as the visual properties of the data surface may be helpful in indicating an appropriate model (moving average or autoregressive, isotropic or anisotropic, first order or higher order). In special instances model selection may be guided by specifying a space-time process and identifying its spatial properties (Bartlett, 1975; Whittle, 1962).

Within the class of variate interaction schemes there is the option of choosing models from the joint or conditional specifications. Some of the theoretical differences between the two specifications are raised in Besag (1974) and in Whittle's contribution to the discussion. An issue that may be of practical significance is that the conditional scheme arises as the equilibrium form for a space–time autoregressive process. Ord (1981) notes that parameter estimation is often easier for CAR rather than SAR models and that boundary information can be more easily incorporated (see Chapter 4). Although the class of SAR schemes are contained within the class of CAR schemes the former may summarise observed variation with fewer parameters. Finally, the symmetry condition that limits the form of W or results in the imposition of conditions on the conditional variance may

Table 3.5. *Modelling strategies for different forms of spatial stationarity and non-stationarity*

Mean	Stationary covariances	Stationary increments
Stationary	Interaction models Covariance models Semi-variograms	Semi-variograms
Non-stationary (order of polynomial $k > 0$)	Remove mean and use interaction models, covariance models or semi-variograms	Use PGCFs or remove mean and use semi-variograms

Interaction models defined on finite lattices do not have stationary covariances nor constant variances across all the sites.

create difficulties for the analysis of irregularly spaced data using the CAR model.

Certain practical points from this chapter deserve emphasis. First, with continuous spatial surfaces the modelling strategy depends on the form of non-stationarity present in the data. Table 3.5 summarises approaches for different surfaces. Second, for partitioned surfaces and processes defined on point sites, the specifications of edge and weighting schemes are important because they affect model properties (including variances and covariances). Third, boundary effects are important. As a starting point the analyst should consider whether the study area is bounded (so that assuming fixed border values is reasonable) or part of a much larger area (in which case locally stationary models might be preferable). It may be particularly important in any statistical analysis to try to assess the influence of boundary assumptions on model fit by considering alternatives, and the same comment applies to the choice of edge and weighting schemes. Fourth, certain models do not appear to have spatially correlated forms that are of practical use (they can only describe negative or competitive relationships between neighbouring sites). In practice, data distributions that closely approximate these forms may be encountered so normalising transformations may be needed. Finally, in the description of some surfaces it may be necessary to accommodate spatial heterogeneity and various forms of spatial discontinuity. Here we shall only attempt to model such features through the mean. Interestingly, however, the models of section 3.3 can display apparent discontinuities including linear and other regular textures when certain orders of model and parameter values are used. Some

114

examples are given in Figure 3.16 but the interested reader should consult Cross and Jain (1983) for a more extensive treatment of possibilities. This is not an issue which will be considered further here although it is an issue of some importance in the reconstruction of distorted or contaminated remotely sensed images where linear or block features are present (Bernstein *et al.*, 1984; Besag, 1986).

In the next chapter we examine how these models are fitted to data: how parameters are estimated, how models are compared, how their fit to any given set of data can be assessed. It is necessary to approach the problem of model selection and statistical analysis with some care for empirical results can prove sensitive to even subtle changes in the way spatial variation is represented. Given that it is often the case that these models provide only very crude approximations to the sorts of dependency relations commonly encountered in spatial data, this is a worrying feature. It seems likely that the only safe program will be to consider many different models in order to gauge the sensitivity of results to different specifications. We shall return to this theme later in the book but it is important to be aware of the range of models available, which has been the purpose of this chapter.

NOTES

1.   There is an important exception of this. Time series modelling often needs to accommodate periodicities of varying length. Economic series, for example, often display seasonal periodicities and longer term cycles associated with fluctuations in the level of business activity and the introduction of new technologies. The case is not so strong for spatial data where it is trend rather than periodicity that is usually most important. There are exceptions of course. Some theories of spatial phenomena suggest the presence of spatial periodicity (Rayner, 1971; Rayner and Golledge, 1972, 1973; Tobler, 1969). Periodicity has appeared as a strong empirical feature in: aeromagnetic maps (Horton, Hempkins and Hoffman, 1964); vegetation maps arising from planting practices (Ford, 1976); variations in soil fertility and crop yields (Ripley, 1981, pp. 81–5; McBratney and Webster, 1981; Burrough *et al.*, 1985). It is perhaps more frequently encountered at relatively small, rather than large, spatial scales.

2.   Length can be defined in several ways but two are of particular importance. First, it may be defined as the straight line distance between sites (Figure F1(*a*)); second, as moves along an imposed edge system. The latter often gives rise to the terms covariance (correlation) at 'lag' one (one edge removed), 'lag' two (two edges removed) and so on. The effect of this is shown in Figure F1(*b*). The first definition appears to give a stronger ordering to the sites – for example, the eight lag two sites in Figure F1(*b*) are divided into two subsets in Figure F1 (*a*) (2 and 2$^{\frac{1}{2}}$). However, the lag two sites can also be divided by noting that the circled lag two sites of point x can be reached in two distinct ways, whereas the other lag two sites can

115

only be reached by one route so that the first group might be considered closer and expected to interact more strongly with x. The other lag orders of three and above can be similarly subdivided.

3.   A Dirichlet partition divides D such that to each point in D we associate a Dirichlet cell (also known as a Voronoi or Thiessen polygon) which is that part of D that is nearer to the point than to any other point on the surface (Ripley, 1981, pp. 39–40). The system of edges formed by joining sites that have a common boundary in the Dirichlet partition is called a Delaunay triangulation.

4.   The model derives from geometry: the volume of intersection of two spheres of diameter α the centres of which are r units apart.

5.   Martin (1987, p. 277) notes that 'although stationarity is often defined in terms of an infinite process, it may be perfectly reasonable in applications to assume the stationary form for V or V^{-1} for a finite realisation, even when an infinite realisation is physically impossible'. On the other hand where a region is physically bounded, areas near the boundary may have different properties from those near the centre.

6.   When restricted to a single row (or column) this model reduces to a first order (time series) autoregression. Other models when restricted may reduce to time series models. The Kiefer and Wynn (1981) nearest neighbour model reduces to a moving average scheme (Martin, 1984).

Appendix: simulating spatial models

It is often of interest to obtain realisations of the models of this chapter. These may be required to investigate the behaviour of inference procedures, particularly in small lattice situations where analytical results are not available. The fit of a model can be assessed by comparing properties of the data with properties derived from realisations of the model. A process model, such as a hydrological model for a river basin, may require spatial inputs (of precipitation for example) to drive it.

Suppose we wish to simulate a drawing (**y**) from a MVN(**µ**,V) distribution. Let

Figure F1. Lattice lengths (a) straight line, (b) edge moves.

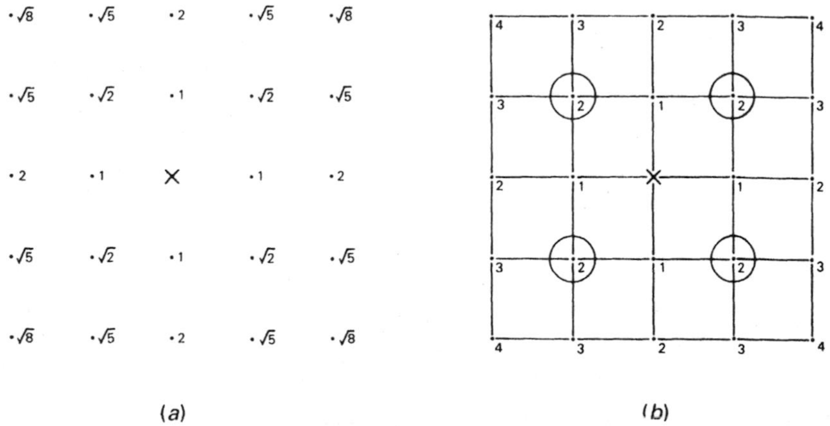

(a)   (b)

$e_1, \ldots, e_n$ be $IN(0,1)$ random variables then $\mathbf{y} = \mathbf{Le}$, where $\mathbf{L}$ is an $n \times n$ matrix, has $\mathbf{V} = \mathbf{LL}^{\mathsf{T}}$. So to generate the required realisation obtain

$$y_i = \mu_i + \Sigma_{j=1,\ldots,n} l_{ij} e_j \qquad (A1)$$

where μ_i is specified for each site (and may be constant) and a set of e_js can be obtained using an algorithm from the NAG or IMSL libraries for example. The first step would be to generate the e_js for all the n locations and then compute (A1). It is often recommended that the sequence of generated values $\{e_j\}$ be checked to ensure they have the required properties for a sample from the normal distribution. In addition the spatial distribution should probably be checked for no autocorrelation. (See Chapter 6 for possible test procedures for normality and autocorrelation.) Where many realisations are required, a certain number would be expected to have e_js that failed one or both of these checks. It may be as undesirable to analyse a set of realisations where all realisations that failed one or both checks have been discarded as to analyse a set of realisations where an excessively large number of realisations that failed one or both checks were retained. Properties of the simulated e_js should be monitored carefully and results for specific cases compared with whether or not these checks were positive or not.

L can be obtained numerically since L is the lower triangular matrix of the Cholesky decomposition of V. If V has a simple decomposition $\mathbf{Q\Lambda Q}^{\mathsf{T}}$ where the columns of the $n \times n$ matrix Q are the normalised eigenvectors of V and $\mathbf{\Lambda}$ is a diagonal $n \times n$ matrix of corresponding eigenvalues then:

$$\mathbf{L} = \mathbf{Q\Lambda}^{\frac{1}{2}}$$

where $\mathbf{\Lambda}^{\frac{1}{2}}$ denotes the square roots of the eigenvalues.

In some cases it may be possible to specify L directly. In the case of example 3.1

$$\mathbf{V} = \sigma^2[(\mathbf{I} - \mathbf{S}^{\mathsf{T}})(\mathbf{I} - \mathbf{S})]^{-1} \text{ and } \mathbf{L} = \sigma[\mathbf{I} - \mathbf{S}]^{-1}$$

and for example 3.1(a) $\mathbf{S} = \rho\mathbf{W}$. The SAR model is simulated using its moving average representation.

For model 3.2(a)

$$\mathbf{V} = \sigma^2[(\mathbf{I} + \nu\mathbf{W})(\mathbf{I} + \nu\mathbf{W})^{\mathsf{T}}] \text{ and } \mathbf{L} = \sigma(\mathbf{I} + \nu\mathbf{W})$$

There is further discussion in Haining, Griffith and Bennett (1983), Clifford *et al.* (1989) and Kindermann and Snell (1980).

Cross and Jain (1983) describe simulation methods for discrete valued interaction models such as the autologistic or autobinomial models. Ripley (1981, pp. 16–18) discusses other simulation and Monte Carlo methods for spatial processes.

4

Statistical analysis of spatial populations

This chapter deals with inference procedures for the models of Chapter 3: model selection and parameter estimation; estimating confidence intervals; tests of hypotheses for choosing between alternative models; evaluating goodness of fit.

Maximum likelihood methods will be emphasised but not to the exclusion of alternatives. Small samples are often encountered in spatial data analysis, boundary effects are severe and the bilateral nature of many spatial interaction models give rise to awkward numerical problems. Maximum likelihood estimation is often very sensitive to correct model specification. For these reasons inference *via* either exact or approximate maximum likelihood may in some cases seem more trouble than it is worth and other methods lacking some of the apparently desirable properties of maximum likelihood estimators but computationally easier may be worth exploring. Further, most of these desirable properties are asymptotic. In the case of spatial data letting n (lattice size) increase to infinity has an ambiguous interpretation: does it imply more points within a fixed area or increasing the size of the study area? The former changes the nature of inter-site relationships, the latter may be inappropriate where the study area is of fixed size with a specified border. Below we refer to 'large' and 'small' sample situations. However, there is no natural continuity between the two as in the case of a time series with an increasing number of observations. Model properties may change for any subset of sites in the spatial case when these sites are embedded in larger lattices of different size (see Chapter 3).

4.1 Model selection

Model selection is based on many criteria (see Chapter 1 and section 3.6). Here we examine several important statistics that are used to characterise spatial variation and to provide an initial model specification, although their importance is not limited only to this stage of analysis. We describe estimators of the spatial covariance function, the periodogram and the semi-variogram. Robust estimators are described in section 6.3.

If $\hat{C}(j,k)$ denotes the estimated spatial covariance at lags j and k in the two directions of a rectangular lattice of dimension $p \times q$ then (Ripley, 1981, p. 79):

$$\hat{C}(j,k) = \sum_{m_1}^{M_1} \sum_{m_2}^{M_2} [(y(s_1,s_2) - \bar{y}(s_1,s_2))$$

$$(y(s_1+j,s_2+k) - \bar{y}(s_1+j,s_2+k))]/pq \tag{4.1}$$

where $m_1 = \max(1, 1-j)$; $m_2 = \max(1, 1-k)$; $M_1 = \min(p, p-j)$; $M_2 = \min(q, q-k)$ and $\bar{y}(s_1,s_2)$ is the mean at site (s_1,s_2). This is sometimes called the Whittle estimator. Note that $C(j,k)$ is half symmetric in that:

$$\hat{C}(j,k) = \hat{C}(-j,-k) \text{ but } \hat{C}(j,k) \neq \hat{C}(-j,k)$$

An alternative estimator is:

$$C^*(j,k) = \hat{C}(j,k)(pq/(p-j)(q-k)) \tag{4.1a}$$

which allows for boundary effects. Dalhaus and Kunsch (1987) have proposed tapered covariance estimates, a procedure similar to smoothing spectral density estimates. For all three covariance estimators different estimators for $\bar{y}$ can be used based on different subsets of the data and different estimation procedures (see section 3.2 and Cox (1983)).

If the mean is known, providing j and k are small relative to p and q then $\hat{C}(j,k)$ is an approximately unbiased estimator of $C(j,k)$ the theoretical spatial covariance. However, the sampling variance depends on $C(j,k)$ and neighbouring values of $\{\hat{C}(j,k)\}$ are substantially correlated because each observation appears in several covariance estimates (Ripley, 1981, p. 80). $C^*(j,k)$ is also an unbiased estimator but does not always result in a positive definite function and has a large variance, particularly for larger lags. It was because of these deficiencies that Dalhaus and Kunsch proposed the tapered covariance estimator which avoids both disadvantages but involves much more computation.

Usually the mean is not known and must be estimated from the data. In this case both $\hat{C}(j,k)$ and $C^*(j,k)$ are biased but the bias in $C^*(j,k)$ is $(pq/(p-j)(q-k))$ times larger than for $\hat{C}(j,k)$. The estimator C^* may be more sensitive to a misspecified trend as well as being less efficient. (See Mardia and Marshall, 1984; Cressie, 1984.) The estimator for spatial correlation, $\hat{R}(j,k) = \hat{C}(j,k)/\hat{C}(0,0)$, with $R^*(j,k)$ defined analogously.

In the case of observations on an irregular lattice, spatial covariance and correlation estimators use pairs of sites separated by a specified number of intervening edges or sites, or by a specified distance band. $C^*(g)$ is obtained by considering all pairs of sites separated by a shortest distance path of g edges (or $(g-1)$ intervening sites) and dividing through by the number of such pairs. If $\mathbf{\Delta}$ describes the edge structure of the graph then an easy way to identify the sites separated by g edges is to obtain $\mathbf{\Delta}^g$ setting to zero all those entries where $\mathbf{\Delta}'(r<g)$ entries are non-zero in order to eliminate g order links that include some backtracking. In the case of an estimator based on distance bands then $C^*(g)$ is computed by taking all pairs of sites that lie within the gth distance band of one another and dividing through by the number of such pairs. Pairs can be further subdivided by direction (see Figure 4.1). These estimators are discussed in Agterberg (1970) and Cliff and Ord (1981, pp. 118–19).

For large regular lattices the periodogram provides a description and model specification statistic with better asymptotic sampling properties than covariance estimators. The periodogram is defined (Ripley, 1981, p. 79):

Figure 4.1. Estimating correlations with non-lattice data. ● = site in distance band g (North-East); × = site in distance band g (South-East).

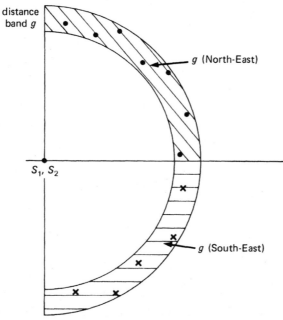

$$I(\lambda,\phi) = (pq/4\pi^2)\,|\,(1/pq)\Sigma_{s_1}\Sigma_{s_2}(y(s_1,s_2) - y(s_1,s_2))$$
$$\times \exp(-i\lambda s_1)\exp(-i\phi s_2)\,|^2$$

and is also called the empirical spectral function $\hat{f}(\lambda,\phi)$ and written

$$\hat{f}(\lambda,\phi) = \Sigma_{j=-m,\ldots,m}\Sigma_{k=-n,\ldots,n}\hat{C}(j,k)\exp(i\lambda j)\exp(i\phi k)$$

Alternatively, f^* is defined by using C^* instead of $\hat{C}$.

For almost all Fourier frequencies

$$E[I(\lambda,\phi)] \approx f(\lambda,\phi) \text{ and } \text{Var}[I(\lambda,\phi)] \approx (f(\lambda,\phi))^2$$

Asymptotic unbiasedness is retained whilst lowering the variance of the estimator by smoothing the periodogram using a window (as in the construction of the Dalhaus and Kunsch estimator for C). Computation of the smoothed estimate of the spectral density is done using the Fast Fourier Transform. Ripley (1981, pp. 80–1) discusses the procedure in more detail and describes the effect of using a bivariate normal density as the smoothing or window function for the periodogram.

Geostatistical analysis of spatial data is usually based on the estimation of $\gamma(j,k)$, the semi-variogram. In the case of a $p \times q$ lattice the estimated or experimental semi-variogram is:

$$\hat{\gamma}(j,k) = \Sigma_{s_1=1,\ldots,p-j}\Sigma_{s_2=1,\ldots,q-k}(y(s_1,s_2)$$
$$- y(s_1+j,s_2+k))^2/2(p-j)(q-k) \qquad (4.2)$$

The estimate is sometimes computed on the original data, but should be computed after the removal of any higher order trend in the data. As with the covariance function:

$$\hat{\gamma}(j,k) = \hat{\gamma}(-j,-k) \text{ but } \hat{\gamma}(j,k) \neq \hat{\gamma}(-j,k)$$

so estimates are computed for lags $j = -m,\ldots,m$ and $k = 0,\ldots,n$ and the rest can be computed by the half symmetry property of the function. Note that division is by the number of paired observations rather than pq. In the case of irregular distributions of observations the semi-variogram is usually estimated by pairing observations in terms of distance and direction (as was used to estimate the spatial covariance function). If $\mathbf{h}$ is a vector of length r and direction α, and if there are n_h pairs of data separated by $\mathbf{h}$ then the experimental semi-variogram in the direction α and for distance r is:

$$\hat{\gamma}(r,\alpha) = (1/2n_h)\Sigma_{i=1,\ldots,n_h}(y(\mathbf{u}_i+\mathbf{h}) - y(\mathbf{u}_i))^2 \qquad (4.3)$$

A series of distances and directions are chosen with the lag distance and direction specified by the band centre. For isotropic surfaces $\hat{\gamma}(r,\alpha)$ is computed only for different distance bands, irrespective of direction. There is further discussion in Journel and Huijbregts (1978, pp. 207–12).

The estimator for the semi-variogram given by (4.2) or (4.3) is optimal when the random variables are normally distributed. On the other hand, these estimators are sensitive to departures from normality and to the presence of outliers. Since the values used to compute $\hat{\gamma}$ are point samples from a continuous surface there are two components of sampling variation: 1. deviation between the experimental semi-variogram and the semi-variogram (called the 'local semi-variogram') that would be obtained if all pairs of points on the surface were included in the estimate and 2. deviation between the local semi-variogram and the theoretical semi-variogram from which the particular realisation was generated. The first component is called the variance of estimation and the second component is called the fluctuation variance. It is the first that is of practical importance in geostatistical applications. The estimation variance is a function of sample size and vanishes as this increases. An important advantage of the semi-variogram relative to the covariance function is that in the case of a constant mean surface, if the mean is not known, $\hat{\gamma}$ is still an unbiased estimator. However, when the mean is not constant and has to be estimated then $\hat{\gamma}$ is also biased (Cressie, 1984). There is further discussion of the properties of $\hat{\gamma}$ in Journel and Huijbregts (1978, pp. 193–5) and Dowd (1984).

When the processes are isotropic the functions of this section can be depicted as graphs. For non-isotropic surfaces isarithmic or contour maps can be used, alternatively block diagrams or stereograms can be drawn to represent the three-dimensional forms. Examples are given in Chapter 7.

The empirical spatial correlation function can be used to select an interaction scheme by comparing the set of observed values with the theoretical spatial correlation structures generated by different models.[1] This stage of analysis is based on judgement rather than any formal criteria. Several models may seem appropriate. In later sections we consider how the models can be compared. Since each interaction scheme generates (for large n) a spectral density function, the periodogram can also be used for model selection. However, the periodogram is of most use in identifying periodicities in the spatial surface and subsequent statistical analysis (such as parameter estimation) is often more easily carried out using estimated spatial covariances rather than periodogram estimates.

Covariance or semi-variogram estimates can also be used to specify a theoretical covariance or semi-variogram model. In both cases it is usually considered better to find a functional representation or summary of these estimates rather than working directly with the empirical values because of sampling variation which can be large. The number of pairs of observations used to obtain the estimates increases with increasing distance, eventually decreasing due to border effects. A consequence of this is that sampling variability tends to be largest for the smallest distances which unfortunately are the estimates of most interest. (Estimates for distances greater than half the dimension of the region or where $n_h < 30$ are usually considered too unreliable to have any useful interpretation.) Further, in the case of irregular site distributions some distance bands may contain no observations.

The selection of one or more functional forms to describe estimated covariances or the semi-variogram is often done by eye, plotting values on a graph and comparing the fit of different models. An interactive graphics terminal to display estimates and superimpose functional forms will be helpful. In three dimensions with non-isotropic forms it is probably easier to work with block diagrams rather than contour maps.

4.2 Statistical inference with interaction schemes

4.2.1 *Parameter estimation: maximum likelihood (ML) methods*

We assume that the n observations $(y_1, \ldots, y_n)$ have been drawn from a MVN(μ, V) distribution. In practice neither μ nor V will be known. However, methods are more easily developed if we first consider cases where either μ or V are known.

4.2.1 (a) μ *unknown; V known*

If $V = \sigma^2 I$ then the ML estimators for the constant mean $(\mu = \mu 1)$, trend surface $(\mu = A\theta)$ and regression $(\mu = X\beta)$ models are:

Constant mean: $\hat{\mu} = (1^T 1)^{-1} 1^T y$ where $\hat{\mu}$ is $N(\mu, \sigma^2 (1^T 1)^{-1})$ (4.4)

Trend surface: $\hat{\theta} = (A^T A)^{-1} A^T y$ where $\hat{\theta}$ is $N(\theta, \sigma^2 (A^T A)^{-1})$ (4.5)

Regression: $\hat{\beta} = (X^T X)^{-1} X^T y$ where $\hat{\beta}$ is $N(\beta, \sigma^2 (X^T X)^{-1})$ (4.6)

These are the best linear unbiased estimators of the unknown parameters, providing the conditions of the Gauss Markov theorem hold.

Now the log likelihood function for the n observations for general V is (assuming an MVN distribution):

$$-(n/2)\ln(2\pi) - \tfrac{1}{2}\ln|V| - \tfrac{1}{2}(y-\mu)^T V^{-1}(y-\mu)$$

and since V is known, the ML estimator of μ is obtained by minimizing with respect to the unknown parameters of μ:

$$(y-\mu)^T V^{-1}(y-\mu)$$

Now the estimators are:

Constant mean: $\tilde{\mu} = (1^{\mathsf{T}}V^{-1}1)^{-1}1^{\mathsf{T}}V^{-1}y$ where $\tilde{\mu}$ is $N(\mu,(1^{\mathsf{T}}V^{-1}1)^{-1})$ (4.7)

Trend surface: $\tilde{\theta} = (A^{\mathsf{T}}V^{-1}A)^{-1}A^{\mathsf{T}}V^{-1}y$ where $\tilde{\theta}$ is $N(\theta,(A^{\mathsf{T}}V^{-1}A)^{-1})$ (4.8)

Regression: $\tilde{\beta} = (X^{\mathsf{T}}V^{-1}X)^{-1}X^{\mathsf{T}}V^{-1}y$ where $\tilde{\beta}$ is $N(\beta,(X^{\mathsf{T}}V^{-1}X)^{-1})$ (4.9)

4.2.1 (b) μ known; V unknown

Since the mean is known we can set it to zero at each site. V is specified by an interaction model so we can take it that V depends on a small number of parameters denoted $z = (z_1, \ldots, z_r)$ where $z_1 = \sigma^2$.

Assuming a MVN(0,V) distribution the log likelihood function for the n observations is

$$-(n/2)\ln(2\pi) - \tfrac{1}{2}\ln|V| - \tfrac{1}{2}(y^{\mathsf{T}}V^{-1}y) \qquad (4.10)$$

where $|V|$ denotes the determinant of V. The exact ML estimate of z is obtained by maximising with respect to z

$$L = -\tfrac{1}{2}[\ln|V| + y^{\mathsf{T}}V^{-1}y] \qquad (4.11)$$

which consists of a determinant term and a sum of squares term. Differentiating we obtain:

$$\partial L/\partial z_i = -\tfrac{1}{2}[\mathrm{tr}(V^{-1}V_i) + y^{\mathsf{T}}V^i y] \qquad (i = 1, \ldots, r)$$

where $V_i = \partial V/\partial z_i$ and $V^i = \partial V^{-1}/\partial z_i = -V^{-1}V_i V^{-1}$ (Mardia and Marshall, 1984). In order to maximise (4.11) numerical methods are required. This raises the problem of repeatedly evaluating $|V|$ and V^{-1} which may present computational difficulties.[2]

We shall discuss exact and approximate ML estimation for V. Later we shall discuss other estimation methods including least squares and moments estimators. The references to examples refer to the examples of Chapter 3.

Simultaneous autoregressive schemes (Example 3.1)
For the SAR model with dispersion matrix given by

$$V = \sigma^2[(I-S)^{\mathsf{T}}(I-S)]^{-1},$$

the exact ML estimators are:

$$\tilde{\sigma}^2 = y^{\mathsf{T}}[(I-\tilde{S})^{\mathsf{T}}(I-\tilde{S})]y/n \qquad (4.12)$$

and $\tilde{S}$ is obtained by minimising with respect to the parameters of S

$$-2n^{-1}\ln|(I-S)| + \ln\tilde{\sigma}^2 \qquad (4.13)$$

The matrix inversion has been avoided but not the need to evaluate the determinant. Further simplification may be possible. Consider the first order symmetric SAR model with $S = \rho W$ (example 4.2.1). Let $\omega_1, \ldots, \omega_n$ denote the n eigenvalues of W which are all real if W is a symmetric matrix. (If W is asymmetric then some of the eigenvalues may be complex.) Now:

$$\ln|(I-S)| = \ln|(I-\rho W)| = \ln\Pi_{i=1,\ldots,n}(1-\rho\omega_i) \qquad (4.14)$$

(Ord, 1975). The eigenvalues usually have to be found by numerical methods but this only needs to be done once. However, if the set of n sites form a $p \times q$ lattice and if $W = \Delta$ (with border sites having only 2 or 3 neighbours) then

$$w_{rs} = 2[\cos(r\pi/(p+1)) + \cos(s\pi/(q+1))] \qquad r = 1, \ldots, p \qquad s = 1, \ldots, q$$

(Ord, 1975, appendix C). For p and q large:

$$-2n^{-1}\ln|(I-S)| \to \Sigma_{j=1,\ldots,\infty}(1/j)\binom{2j}{j}^2\rho^{2j} \qquad (4.15)$$

the infinite sum (on j) converging rapidly providing $|\rho|$ is not too close to the maximum value 0.25 (Whittle, 1954).

Where the number of sites is large (but not forming a lattice) Cliff and Ord (1981, p. 156) suggest forcing W into a block diagonal structure thus creating disconnected subsets of sites. They suggest that the final estimate should be quite good since only the largest eigenvalues have any effect on the value of the function to be minimised.

Expression (4.12) also simplifies. For general W matrices:

$$\tilde{\sigma}^2 = (y^T y - \tilde{\rho}y^T W^T y - \tilde{\rho}y^T Wy + \tilde{\rho}y^T W^T Wy)/n$$

and in the case of large rectangular lattices (neglecting border effects) the Whittle estimator is:

$$\tilde{\sigma}^2 = (1 + 4\tilde{\rho}^2)\hat{C}(0,0) - 4\tilde{\rho}[\hat{C}(1,0) + \hat{C}(0,1)] + 4\tilde{\rho}^2[\hat{C}(1,-1)$$
$$+ \hat{C}(1,1)] + 2\tilde{\rho}^2[\hat{C}(2,0) + \hat{C}(0,2)] \qquad (4.16)$$

where $\hat{C}(i,j)$ is the estimated spatial covariance at lag (i,j) (see section 4.1). Guyon (1982) has suggested using C^* and Dalhaus and Kunsch (1987) have suggested tapered estimates for reasons given in section 4.1.

Although asymmetric and higher order SAR models may be defined, estimating parameters by exact ML involves considerable computation and there appear to be few shortcuts. In the case of irregular lattices there appears to be no alternative to evaluating the matrix determinant. Even a second order model, where $S = \rho_1 W_1 + \rho_2 W_2$ (example 4.2.1(c)) involves substantial computation for even relatively small sample sizes. It is only in the case of unilateral SAR models that the determinant disappears (Ord,

1975).[3] Approximations analagous to (4.15) and (4.16) for asymmetric and higher order SAR models on large rectangular lattices are available, however, and the interested reader should consult Whittle (1954) and Larimore (1977). These methods can be implemented rapidly by fast Fourier methods.

Since parameters are often estimated by numerical methods starting values are needed. Ord (1975, Appendix A) recommends using several different starting values and notes that the least squares estimator may give values that lie outside the permissible range.

Moving average model (Example 3.2)
For the MA model with dispersion matrix given by

$$V = \sigma^2[(I+M)(I+M)^\top]$$

the exact ML estimators are

$$\tilde{\sigma}^2 = y^\top[(I+\tilde{M})(I+\tilde{M})^\top]^{-1}y/n$$

and $\tilde{M}$ is obtained by minimising with respect to the parameters of M:

$$2n^{-1}\ln|(I+M)| + \ln\tilde{\sigma}^2$$

This model raises both the problem of matrix inversion and evaluation of the determinant.

If we consider the first order MA model where $M = vW$ then the determinant term can be simplified along the same lines as for the SAR model replacing $+v$ for $-\rho$ in (4.14). Also:

$$y^\top[(I+vW)(I+vW)^\top]^{-1}y = x^\top(I-H)U^\top U(I-H)x$$

where $x = Qy$ and $Q^{-1} = U$ where U is the matrix of column eigenvectors of W, and H is a diagonal matrix with elements $h_i = 1 - (1 + v\omega_i)^{-1}$ where ω_i is the ith eigenvalue of W. If W is symmetric then

$$x^\top(I-H)U^\top U(I-H)x = \Sigma_{i=1,\ldots,n}[x_i(1+v\omega_i)^{-1}]^2$$

The estimation of this model is discussed in Cliff and Ord (1981) and in Haining (1978d). While higher order and asymmetric versions can be defined the computational problems of obtaining exact ML estimates become severe. The same approximations to the determinant term can be used as were used for the SAR model. The term $\tilde{\sigma}^2$ raises difficulties although it might be approximated by some small number of low order estimated covariances.

Conditional autoregressive schemes (Example (3.4))
For the CAR model with dispersion matrix given by:

$$V = \sigma^2 [I - K]^{-1}$$

the exact ML estimators are

$$\tilde{\sigma}^2 = y^\top [I - \tilde{K}] y / n \tag{4.17}$$

and $\tilde{K}$ is obtained by minimising with respect to the unknown parameters

$$-n^{-1} \ln |I - K| + \ln \tilde{\sigma}^2 \tag{4.18}$$

As with the SAR model the matrix inversion has been avoided but not the need to evaluate the determinant. When $K = \tau W$ (example 3.4a) then the determinant is evaluated using the simplification suggested for the first order SAR model and:

$$\begin{aligned}
\tilde{\sigma}^2 &= (y^\top y - \tilde{\tau} y^\top W y)/n \\
&= \hat{C}(0,0) - 2\tilde{\tau}[\hat{C}(1,0) + \hat{C}(0,1)]
\end{aligned}$$

for large lattices (neglecting border effects) where $\hat{C}(i,j)$ is the estimated spatial covariance at lag (i,j). Note that this model is fitted on $\hat{C}(0,0)$, $\hat{C}(1,0)$ and $\hat{C}(0,1)$ whereas the first order SAR model also fits on $\hat{C}(2,0)$, $\hat{C}(0,2)$, $\hat{C}(1,1)$ and $\hat{C}(-1,1)$ reflecting its higher order status (see section 3.2).

For asymmetric and higher order schemes approximations are available for CAR models on large rectangular matrices and the interested reader should consult Besag (1974), and Besag and Moran (1975). Estimation of errors in variables schemes are described in Besag (1977b). As noted for the SAR model the least squares estimator is not recommended as a starting value for numerical procedures since it may lie outside the permissible range.

4.2.1   (c) μ *and* V *unknown*

In most practical situations both μ and V have to be estimated. In this case we consider simultaneous ML estimation where the log likelihood function to be maximised is:

$$= -\tfrac{1}{2} [\ln |V| + (y - \mu)^\top V^{-1} (y - \mu)^\top] \tag{4.19}$$

An iterative cycle is usually recommended in which an initial estimate of μ is obtained, this is then used to obtain an initial estimate of V which is in turn used to update the estimate of μ and so on. Figure 4.2 describes the procedure in Ord (1975) which is based on the Cochrane–Orcutt procedure.

127

It involves maximising the conditional likelihood at each stage. We include an optional test for spatial autocorrelation (see Chapter 6) at the first round based on the least squares residuals. It is not necessary to start the cycle using the least squares estimate of μ but this appears to be widely used. At the first iteration the least squares residuals may be used to specify an appropriate model for V based on the data (see section 4.1).

This iterative procedure converges to a local minimum when used in time series analysis but no proof exists that this will occur for spatial analysis. Furthermore, several local minima may exist in the case of small data sets and for certain models (Ripley, 1988). Mardia and Marshall (1984) discuss numerical procedures for simultaneously estimating μ and V and suggest a

Figure 4.2. Iterative estimation of V and μ.

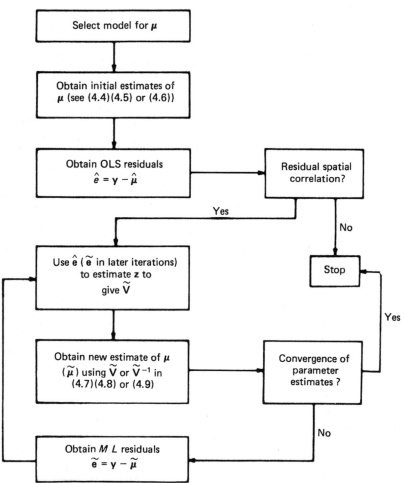

procedure that does not involve finding the conditional maximum of the likelihood at each step but requires only an increase in the likelihood at each step. They comment: 'the extra iterations in finding the conditional maximum likelihood do not seem to speed the overall rate of convergence' (p. 137). Given earlier comments on computational problems, the evaluation of alternative computational procedures deserves closer attention.[4]

In many situations ML estimation of μ and V will be difficult because of the problems of ML estimation of V. For this reason we shall look later at other estimators for the unknown parameters of V.

4.2.1 (d) Models with non-constant variance

In the case of the CAR model (example 3.4) let:

$$V = (I - K)^{-1}M$$

where $M = \sigma^2 \text{Diag}[d_1, \ldots, d_n] = \sigma^2 D$ where the $\{d_i\}$ are known and K must satisfy (3.11). The term $\sigma^2 d_i$ is the error variance associated with the ith observation. Note that M does not depend on the unknown parameters of K. If this specification is part of a regression model then:

$$\tilde{\beta} = (X^T D^{-1}(I - \tilde{K})X)^{-1}(X^T D^{-1}(I - \tilde{K})y)$$
$$\tilde{\sigma} = (y - X\tilde{\beta})^T D^{-1}(I - \tilde{K})(y - X\tilde{\beta})/n$$

and the spatial parameters in K are obtained by minimising

$$-n^{-1}\ln|I - K| + \ln\tilde{\sigma}^2$$

In the case of the SAR model (example 3.1) a model with non-constant error variance (in (3.3) $V_e = D$) is given by:

$$V^{-1} = \sigma^{-2}[(I - S)^T D^{-1}(I - S)]$$

and the ML estimators for the unknown parameters are obtained from (4.7), (4.8) or (4.9) and (4.12) and (4.13).

The usual CAR and SAR models (examples 3.4 and 3.1) can have non-constant variances (on Y) depending on the specification of W (see section 3.5). The models here, however, allow for more severe forms of non-constant variance that can be specified a priori, or empirically from model residuals, and then incorporated into the matrix D. (See Section 2.3.3(h) and for an example Cressie and Chan, 1989.)

4.2.2 Parameter estimation: other methods

4.2.2 (a) Ordinary least squares (OLS) and pseudo-likelihood estimators

We consider the case where μ is known. When μ is not known replace y with $(y - \mu)$ in what follows and the parameters are estimated iteratively (section 4.2.1(c)). OLS estimators for the unknown parameters of a general SAR model are obtained by minimising with respect to S the expression

$$\mathbf{e}^\top\mathbf{e} = \mathbf{y}^\top(\mathbf{I}-\mathbf{S})^\top(\mathbf{I}-\mathbf{S})\mathbf{y} = n\sigma^2$$

which is the error sum of squares. Comparing this with (4.12), OLS and ML estimators are seen to be identical only when $\mathbf{V} = \sigma^2\mathbf{I}$. This occurs when $\mathbf{S}$ is upper or lower triangular, that is if the autoregressive scheme is defined on a directional edge system (Ord, 1975).

OLS estimators in bilateral SAR models are inconsistent since $\mathbf{e}$ and $\mathbf{Y}$ are not independent as shown in example 3.1 and discussed in Whittle (1954) and Ord (1975). However, a modified least squares estimator developed by Ord (1975) may be useful for the estimation of parameters of high order models. If $\mathbf{S} = \Sigma_{j=1,\ldots,k}\,\rho_j\mathbf{W}^j$ (see example 3.1c) then the estimating equation is:

$$(\mathbf{y}-\rho_1\mathbf{W}_1\mathbf{y}-,\ldots.-\rho_k\mathbf{W}_k\mathbf{y})^\top\mathbf{W}_j^2\mathbf{y} = 0 \qquad j=1,\ldots,k$$

(Cliff and Ord, 1981, p. 160). For the first order symmetric case then this least squares estimator gives

$$\hat{\rho} = \mathbf{y}^\top\mathbf{W}^\top\mathbf{W}\mathbf{y}/\mathbf{y}^\top\mathbf{W}^\top\mathbf{W}^2\mathbf{y}$$

However, OLS estimators of the parameters of CAR models are consistent since the errors and $\mathbf{Y}$ are independent, that is $\mathrm{Cov}(\mathbf{e},\mathbf{Y}) = \sigma^2\mathbf{I}$ where $\mathbf{e} = (\mathbf{I}-\mathbf{K})(\mathbf{Y}-\mu)$. This 'innovations' property underlies the reason why OLS is feasible (Besag, 1975, p. 191). For the first order symmetric scheme ($\mathbf{K} = \tau\mathbf{W}$ and $\mathbf{W} = \mathbf{W}^\top$) the OLS estimator for τ ($\hat{\tau}$) is obtained by minimising (Besag, 1975, p. 191):

$$\mathbf{e}^\top\mathbf{e} = \mathbf{y}^\top(\mathbf{I}-\tau\mathbf{W})^2\mathbf{y}$$

and

$$\hat{\tau} = \mathbf{y}^\top\mathbf{W}^\top\mathbf{y}/\mathbf{y}^\top\mathbf{W}^2\mathbf{y}$$

which is also the (inconsistent) OLS estimator for ρ in the first order symmetric SAR model ($\mathbf{W} = \mathbf{W}^\top$). OLS estimates for the parameters of general CAR models are obtained by minimising with respect to the unknown parameters of **K**:

$$\mathbf{y}^T(\mathbf{I}-\mathbf{K})^2\mathbf{y} \qquad (4.20)$$

The least squares estimator for σ^2 is:

$$\hat{\sigma}^2 = n^{-1}\mathbf{y}^T(\mathbf{I}-\hat{\mathbf{K}})^2\mathbf{y}$$

and if a mean (μ) is to be estimated the least squares estimator is:

$$\hat{\mu} = [\mathbf{X}^T(\mathbf{I}-\hat{\mathbf{K}})^2\mathbf{X}]^{-1}[\mathbf{X}^T(\mathbf{I}-\hat{\mathbf{K}})^2\mathbf{y}]$$

substituting **A** for **X** in the case of a trend surface model.

If $\mathbf{K}=\Sigma_{j=1,\ldots,k}\ \tau_j\mathbf{W}_j$, the least squares estimates are obtained by solving the equations:

$$(\mathbf{y}-\tau_1\mathbf{W}_1\mathbf{y}-\ldots-\tau_k\mathbf{W}_k\mathbf{y})^T\mathbf{W}_j\mathbf{y}=0 \qquad j=1,\ldots,k$$

The OLS estimators for τ^2, μ and **K** correspond to pseudo-likelihood estimators in the case of normal variables (Besag, 1975). Pseudo-likelihood estimates are obtained by maximising with respect to the unknown parameters

$$\Sigma_{i=1,\ldots,n}\ \ln p_i(\sigma^2,\mu,\mathbf{K})$$

where $p_i(\sigma^2,\mu,\mathbf{K})$ is the conditional probability of observing y_i given the values at all the other sites. For normal variables:

$$p_i(\sigma^2,\mu,\mathbf{K})=(2\pi\sigma^2)^{-1}\exp[-(1/2\sigma^2)\{y_i-\mu_i-\Sigma_{j\epsilon N(i)}k_{ij}(y_j-\mu_j)\}^2]$$

For large n the method provides estimates which are consistent in the case of the CAR model (Besag, 1977a), however, there is a risk that estimates will lie outside the permissible range. There is further discussion of this estimation procedure in section 4.2.3 below since it may be particularly useful for estimating the parameters of discrete valued interaction schemes.

4.2.2 (b) Coding estimators

Coding estimation requires that the sites are partitioned (or coded) into two groups S_0 and S_1 where no two sites in S_0 are neighbours of one another in the sense described in Chapter 3. Parameter estimates for the CAR model using the observations on the sites S_0 can be obtained by maximising with respect to the unknown parameters the log of the conditional probability of observing each y_i in the coded set S_0 given the values in the set S_1; that is:

$$\Sigma_{i\epsilon s_0}\ln p_i(\sigma^2,\mu,\mathbf{K}) \qquad (4.21)$$

The similarity with the pseudo-likelihood expression will be evident. The only difference here is that summation is only over those sites in the set S_0. The nature of the coding and the number of possible codings depends on the order of the model being fit. Figure 4.3 shows coding sets for two lattices where neighbours have been defined differently to reflect the order of models being fit, and an irregular pattern of sites. For the first order model there are two possible codings on the lattice, for the second order model there are four possible codings, third and fourth order models generate nine codings. The method is based on the fact that the sites in S_0 are conditionally independent given the sites in S_1 so that (4.21) is the conditional log likelihood function. The parameters are estimated by regressing the

Figure 4.3. Coding schemes for different lattice and non-lattice situations.

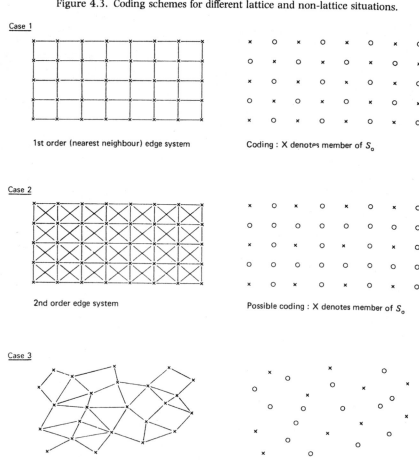

Case 1

1st order (nearest neighbour) edge system

Coding : X denotes member of S_0

Case 2

2nd order edge system

Possible coding : X denotes member of S_0

Case 3

Edge system for irregular lattice

Possible coding : X denotes member of S_0

observed values in S_0 on the observed values in S_1 and the need to evaluate any awkward determinant has been avoided. A second set of estimates can be obtained by regressing the sites S_1 on S_0 but then the problem is to reconcile the two sets of results. Evidence suggests that differences may not be large and that a simple averaging of parameter estimates over the codings may be adequate (Cross and Jain, 1983). The approach may be less easy to justify when there are several codings and (or) estimates differ substantially.

Besag and Moran (1975) and Besag (1977a) evaluated the efficiency of the coding technique for Gaussian conditional schemes. For first order lattice schemes the coding technique is reasonably efficient only when first order correlation is small and it was less efficient than the pseudo-likelihood procedure in the cases examined by Besag (1977a). The importance of this estimator is that it is unbiased and provides exact tests of significance using classical maximum likelihood theory to obtain the approximate conditional distribution of the estimators and to construct conditional likelihood ratio tests (Besag, 1975). However, like the least squares estimator, the coding estimator may give parameter estimates that lie outside the feasible region. Moreover, it may be wasteful of data, on irregular lattices using perhaps as little as 30% of the observations.

For irregular distributions of sites, coding may produce a large number of possible codings each with a small number of observations. For such systems the least squares or pseudo-likelihood estimator seems preferable, which in any case is likely to produce better point estimates (Besag, 1975). Nor does the coding estimator seem a practical proposition for SAR models. Even a first order model would require a third order coding to ensure conditional independence (see Chapter 3) producing nine possible codings.

4.2.2 (c) Moment estimators

In Chapter 3 the lag one theoretical spatial correlations for a first order moving average with parameter v were:

$$R(0,1) = R(1,0) = 2v/(1 + 4v^2)$$

The moments estimator for v (Haining, 1978d) is the solution to:

$$2v/(1 + 4v^2) = \tfrac{1}{2}(\hat{R}(0,1) + \hat{R}(1,0))$$

where $\hat{R}$ denotes the estimated spatial correlation. The estimator is inefficient and moreover the equation cannot be solved if

$$(\hat{R}(0,1) + \hat{R}(1,0)) > 1.0$$

4.2.3 Parameter estimation: discrete valued interaction models

For these models ML estimation is often ruled out because of the difficulty in writing down the joint probability distribution. Both the pseudo-likelihood and coding techniques offer simple alternatives, the former providing better point estimates, the coding technique allowing probability statements about the parameters. Besag (1974) describes coding estimation of the parameters of autologistic schemes on a lattice. For the first order isotropic scheme this requires the maximisation of the conditional log likelihood with respect to α and β:

$$\Sigma_{(s_1,s_2)\varepsilon S_0} \ln p_{(s_1,s_2)}(\alpha,\beta) = \Sigma_{(s_1,s_2)\varepsilon S_0}$$

$$\ln\{\exp[(\alpha + \beta x(s_1,s_2))y(s_1,s_2)]/(1 + \exp[\alpha + \beta x(s_1,s_2)])\}$$

where $p_{(s_1,s_2)}(\alpha,\beta)$ is the conditional probability and

$$x(s_1,s_2) = (y(s_1 - 1,s_2) + y(s_1 + 1,s_2) + y(s_1,s_2 - 1) + y(s_1,s_2 + 1))$$

Besag (1978a) gives an example of fitting the same models using pseudo-likelihood. Cross and Jain (1983) discuss the use of the coding estimator for fitting the autobinomial model.

4.2.4 Properties of ML estimators

4.2.4 (a) Large sample properties

The asymptotic variance covariance matrix of the exact ML estimators for $(z_1 = \sigma^2, z_2, \ldots, z_r, \mu)$ is (Mardia and Marshall, 1984):

$$
V(z,\mu) =
\begin{bmatrix}
t_{1,1} \cdots\cdots\cdots t_{1,r} & \\
t_{2,1} \cdots\cdots\cdots t_{2,r} & \\
\vdots \qquad\qquad \vdots & \mathbf{0} \\
t_{r,1} \cdots\cdots\cdots t_{r,r} & \\
\mathbf{0} & \mathbf{B}
\end{bmatrix}^{-1}
$$

where

$\mathbf{B} = \mathbf{A}^{\mathrm{T}} \mathbf{V}^{-1} \mathbf{A}$ if μ is a trend surface model

$\quad = \mathbf{X}^{\mathrm{T}} \mathbf{V}^{-1} \mathbf{X}$ if μ is a fixed effects regression model and

$$t_{ij} = \tfrac{1}{2}tr(\mathbf{VV}^i\mathbf{VV}^j); \quad t_{ij} = t_{ji}$$

Note that the asymptotic variance of the ML estimator of $\boldsymbol{\mu}$ can be examined independently of the other parameters.

This result generalises results given in Ord (1975) where for the case of a first order Gaussian SAR model with $\boldsymbol{\mu} = \mathbf{X}\boldsymbol{\beta}$ then:

$$\mathbf{V}(\sigma^2, \rho, \boldsymbol{\beta}) = \sigma^4 \begin{bmatrix} n/2 & \sigma^2 tr(\mathbf{PW}) & 0 \\ & \sigma^4[tr(\mathbf{PW})^{\mathsf{T}}(\mathbf{PW}) - \alpha] & 0 \\ & & \sigma^2 \mathbf{X}^{\mathsf{T}}(\mathbf{P}^{\mathsf{T}}\mathbf{P})^{-1}\mathbf{X} \end{bmatrix}^{-1}$$

where $\mathbf{P} = (\mathbf{I} - \rho\mathbf{W})^{-1}$; $\alpha = \partial^2 \ln|\mathbf{I} - \rho\mathbf{W}| / \partial\rho^2 = -\Sigma_{i=1,\ldots,n}\{\omega_i/(1 - \rho\omega_i)\}^2$, and $(\omega_1, \ldots, \omega_n)$ are the n eigenvalues of $\mathbf{W}$.

For the case of a first order symmetric CAR model with $\boldsymbol{\mu} = \mathbf{X}\boldsymbol{\beta}$:

$$\mathbf{V}(\sigma^2, \tau, \boldsymbol{\beta}) = \sigma^4 \begin{bmatrix} n/2 & \sigma^2 tr(\mathbf{L}^{\mathsf{T}}\mathbf{WL})/2 & 0 \\ & \sigma^4\Sigma_{i=1,\ldots,n}(\omega_i/(1 - \tau\omega_i))^2/2 & 0 \\ & & \sigma^2 \mathbf{X}^{\mathsf{T}}(\mathbf{L}^{\mathsf{T}}\mathbf{L})^{-1}\mathbf{X} \end{bmatrix}^{-1}$$

where $\mathbf{L}^{\mathsf{T}}\mathbf{L} = (\mathbf{I} - \tau\mathbf{W})^{-1}$.

Mardia and Marshall (1984) have given sufficient conditions in the case of Gaussian processes for the consistency and asymptotic normality of the exact ML estimators. The Whittle estimator using $\{\hat{C}(j,k)\}$ is not consistent but if $\{C^*(j,k)\}$ are used instead the estimator is consistent and asymptotically normal for second order stationary (zero mean) processes and in addition if the processes are Gaussian the estimators are asymptotically efficient (Guyon, 1982; Dalhaus and Kunsch, 1987).

4.2.4   (b) Small sample properties

Some simulation experiments have been reported on the behaviour of ML estimators in different small lattice situations. Haining (1978c) studied the behaviour of the exact ML estimator of ρ, the parameter of a symmetric first order SAR model with the mean assumed known. Figure 4.4 shows the behaviour for 25×1, 5×5 and 7×7 lattices and Table 4.1 shows the mean, variance and mean square error of the estimator for the last two cases. The estimator is generally biased downwards with a variance that decreases as ρ increases. Figure 4.5 contrasts the ML and OLS estimators for the 5×5 case. The OLS estimator produces some inadmissible estimates and is biased upwards. The effect of the determinant term which is symmetric about $\rho = 0,0$ and present in the ML estimator is to depress the ML estimate

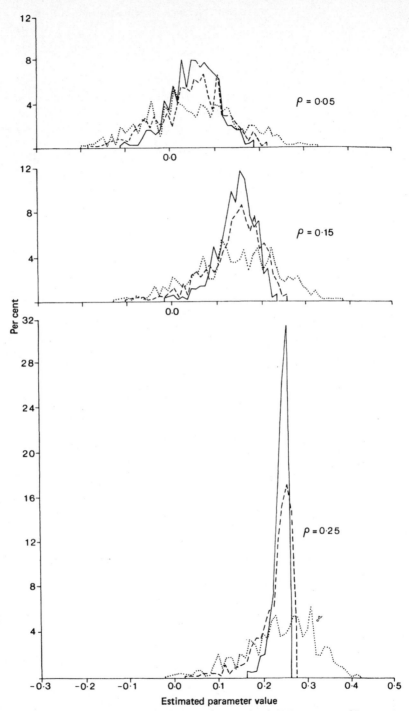

Figure 4.4. Simulated sampling distributions for the ML estimator of SAR models (single parameter) on small lattices and with different ρ values. $\cdots\cdots$ 25 × 1, --- 5 × 5, —— 7 × 7. After Haining (1978c).

136

Table 4.1. *Arithmetic mean, variance and mean square error of the exact ML estimator of the parameter of a first order SAR model. ($\mu=0.0$ and known; border values set to zero; max $\rho \simeq 0.25$; 500 simulations.)*

Parameter value (ρ)	5 × 5 lattice			7 × 7 lattice		
	$E(\hat{\rho})$	$Var(\hat{\rho})$	$MSE(\hat{\rho})$	$E(\hat{\rho})$	$Var(\hat{\rho})$	$MSE(\hat{\rho})$
0.00	0.004	0.001	0.001	0.005	0.003	0.003
0.05	0.048	0.005	0.005	0.051	0.003	0.003
0.10	0.093	0.004	0.004	0.098	0.002	0.002
0.15	0.136	0.003	0.003	0.145	0.002	0.002
0.20	0.182	0.002	0.003	0.192	0.001	0.001
0.25	0.230	0.002	0.002	0.242	0.000	0.000

Haining (1978c)

below the OLS estimate (when $\hat{\rho} > 0.0$). The shape of this function is shown in Figure 4.6. There is further evidence in Dow *et al.* (1982).

Simulation results for the symmetric first order CAR model indicate that for 9 × 9 lattices the ML estimator for τ is biased downwards and the variance of the estimator decreases as τ increases. Asymptotic results provide a poor estimate of sampling variances for τ when $n \leqslant 81$ and when τ is large. The ML estimator for the mean is biased downward and the asymptotic variance estimate is again poor when τ is large (Table 4.2). The downward bias of the ML estimator is also evident in the results given in Kunsch (1983) on the Whittle estimator for 5 × 5 and 10 × 10 lattice realisations of a stationary symmetric first order CAR model (Table 4.3). The use of the Whittle approximation to the determinant term for small lattices will have a strongly depressing effect, as evident from Figure 4.6.

Haining (1978d) reports results on the behaviour of the exact ML estimator for moving average schemes with the mean assumed known. Figure 4.7 shows the behaviour of the estimator for different values of the parameter θ. The ML estimator is biased upwards, the estimator variance declining as θ increases.

There are few reported studies on small sample distributional behaviour. Some results are given in Cliff and Ord (1975).

4.2.4   (c) A note on boundary effects

The earlier discussion assumed that unobserved boundary values are equal to the mean of the process. Expressions for $\mathbf{V}^{-1}$ are greatly complicated

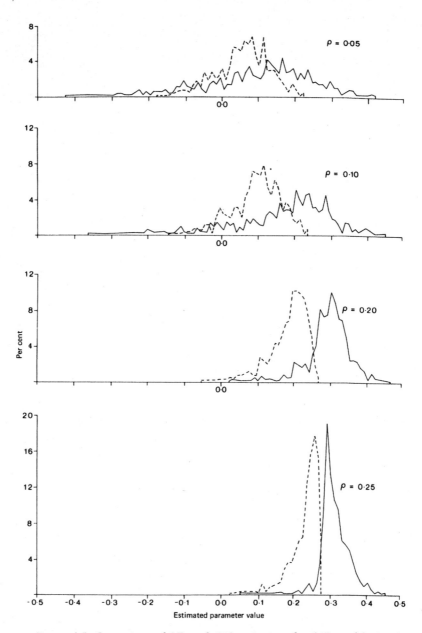

Figure 4.5. Comparison of ML and OLS estimators for SAR models (single parameter). --- ML, —— OLS. After Haining (1978c).

Table 4.2. *Arithmetic mean and standard deviation of the exact ML estimators of the parameter of a first order CAR model and the surface mean.* ($\mu = 0.0$; *border values set to zero; max* $\tau \simeq 0.25$; *300 simulations;* 9×9 *lattice.*)

Parameter value (τ)	$E(\bar\tau)$	$SD(\bar\tau)$		Parameter value (μ)	$E(\bar\mu)$	$SD(\bar\mu)$	
0.075	0.053	0.079	(0.078)	0.0	−0.0010	0.127	(0.129)
0.150	0.124	0.073	(0.064)	0.0	−0.0008	0.165	(0.162)
0.225	0.222	0.051	(0.034)	0.0	−0.0042	0.275	(0.248)

The figures in brackets are the asymptotic estimate of the standard deviation. Haining *et al.* (1988)

Figure 4.6. The large and small lattice weighting function for the SAR model ($\rho > 0$).

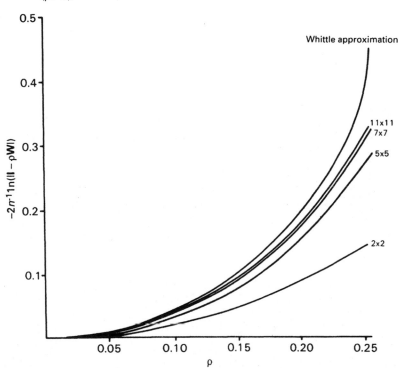

Table 4.3. *Arithmetic mean and standard deviation of the Whittle estimator of the parameter of a first order stationary CAR model. ($\mu=0.0$ and estimated; max $\tau \simeq 0.25$; 100 simulations; 5×5 and 10×10 lattices.)*

5 × 5 case			10 × 10 case		
Parameter value (τ)	E($\bar{\tau}$)	SD($\bar{\tau}$)	Parameter value (τ)	E($\bar{\tau}$)	SD($\bar{\tau}$)
0.20	0.110	0.106	0.20	0.174	0.043
0.24	0.149	0.102	0.24	0.217	0.030

Kunsch (1983)

when other boundary assumptions are made (see Chapter 3). For the CAR model for example, if the region is a finite lattice realisation of a stationary process then $\mathbf{V}^{-1}$ contains additional terms which should appear in the *exact* likelihood expression. In particular (see page 103)

$$\mathbf{V}^{-1} = \sigma^{-2}[\mathbf{I} - (\mathbf{K} + \mathbf{\Gamma})]$$

(Kunsch, 1983) and this alters both (4.17) and (4.18). Other boundary assumptions introduce other modifications to the likelihood equations (see for example Ripley, 1982).

There appear to be few comparative results in this area. Kunsch (1983) compared the previously defined Whittle estimator with the exact ML estimator on 5×5 and 10×10 lattice realisations of a stationary CAR process and concluded that the simpler (Whittle) estimator did not perform any worse than the exact ML estimator when interaction was weak and lost only a little in efficiency when interaction was strong. In addition the Whittle estimator has good asymptotic properties (see section 4.1). These issues are discussed further in Martin (1987).

Although incorporating boundary effects is awkward in the case of ML estimation, in the case of other estimation procedures for conditional schemes boundary assumptions can be introduced rather more easily. For example, suppose unobserved boundary values are assigned specific values. In the case of least squares, pseudo-likelihood and coding estimators these values can be treated as additional 'data' relating to the neighbourhoods of the observed boundary sites and included directly in the computations.

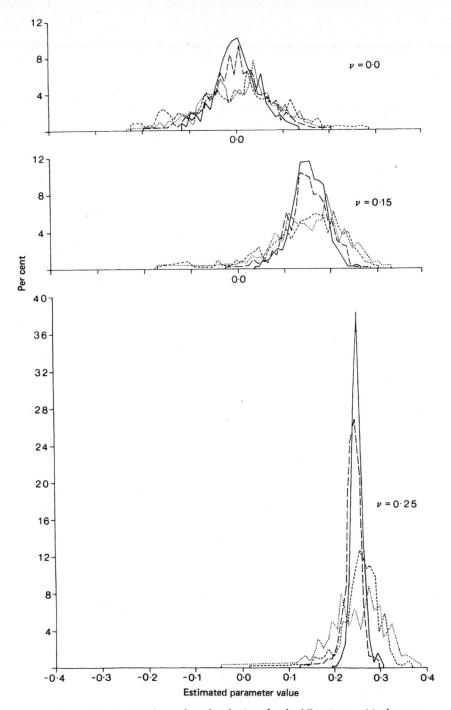

Figure 4.7. Simulated sampling distributions for the ML estimator (single parameter) on small lattice realisations of the MA model and with different v values. ——— 9×9, ––– 7×7, ‑‑‑‑ 5×5, ⋯⋯⋯ 49×1 lattice. After Haining (1978d).

4.2.5 Hypothesis testing for interaction schemes

4.2.5 (a) Likelihood ratio tests

In this section we examine the problem of comparing the relative fit of two or more models to a set of data using the likelihood ratio (LR) test. The question of how well any chosen model describes the data is left until section 4.4.

Let $p(\mathbf{y},\boldsymbol{\mu},\mathbf{V})$ denote the joint density of the n observations. For samples from a MVN distribution

$$p(\mathbf{y},\boldsymbol{\mu},\mathbf{V}) = (2\pi)^{-(n/2)} |\mathbf{V}|^{-\frac{1}{2}} \exp(-\tfrac{1}{2}(\mathbf{y}-\boldsymbol{\mu})^{\top}\mathbf{V}^{-1}(\mathbf{y}-\boldsymbol{\mu}))$$

Let H_0 and H_1 denote two hypotheses where H_0 is included in H_1, that is the two hypotheses are nested and H_0 is a special case of H_1. In this case if λ denotes the LR statistic:

$$\lambda = \max_{H_0} p(\mathbf{y},\boldsymbol{\mu},\mathbf{V}) / \max_{H_1} p(\mathbf{y},\boldsymbol{\mu},\mathbf{V})$$

Let $\tilde{\boldsymbol{\mu}}_0, \tilde{\mathbf{V}}_0$ denote the ML estimators of $\boldsymbol{\mu}$ and $\mathbf{V}$ under H_0 and let $\tilde{\boldsymbol{\mu}}_1$ and $\tilde{\mathbf{V}}_1$ denote the ML estimators under H_1, then the maximised log likelihood functions corresponding to the two densities are (ignoring constants)

$$\mathscr{L}_q(\tilde{\boldsymbol{\mu}}_0,\tilde{\mathbf{V}}_0) = -\tfrac{1}{2}\ln|\tilde{\mathbf{V}}_0| - \tfrac{1}{2}((\mathbf{y}-\tilde{\boldsymbol{\mu}}_0)^{\top}\tilde{\mathbf{V}}_0^{-1}(\mathbf{y}-\tilde{\boldsymbol{\mu}}_0)) = -\tfrac{1}{2}\ln|\tilde{\mathbf{V}}_0| - (n/2)$$
$$\mathscr{L}_{p+q}(\tilde{\boldsymbol{\mu}}_1,\tilde{\mathbf{V}}_1) = -\tfrac{1}{2}\ln|\tilde{\mathbf{V}}_1| - \tfrac{1}{2}((\mathbf{y}-\tilde{\boldsymbol{\mu}}_1)^{\top}\tilde{\mathbf{V}}_1^{-1}(\mathbf{y}-\tilde{\boldsymbol{\mu}}_1))$$
$$= -\tfrac{1}{2}\ln|\tilde{\mathbf{V}}_1| - (n/2)$$

where q and $p+q$ denote the number of parameters estimated under H_0 and H_1 respectively. Then

$$\ln(\tilde{\lambda}) = \mathscr{L}_q(\tilde{\boldsymbol{\mu}}_0,\tilde{\mathbf{V}}_0) - \mathscr{L}_{p+q}(\tilde{\boldsymbol{\mu}}_1,\tilde{\mathbf{V}}_1)$$

and H_0 is rejected if $\ln(\tilde{\lambda}) \leqslant K$ where K is a constant that depends on the significance level of the test.

This test can be used to choose between two models that might differ in terms of their order or degree of anisotropy.

Example 4.1 (Haining, 1977)
Let $\boldsymbol{\mu}=0$ under H_0 and H_1 and let $\mathbf{V} = \sigma^2[(\mathbf{I}-\rho\mathbf{W})^{\top}(\mathbf{I}-\rho\mathbf{W})]^{-1}$. Under $H_0:\rho=0$ and under $H_1:\rho\neq0$. This test establishes whether a single parameter first order SAR model provides a better description than a model of independence.

142

Under H_0, the estimate of σ^2 is $\mathbf{y}^\top\mathbf{y}/n$ and under H_1 the estimate is given by (4.12) with $\mathbf{S}=\rho\mathbf{W}$. Since under H_0, $\rho=0$ then using (4.12):

$$\tilde{\lambda}=|\tilde{\mathbf{V}}_1|^{\frac12}/|\tilde{\mathbf{V}}_0|^{\frac12}=((\mathbf{y}^\top(\mathbf{I}-\tilde{\rho}\mathbf{W})^\top(\mathbf{I}-\tilde{\rho}\mathbf{W})\mathbf{y})/\mathbf{y}^\top\mathbf{y})^{n/2}|\mathbf{I}-\tilde{\rho}\mathbf{W}|^{-1}$$

where $\tilde{\rho}$ is the ML estimate of ρ under H_1.

For the first order single parameter CAR model ($\boldsymbol{\mu}=0$)

$$\tilde{\lambda}=((\mathbf{y}^\top(\mathbf{I}-\tilde{\tau}\mathbf{W})\mathbf{y})/\mathbf{y}^\top\mathbf{y})^{n/2}|\mathbf{I}-\tilde{\tau}\mathbf{W}|^{-\frac12}$$

where $\tilde{\tau}$ is the ML estimate of τ under H_1.

The next step might be to decide whether an anisotropic form would fit significantly better. However, for hypothesis tests where more than two hypotheses are being compared, Akaike's information criterion (AIC) may be better suited than multiple pairwise LR tests. Such tests arise for example where model order is being decided and there is therefore a sequence of nested hypotheses $(H_0,\ldots,H_k)$ ranging from independence to some possibly high order interaction scheme. These tests have been discussed by Larimore (1977) for large lattice systems. To test amongst $k+1$ possible models, minimise

$$\text{AIC}(i)=-2\ln p(\mathbf{y},\tilde{\boldsymbol{\mu}}_i,\tilde{\mathbf{V}}_i)+2|\tilde{\boldsymbol{\mu}}_i,\tilde{\mathbf{V}}_i| \qquad i=0,1,\ldots,k$$

where $|\tilde{\boldsymbol{\mu}}_i,\tilde{\mathbf{V}}_i|$ denotes the number of parameters estimated (by ML methods) under hypothesis i.

If the null hypothesis for the LR test is a model of independence then $-2\ln\tilde{\lambda}$ is χ^2 distributed with the number of degrees of freedom given by the number of additional parameters estimated under H_1. This result also holds where two (nested) models of dependence are being compared (Whittle, 1954). Where the number of parameters is large relative to n the term $-2((n-p-q)/n)$ rather than -2 may give a better approximation in small samples. The main result also holds using $\hat{C}$ in the Whittle expression for the sums of squares term (see Guyon, 1982).

Example 4.2
Let $\boldsymbol{\mu}=0$ under H_0 and H_1 and let:

$$\mathbf{V}=\sigma^2[(\mathbf{I}-\rho_1\mathbf{W}_1-\rho_2\mathbf{W}_2)^\top(\mathbf{I}-\rho_1\mathbf{W}_1-\rho_2\mathbf{W}_2)]^{-1}$$

where under H_0: $\rho_1\neq0$, $\rho_2=0$ and under H_1: $\rho_1\neq0$, $\rho_2\neq0$. Let:

$$\rho_1\mathbf{W}_1=\mathbf{S}_0; \; \rho_1\mathbf{W}_1+\rho_2\mathbf{W}_2=\mathbf{S}_1;$$

$$\tilde{\sigma}_0^2=\mathbf{y}^\top(\mathbf{I}-\tilde{\mathbf{S}}_0)^\top(\mathbf{I}-\tilde{\mathbf{S}}_0)\mathbf{y}/n; \; \tilde{\sigma}_1^2=\mathbf{y}^\top(\mathbf{I}-\tilde{\mathbf{S}}_1)^\top(\mathbf{I}-\tilde{\mathbf{S}}_1)\mathbf{y}/n$$

143

Table 4.4. 5% *critical values for* $-2\ln\lambda$ *obtained by simulation methods*

(a) SAR model ($\mu=0$; σ^2 estimated under H_0; σ^2 & ρ estimated under H_1)

7×7	5×5	25×1
4.395	3.896	3.998

(b) MA model ($\mu=0$; σ^2 estimated under H_0; σ^2 & v estimated under H_1)

9×9	49×1	7×7	7×6	25×1	5×5	3×3
4.582	4.729	4.085	4.800	4.329	4.538	5.713

then

$$\tilde{\lambda}=[(|\mathbf{I}-\tilde{\mathbf{S}}_1|^{-(2/n)}\tilde{\sigma}_1{}^2)/(|\mathbf{I}-\tilde{\mathbf{S}}_0|^{-(2/n)}\tilde{\sigma}_0{}^2)]^{(n/2)}$$

and $-2\ln\tilde{\lambda}$ is χ^2 distributed with one degree of freedom under the null hypothesis for large n. This is a test to establish whether a second order symmetric SAR model provides a significantly better fit than a first order version.

There is some evidence on the validity of the approximation for small lattice and non-lattice situations under different models. Haining (1977), using simulation methods, contrasts a first order symmetric SAR model (H_1) against a model of independence ($\rho=0$). Haining (1978d) compares a moving average scheme (H_1) against a model of independence ($v=0$). Evaluating $-2\ln\tilde{\lambda}$ for n large in both cases gives a 5% critical value of 3.841. A summary of the simulation results is given in Table 4.4. Of course these simulation results are very sensitive to the lumpiness of the simulated distributions at the tail but the results suggest that the χ^2 approximation provides a reasonable guide for lattice situations. Results in Brandsma and Ketellapper (1979b), however, suggest that the χ^2 approximation is not satisfactory for small non-lattice situations.

In the case of non-nested hypotheses these distributional results may no longer hold. For two non-nested hypotheses, the LR test under H_1 converges to a non-central χ^2 distribution with the same degrees of freedom as previously (Larimore, 1977). For comparing many non-nested hypotheses, Larimore suggests that whilst the AIC statistic may be computed, a different correction to $2|\mathbf{\mu}_i,\mathbf{V}_i|$ would probably give better results, but does not say what correction is required. Where all the models have the same number of parameters it seems sensible to just select the model that maximises the likelihood function. Haining (1979, p. 53) suggests how the LR statistic might be used to test the fit of a moving average scheme against the null hypothesis of a first order symmetric SAR scheme.

Hypothesis testing using the coding procedure on normal variables can be based on classical maximum likelihood theory providing the sample size is large enough to justify the asymptotic theory (Besag, 1975). Coding is of most value in fitting models arising from the conditional specification on large lattices. When models of different order are compared, the maximised likelihood functions must be comparable – i.e. when comparing first and second order models, the first order model must be fit on the set of second order codings. Further, the results from the different codings must be reconciled. It is quite possible, for example, for different codings to give conflicting results. The tests are not independent. Cross and Jain (1983) suggest a conservative strategy, taking the overall significance level of the test to be $r \times p$ where r is the number of codings and p is the level at which the most significant result was obtained. With lower order schemes there are fewer codings and hence fewer opportunities for rejection.

For a likelihood ratio test on a suitable coding set S_0 and two nested hypotheses H_0 and H_1:

$$\ln(\hat{\lambda}) = \Sigma_{i \varepsilon S_0}[\ln p_i(\mathbf{y}, \hat{\boldsymbol{\mu}}_0, \overset{+}{\mathbf{V}}_0) - \ln p_i(\mathbf{y}, \hat{\boldsymbol{\mu}}_1, \overset{+}{\mathbf{V}}_1)]$$

where p_i is the conditional probability and where $+$ represents the coding estimates of the parameters under the two hypotheses. From classical theory $-2\ln(\hat{\lambda})$ converges to a χ^2 distribution with the number of degrees of freedom equal to the number of parameters constrained under H_0.

4.2.5 (b) Lagrange multiplier tests

An objection to the LR test is that it requires the model under H_1 to be estimated. It would be better if this could be avoided, particularly if H_0 was retained. (Where H_0 is a model of independence this has considerable computational advantages.) One approach suggested by Cliff and Ord (1981, p. 134) is based on estimating partial autocorrelations. More recently there has been interest in applying Lagrange Multiplier (LM) tests (Burridge, 1981; Anselin, 1988a) for nested spatial hypothesis testing, treating estimation under the null hypothesis as a constrained maximisation problem.

The constrained log likelihood function is formed:

$$\mathscr{L}^*(\boldsymbol{\varphi}) = \mathscr{L}(\boldsymbol{\varphi}) - \boldsymbol{\lambda}^\mathsf{T} H(\boldsymbol{\varphi})$$

where $H(\boldsymbol{\varphi})$ specifies the p independent restrictions on the parameter vector $\boldsymbol{\varphi}$ under H_0 relative to H_1, and $\mathscr{L}(\boldsymbol{\varphi})$ is the log likelihood under H_0 and

$\boldsymbol{\varphi} = (\phi_1, \ldots, \phi_q)$. $\boldsymbol{\lambda}$ is the vector of Lagrange multipliers. Then

$$\text{LM} = \tilde{\boldsymbol{\lambda}}^\mathsf{T} H'(\tilde{\boldsymbol{\varphi}})^\mathsf{T} I^{-1}(\tilde{\boldsymbol{\varphi}}) H'(\tilde{\boldsymbol{\varphi}}) \tilde{\boldsymbol{\lambda}}$$

where $\tilde{\boldsymbol{\lambda}}$ and $\tilde{\boldsymbol{\varphi}}$ are the solutions of the $p+q$ equations:

$$\partial L(\boldsymbol{\varphi})/\partial\boldsymbol{\varphi} - H'(\boldsymbol{\varphi})\boldsymbol{\lambda} = 0$$
$$H(\boldsymbol{\varphi}) = 0$$

where $H'(\boldsymbol{\varphi})$ is the matrix of derivatives of H with respect to $\boldsymbol{\varphi}$ and $I(\boldsymbol{\varphi})$ is Fisher's information matrix. Under H_0 LM is χ^2 distributed with p degrees of freedom and is asymptotically equivalent to the likelihood ratio test. We now consider an important special case.

Let H_0 denote a model of independence (mean not necessarily known) and H_1 a model of spatial dependence. Define

$$\text{GMC} = \hat{\mathbf{e}}^\mathsf{T} \mathbf{W} \hat{\mathbf{e}} / \hat{\mathbf{e}}^\mathsf{T} \hat{\mathbf{e}} \qquad (4.22)$$

where $\hat{\mathbf{e}}$ is the vector of ordinary least squares residuals obtained from estimating the mean of the surface under H_0. The matrix W is scaled and defines the connectivity (edge) structure. The statistic (4.22) is called the generalised Moran coefficient and is analogous to the Moran (I) statistic to be described in Chapter 6. In situations where H_1 denotes an SAR or MA model of spatial dependence then the GMC test is the LM statistic (Burridge, 1980). The GMC statistic also corresponds to the LR test for fixed τ in the case of a symmetric first order CAR model (see for example Cliff and Ord, 1981, p. 166; Besag and Moran, 1975) and to the LR test for fixed ρ in the case of a symmetric first order SAR model if terms in ρ^2 are ignored (Cliff and Ord, 1981). Haining (1977) reports simulation results on the power curves of the Moran and LR tests for the SAR model ($\rho > 0$) for different sized small lattices. Results are summarised in Table 4.5. As expected, although the LR test is more powerful the differences are small. In the case of a moving average model, however, where the Moran test is not sufficient for θ, the LR statistic is rather more powerful, although again for moderately large samples the differences are small. Power curves are shown in Figure 4.8 (Haining, 1978d).

A test procedure using the GMC test is constructed as follows. If $\hat{\mathbf{e}}$ is the vector of least squares residuals from a regression model then (Brandsma and Ketellapper, 1979b):

$$E[\text{GMC}] = \text{trace}[\mathbf{W} - \mathbf{X}(\mathbf{X}^\mathsf{T}\mathbf{X})^{-1}\mathbf{X}^\mathsf{T}\mathbf{W}]/(n-k)$$

where $(k-1)$ is the number of explanatory variables and X is the $n \times k$ matrix of observations on these variables (the first column consisting of 1s).

146

Table 4.5. *Percent power for likelihood ratio (LR) and Moran tests for a first order SAR model*

	\multicolumn{6}{c}{Parameter value (ρ)}					
	0.0	0.05	0.10	0.15	0.20	0.25
5 × 5						
LR	5.0	10.8	25.4	54.6	79.8	95.0
Moran	5.0	9.0	21.8	43.6	69.2	88.4
25 × 1						
LR	5.0	9.2	15.8	31.2	52.0	73.6
Moran	5.0	9.0	14.6	26.2	43.2	62.6
7 × 7						
LR	5.0	13.6	43.2	82.0	97.2	100.0
Moran	5.0	10.4	35.8	72.4	95.2	100.0

Haining (1977)

$$E[\text{GMC}^2] = \{\text{trace}[\mathbf{MWMW}^\top] + \text{trace}[(\mathbf{MW})^2] + [\text{trace}(\mathbf{MW})]^2\}/(n-k)(n-k+2)$$

where $\mathbf{M} = (\mathbf{I} - \mathbf{X}(\mathbf{X}^\top\mathbf{X})^{-1}\mathbf{X}^\top)$. The evidence from simulation experiments in Cliff and Ord (1981) indicates that unless n is very small ($n \leqslant 10$) or $\mathbf{W}$ is very sparse the normal approximation can be used for the test. Note that the test assumes constant variance of the residuals and that if tests are carried out for a succession of distances or lags they are not independent.

When H_0 is a model of dependence the computational advantages of the LM test relative to the likelihood ratio test may disappear. Anselin (1988a) has examined the use of LM test procedures for the residuals of models such as (2.4) and (2.5) where the GMC test is no longer valid.

For binary data the 'black/white' (BW) join-count statistic provides a simple test for independence but is less powerful than the Moran test in the case of interval data (Cliff and Ord, 1981, Chapter 6). The inference problems associated with this and other 'spatial autocorrelation' tests are discussed in Chapter 6.

4.3   Statistical inference with covariance functions and intrinsic random functions

This section examines the problems of statistical analysis when covariance functions, semi-variograms or polynomial generalised covariance functions (PGCFs) are used to model variation in spatial data. As in section 4.2 we

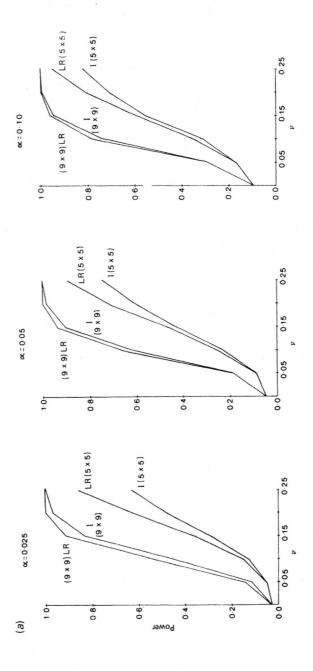

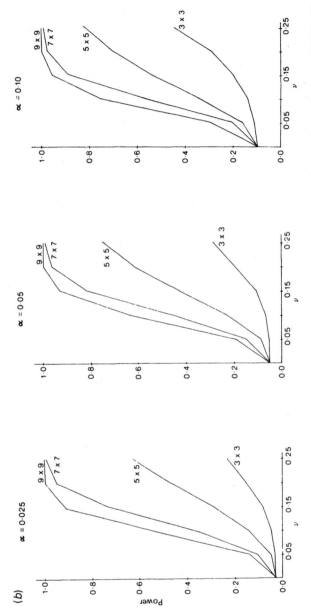

Figure 4.8. Power curves for (a) the I and LR statistics for the single parameter MA model for 5 × 5 and 9 × 9 lattices and (b) the LR statistic for the single parameter MA model for different sized lattices. After Haining (1978d).

distinguish between ML and other methods. Later in this section we describe properties of the procedures and make some comments on hypothesis testing.

4.3.1 Parameter estimation: maximum likelihood (ML) methods

We assume that the vector of n observations $y^T = (y_1, \ldots, y_n)$ are from a multivariate normal distribution with mean μ. Usually these n observations are point samples from a continuous surface or they may be areal aggregates again from a continuous surface.

If the mean is known (and set to zero) and y is MVN(0,V), where V is a positive definite stationary covariance function, then the log-likelihood function for the n observations is given by (4.10) and exact ML estimates of the unknown parameters of V are obtained by maximising (4.11) with respect to the unknown parameters. If $V = \{\sigma(s_i, s_j; \theta)\}$ where s_i and s_j are sites in the region and θ denotes the set of unknown parameters then in the case of the spherical model (page 91):

$$\sigma(s_i, s_j; \theta) = \sigma(r; \sigma_1{}^2, \alpha) = \sigma_1{}^2 C(r; \alpha)$$

where

$$C(r; \alpha) = 1 - (3|r|/2\alpha) + (|r|^3/2\alpha^3) \qquad |r| \leqslant \alpha$$
$$= 0 \qquad |r| > \alpha$$

For the Whittle model:

$$\sigma(s_i, s_j; \theta) = \sigma(r; \sigma_1{}^2, \eta) = \sigma_1{}^2 C(r; \eta)$$

where

$$C(r; \eta) = |r| \eta K_1(|r| \eta) \qquad \eta > 0$$

and K_1 is a modified Bessel function of the second kind.

Mardia and Marshall (1984) discuss numerical methods for estimating the parameters of spatial covariance functions and warn that convergence can be very slow. There is the further difficulty of estimating the mean of the surface which ought to be tackled as part of an overall iterative cycle as described in section 4.2.

Kitanidis (1983) describes ML estimation for the unknown parameters $(\psi_1, \ldots, \psi_k)$ of PGCFs of IRF-k models which are linear in the parameters. Estimation requires correct specification of the order (k) of the IRF and also prior estimates of the $\{\lambda_i\}$ – see section 3.1. These $\{\lambda_i\}$ are the coefficients of a linear transformation that filter out the kth order polynomial (the mean of the surface). If T denotes the required transformation matrix so that $z = Ty$ is a vector of stationary increments then the usual recommendation is to represent T by a trend surface and estimate trend surface parameters by (ordinary) least squares. The PGCF is fit to the residuals from this estimated surface. The linear form does lead to some simplification of the ML procedure particularly if there is only one parameter to estimate in the PGCF (Kitanidis, 1983).

In the light of comments in section 4.2 (see also section 4.5) the use of ordinary least squares to filter out the polynomial trend might seem unsatisfactory, the iterative procedure, outlined above, being preferable. Kitanidis remarks that because of the linearity of the PGCF the numerical values of the estimates of $\psi_1, \ldots, \psi_k$ are not affected by the addition to y of any arbitrary polynomial trend up to the same order as that filtered out from the data (the invariance property) so in this case ML estimation is robust with respect to trend surface parameter estimation. (An improved estimate of the trend can be obtained, if needed, using weighted least squares with the estimate of the PGCF but it is not then necessary to re-estimate $(\psi_1, \ldots, \psi_k)$.) However, the procedure is not robust to order misspecification and some anxiety must remain since least squares trend surface modelling in this situation does tend to lead to the fitting of a surface of too high an order (Ripley, 1981, p. 34).

4.3.2 *Parameter estimation: other methods*

One simple approach is to fit by eye, varying model parameters until a good visual fit is observed. Despite the obvious subjectivity of such an approach it does allow the analyst to introduce non-statistical information into the fitting procedure or to preserve values of the function at certain important distances or where estimates are known to be more reliable. Kitanidis (1983) discusses some rules of thumb, Ripley (1981, p. 57) gives an example and comments on some of the problems.

A second approach is to fit by ordinary least squares using the estimates of the function at different distances. In the case of a semi-variogram model the approach is:

$$\min_{\psi} \Sigma_i (2\hat{\gamma}(h_i) - 2\gamma(h_i; \psi))^2$$

where $\hat{\gamma}(h_i)$ is the estimated semi-variogram at distance h_i and $\gamma(h_i; \psi)$ is the model. The method is easy to apply when the models are linear or can be linearised by some simple transformation, but the assumptions of least squares regression are violated since neighbouring values of $\hat{\gamma}(h_i)$ are correlated. A solution to this difficulty is to estimate by generalised least squares, that is:

$$\min_{\psi}(2\hat{\gamma}(h_i)-2\gamma(h_i;\psi))^\top\Sigma^{-1}(2\hat{\gamma}(h_i)-2\gamma(h_i;\psi))$$

where Σ is the covariance matrix of $\{2\hat{\gamma}(h_i)\}$. The form of Σ, which contains diagonal and off-diagonal terms, makes implementation of this method difficult (Cressie, 1985). A more manageable approach is provided by weighted least squares that requires:

$$\min_{\psi}\Sigma_i w_i(2\hat{\gamma}(h_i)-2\gamma(h_i;\psi))^2$$

where $\{w_i\}$ are a set of weights. If the $\{w_i\}$ are interpreted as the diagonal elements of Σ, then (Cressie, 1985):

$$w_i = \mathrm{var}(2\hat{\gamma}(h_i)) \approx 2[2\gamma(h_i;\psi)]^2/n_i$$

where n_i is the number of observations used to estimate $\hat{\gamma}(h_i)$. The parameter estimates are obtained by minimising

$$\Sigma_i n_i\{[\hat{\gamma}(h_i)/\gamma(h_i;\psi)]-1\}^2$$

Note that n_i will give a greater weighting to semi-variogram estimates at small distances. Other weighting schemes have also been proposed and there is further discussion in McBratney and Webster (1986).

An early method used to fit PGCFs is weighted least squares using authorised combinations of the original data (Delfiner, 1976). The coefficients are estimated by minimising with respect to the parameters (ψ):

$$\Sigma_i w_i^2[z_i^2 - E(z_i^2 \mid \psi)]^2$$

where z_i^2 is a squared generalised increment and $E(z_i^2 \mid \psi)$ is the expected value. For the ith generalised increment

$$z_i = \Sigma_j \lambda_{ij} y_j \text{ and } z_i^2 = \Sigma_j \Sigma_k \lambda_{ij}\lambda_{ik} y_j y_k$$

where $\{\lambda_{ij}\}$ are the coefficients of the authorised combination and $(y_1, \ldots, y_n)$ are the original measurements. Now if the Y process is IRF-k with nugget variance then:

$$K(r;\psi) = \Sigma_{d=0,\ldots,k} \psi_d \mid r \mid^{2d+1} + c_0\delta(\mid r \mid)$$

(see section 3.2) and

$$E[z_i^2 \mid \psi] = \Sigma_j \Sigma_k \lambda_{ij}\lambda_{ik} K(r;\psi)$$
$$= \Sigma_{d=0,\ldots,k} \psi_d \Sigma_j \Sigma_k \lambda_{ij}\lambda_{ik} \mid r_{jk}\mid^{2d+1} + c_0\Sigma_j \lambda_{ij}^2$$

Statistical inference with covariance functions

where r_{jk} is the distance between observation site j and k. Given the linearity of the PGCF the estimation problem is equivalent to estimating the parameters of a linear regression model. That is, the coefficients $\{\psi_d\}$ and c_0 are estimated by regressing z_i^2 on the independent variables $\Sigma\lambda_{ij}^2$ and $T_{d,i}$ where

$$T_{di} = \Sigma\Sigma\lambda_{ij}\lambda_{ik}\,|\,r_{jk}\,|^{2d+1}$$

Delfiner further recommends weighting with $w_i^2 = T_{0,i}^{-2}$ or $T_{1,i}^{-2}$. Since Var(z_i^2) is proportional to $T_{0,i}^2$ or $T_{1,i}^2$ this weighting gives an improvement over ordinary (unweighted) least squares.

The method is applicable to estimating the parameters of decomposable covariance functions (Streitberg, 1979) and an early paper by Agterberg (1970) suggested fitting covariance functions of the form

$$C(r) = a + br + cr^2$$

where a, b and c are unknown constants by (weighted) least squares. However, for non-linear models the method is frequently unworkable although sometimes a transformation may help matters. There is discussion of these issues in Armstrong (1984). There are a number of problems with least squares methods. Since squared differences are used the assumption of independent errors is unreasonable, also it can lead to inadmissable models (with parameter estimates outside the permissible range).

Davis and David (1978), Hughes and Lettenmaier (1981) and Wahba and Wendelberger (1980) use jacknifing to estimate the parameters of the covariance function. The aim is to minimise the sum of squared prediction errors and the procedure deletes individual observations in turn and uses the remaining observations to predict the deleted value. The parameter estimates are those values that minimise the prediction errors. Ripley (1981, p. 57) comments on the use of this approach. One or two extreme data values that are difficult to predict may dominate the results. The robustness of the procedure might be improved by using a different numerical summary (such as mean absolute deviation) but in any case the approach is likely to be computationally very expensive for all but the smallest sets of regularly spaced data.

When the covariance model is linear in the parameters (e.g. PGCF models or the decomposable covariance structures of Streitberg, 1979) the parameters can be estimated by minimum norm unbiased quadratic estimation (MINQUE) or some variant (Rao, 1971, 1972). Suppose we define the usual regression model where $Y = X\beta + e$ and specify

$$E(ee^T) = V(\psi) = \Sigma_{i=1,\ldots,k}\,\psi_i V_i$$

153

where V_i is a known $n \times n$ matrix and $\boldsymbol{\psi}^\top = (\psi_i, \ldots, \psi_k)$ is a vector of unknown coefficients. Rao considers the estimation of a linear function of $\{\psi_i\}$ by the quadratic form $\mathbf{y}^\top \mathbf{A} \mathbf{y}$ where $\mathbf{A}$ is chosen to satisfy invariance, unbiasedness and minimum norm criteria. The MINQUE for $\boldsymbol{\psi}$ is given by

$$\boldsymbol{\psi} = \mathbf{F}^{-1} \mathbf{v} \tag{4.23}$$

where $\mathbf{F} = \{f_{ij}\}$ and

$$f_{ij} = \mathrm{trace}(\mathbf{R} \mathbf{V}_i \mathbf{R} \mathbf{V}_j) \qquad \mathbf{R} = \mathbf{V}_*^{-1} \mathbf{Q}$$
$$\mathbf{Q} = (\mathbf{I} - \mathbf{X}(\mathbf{X}^\top \mathbf{V}_*^{-1} \mathbf{X})^{-1} \mathbf{X}^\top \mathbf{V}_*^{-1})$$
$$\mathbf{v}^\top = (v_1, \ldots, v_n) \text{ with } v_i = \mathbf{e}^\top \mathbf{V}_i \mathbf{e} \text{ and } \mathbf{e} = \mathbf{R} \mathbf{y}$$

where $\mathbf{V}_*$ is $\mathbf{V}(\boldsymbol{\psi}_0)$ where $\boldsymbol{\psi}_0$ denotes a vector of provisional or prior estimates of $\boldsymbol{\psi}$. The MINQUE of $\boldsymbol{\psi}$ is a linear combination of the quadratic forms in the residuals $\mathbf{Q}\mathbf{y}$, which are obtained by regressing $\mathbf{V}_*^{-\frac{1}{2}} \mathbf{y}$ on $\mathbf{V}_*^{-\frac{1}{2}} \mathbf{X}$. The theory is overviewed by P. Rao (1977).

There are several variants of this estimation procedure that differ in their large and small sample properties. For example, the above procedure can be iterated with the estimate of $\boldsymbol{\psi}$ from the previous round used to provide provisional estimates at the next. The cycle is continued until convergence. (Initial estimates, $\boldsymbol{\psi}_0$, might be based on a visual inspection of the empirical covariances. Kitanidis (1983) remarks that $\boldsymbol{\psi}$ is not very sensitive to the choice of $\boldsymbol{\psi}_0$.) Kitanidis (1983) suggests a minimum norm estimator which is given by (4.23) with $\mathbf{V}^{*-1} = \mathbf{I}$. Stein (1987) describes MINQUE (where the invariance but not the unbiasedness condition is fulfilled) for the parameters of semi-variograms. A limitation with these procedures is their restriction to linear models. Stein (1987) discusses the extension of MINQUE to non-linear covariance models.

The invariance property of these estimators means that estimates of $\boldsymbol{\psi}$ will be identical if $\boldsymbol{\beta}$ is replaced by $(\boldsymbol{\beta} - \boldsymbol{\beta}_0)$. This implies that the iterative procedure of earlier sections (involving $\boldsymbol{\beta}$ and $\mathbf{V}$) can be avoided. An initial estimate of $\boldsymbol{\beta}$ is obtained by ordinary least squares. MINQUE, or some variant, is used to obtain $\boldsymbol{\psi}$ and then finally $\boldsymbol{\beta}$ can be re-estimated by generalised least squares using $\mathbf{V}(\boldsymbol{\psi})$. (However MINQUE is not robust with respect to the specification of the order of trend surface or the set of independent variables in a regression model.)

4.3.3 *Properties of estimators and hypothesis testing*

The large lattice properties of ML estimators for spatial processes have been reviewed in section 4.2. The conditions for the asymptotic normality and weak consistency of the ML estimators arc slightly weaker in the case of covariance stationary processes (Mardia and Marshall, 1984, theorem 3). Kitanidis (1983) discusses the properties of the ML estimator for estimating the parameters of PGCFs. ML estimation does not guarantee conditional positive definiteness unless constrained, although if this condition is not satisfied this could be taken to imply an incorrectly specified model. For the case of a single unknown parameter, the ML estimator is the minimum variance unbiased estimator and it is consistent. These results together with the results of section 4.2 allow the construction of confidence intervals and hypothesis tests on covariance model parameters.[5] The nested hypothesis tests of section 4.2 are appropriate for testing covariance models against the $V = \sigma^2 I$ case but it would be useful to be able to compare alternative covariance models and for this non-nested tests are required. Akaike's information criterion provides a method of comparison. Where two covariance models have the same number of parameters a straightforward comparison of the maximised likelihood functions is another possibility. Maps of prediction errors for the different models obtained by cross validation should be considered. Each data point is deleted in turn and its value predicted from the remaining observations. The model is not re-estimated each time. This amounts to kriging the observations (see Chapter 7). Prediction errors could then be summed for each of the different models and results compared. But such measures could be distorted by one or two values that are difficult to predict. If several models are to be compared this could be time consuming to implement.

Mardia and Marshall (1984) have examined the accuracy of asymptotic results for finite lattice schemes. They examined square (6×6, 8×8 and 10×10) lattices, simulating zero mean processes with different covariance stationary functions. They simulated the spherical and Whittle models both with nugget effects. For the spherical model the agreement with asymptotic theory was generally quite good for all the parameters for $n = 10 \times 10$ although significant bias was evident for the 6×6 case. The empirical distribution of the estimate of the mean agreed closely with asymptotic theory even for the 6×6 case. The estimate for η in the Whittle model is significantly biased to the right even for the 10×10 case. Figure 4.9 shows a selection of these results.

MINQUE was constructed to ensure unbiasedness and invariance of the estimators. If Y is normal and prior estimates of ψ are used then MINQUE is also minimum variance (over the class of quadratic estimators) or MIVQUE (Rao, 1971). There is further discussion of the properties of MIVQUE and 'minimum norm' estimators for estimating the parameters of PGCFs in Kitanidis (1983) who also gives the standard errors for the parameters so that confidence intervals and approximate tests of significance can be constructed. Iterated MIVQUE yields identical estimates to ML estimation and in the case of a single parameter model the two procedures yield identical estimates without iteration. Rao (1972) remarks that MIVQUE usually loses its unbiasedness property if the procedure is iterated so a small sample comparison of these two procedures would be useful to determine the relative merits of (non-iterated) MIVQUE and ML estimation. The much simpler minimum norm estimator which avoids matrix inversion is also worth exploring since it is also unbiased and invariant. MINQUE ensures that parameter estimates are non-negative and asymptotic properties are discussed in Stein (1987).

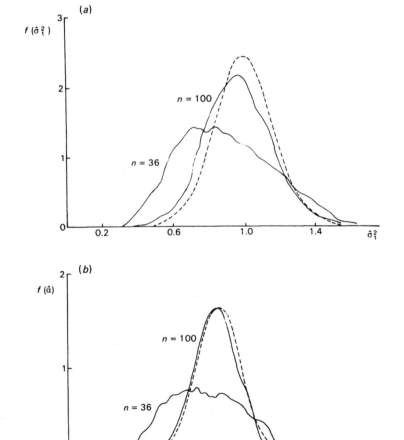

Figure 4.9. Simulated sampling distributions for the ML estimator of the parameters of models with different covariance structures. (a) Spherical model, σ_1^2, $\sigma = 1$, (b) spherical model, $\hat{\alpha}$, $\alpha = 3$.

For most of the estimation procedures reviewed in section 4.3.2 it appears that little is known concerning their properties and it is not possible to construct confidence intervals for the parameters (Kitanidis, 1983). In the case of semi-variogram modelling, Cressie (1985) argues that the method of weighted least squares provides a statistically sensible estimator which can be improved if necessary by using iterated generalised least squares although this involves much more computation. The efficiency of the estimator is further improved by using the Cressie and Hawkins (1980) robust estimator of the semi-variogram (see Chapter 6).

When dealing with surfaces with trend, PGCF models can be fitted but estimation methods, although robust to estimation of $\{\lambda_i\}$, are not robust to order misspecification. Properties of Delfiner's weighted least squares method are examined by Starks and Fang (1982). The distribution of z_i^2 is usually highly skewed with a few large outliers that have a strong influence on parameter estimates. Confidence intervals are usually large. Further, multicollinearity amongst the T_{di} means that estimates will be unstable. Starks and Fang conclude that PGCF models can only be estimated satisfactorily when the order of trend is small. Perhaps for these reasons, rather than dealing with PGCF models, it is more common to detrend the data and then fit a semi-variogram or covariance function. However, here too serious bias in the parameter estimates may arise if the order of trend surface is misspecified (Cressie, 1985). In the case of semi-variogram modelling Cressie recommends robust procedures (such as median polish) to remove non-stationarity. Models are fit to the residuals using robust estimators of the semi-variogram.

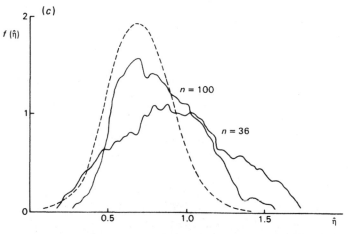

(*c*) Whittle model, $\hat{\eta}$, $\eta = 0.7$. After Mardia and Marshall (1984).

Covariance models should be positive definite and variogram models conditionally nonpositive definite. None of these procedures guarantees this unless constrained.

4.4   Validation in spatial models

Validation is concerned with assessing the fit of a model to the data. Several issues are of importance: how well does the model describe the data; are the statistical assumptions of the model satisfied; how sensitive is the fit to specific observations? Here methods of assessing the fit of any permissible model to the *spatial* properties of the data are considered. Wider issues of model validation are treated in Part 4.

One assessment criterion is to establish how well the spatial correlation properties of the data have been described by the chosen model. Figure 4.10 shows a plot of empirical and theoretical spatial correlations arising from two models used to represent aerial survey data. One of the models is a symmetric first order SAR model ($\rho = 0.225$) whilst the other is an autocorrelation function fitted using a method suggested by Agterberg (1970, p. 129). The Mercer and Hall wheat yield data has been the subject of repeated statistical analysis and was one of the data sets analysed by Whittle (1954). Besag (1974) fitted various CAR models and an errors in variables scheme (Besag, 1977b) and reported the pattern of theoretical autocorrelations generated by each scheme and compared them with the empirical estimates. Haining (1978b) compared empirical autocorrelations for county wheat yield data with those derived from different theoretical models. The approach is subjective and assumes stationarity. Larimore (1977) gives an example comparing the true spectral density and sample spectral density for a large set of simulated SAR data. Even with large data sets sampling variation can make this a hazardous procedure.

For interaction models maps of prediction errors or residuals can be used to identify those areas that are not well summarised by the model and these are usually easy to obtain. Poorly modelled areas may be evident in the form of isolated outlier values or clusters of positive (or negative) values. Graph plots of residual pairs that are separated by a fixed distance or correlation estimates at different distance lags should help to indicate whether there is any residual structure unaccounted for by the model. There appear to be few formal tests but graphical evidence can be highly informative. For a wide range of problems these considerations are likely to be useful and examples will be given in Part 4. For assessing the fit of covariance functions or semi-variogram models cross validation can be used and the map of prediction errors obtained.[6]

158

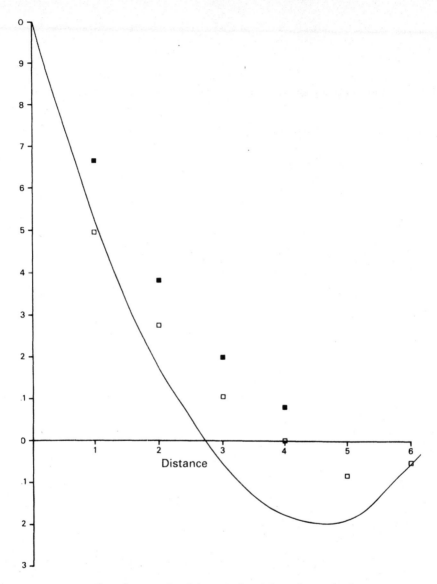

Figure 4.10. Plots of empirical and theoretical spatial correlations for data arising from an aerial survey. ■ First four spatial correlations for model 1 ($\rho = 0.225$), □ Empirical spatial correlations, —— Agterberg's function ($\hat{a} = 0.921$, $b = -4.524$, $c = 4.711$). After Haining (1987c).

Statistical analysis of spatial populations

Another informal assessment criterion which may be informative is to visually compare realisations of the chosen model and the observed map. This approach can be made more rigorous if we wish to make comparisons in terms of specific statistics. Simulating the model m times and then comparing observed statistics (such as low order correlations) with the simulated distribution of the statistics allows some assessment. A problem here is that we are usually interested in the tails of the distributions and most simulations do not provide much relevant information on the tails without large numbers of simulations. These may be expensive or time consuming to compute. Ripley (1981, p. 16) discusses this.

For models for discrete valued data on lattices, chi-square goodness of fit tests can be performed comparing observed and expected frequencies under the chosen model. Suppose a first order isotropic autologistic model is fit to lattice data where

$$p(y(s_1,s_2)\,|\,\text{all other values}) = \exp(y(s_1,s_2)(\tilde{\alpha}+\tilde{\beta}x(s_1,s_2)))/$$
$$(1+\exp(\tilde{\alpha}+\tilde{\beta}x(s_1,s_2))) \quad (4.24)$$

where $x(s_1,s_2) = (y(s_1-1,s_2)+y(s_1+1,s_2)+y(s_1,s_2-1)+y(s_1,s_2+1))$ and $\tilde{\alpha}$ and $\tilde{\beta}$ are the estimates of α and β. Under this model $y(s_1,s_2)=0$ or 1 and $x(s_1,s_2)=0,1,\ldots,4$ so for any coding a 5×2 table of observed and theoretical frequencies (based on (4.24)) can be constructed. A simple χ^2 goodness of fit test can be computed where the null hypothesis is that the observed frequencies follow the model given by (4.24) with estimated parameters α and β. The degrees of freedom for the test are

$$df = (G-1) \times N_c - E$$

where G is the number of possible values at a site, N_c is the number of neighbourhood configurations and E is the number of estimated parameters (in the above case: $G=2$, $N_c=5$ and $E=2$). There must be at least one expected observation per cell. Besag (1974) and Cross and Jain (1983) give examples of this type of analysis. Since there are overlaps in the set of neighbours for close members of any coding set the columns of the table are not independent. The χ^2 test requires independence, however, and for this reason the test may be unreliable.

A more difficult area in which to make recommendations is that concerned with assessing the influence of observations (single or in clusters) on model fit. In the case of data arising from sampling a continuous spatial surface each observation could be deleted in turn and model parameters re-estimated. Since some of the models may be quite expensive to fit (or time consuming) repeating the exercises deleting one observation at a time could be prohibitive. Some of the methods discussed in Chapter 6 have been

160

developed to try and identify values that may have a disproportionate influence on model fit. In the case of data for an areal partition, deleting values may create awkward boundary problems due to data dependency. This has been mentioned in Chapter 2 and will be considered further in Chapter 7.

4.5   The consequences of ignoring spatial correlation in estimating the mean

Often the main objective in data analysis, particularly in the social sciences where these methods are employed, is to get a good estimate of the mean (μ). Estimation and analysis of second order properties are only of interest inasmuch as they improve estimation of the mean. This final section reviews the problems that arise in the estimation of μ if it is assumed that $V = \sigma^2 I$ when this is not true. The estimators for μ in this case for a constant mean ($\mu = \mu 1$) a trend surface mean ($\mu = A\theta$) and a regression mean ($\mu = X\beta$) are given by (4.4), (4.5) and (4.6) respectively. We shall also consider the effect of increasing levels of dependency on the properties of the ML estimators given by (4.7), (4.8) and (4.9) respectively.

Consider the constant mean case first. If observations are not independent (4.4) will still be unbiased but

$$\begin{aligned}\text{Var}(\hat{\mu}) &= E[\hat{\mu} - \mu]^2 \\ &= (1/n^2)[\Sigma_{i=1,\dots,n}\text{Var}(y_i) + \Sigma\Sigma_{i\neq j}\text{Cov}(y_i,y_j)]\end{aligned} \qquad (4.25)$$

where $\text{Cov}(y_i,y_j)$ denotes the covariance between y_i and y_j. If $\text{Var}(y_i) = \sigma^2$ for all i then

$$\text{Var}(\hat{\mu}) = \sigma^2/n + (1/n^2)\Sigma\Sigma_{i\neq j}\text{Cov}(y_iy_j) = (1^T V 1)/n^2$$

So σ^2/n underestimates the true sampling variance of $\hat{\mu}$ if there is positive spatial correlation in the data and overestimates it if there is negative spatial correlation. In addition the usual estimator for σ^2 is given by

$$s^2 = (1/(n-1))\Sigma_{i=1,\dots,n}(y_i - \bar{y})^2$$

This estimator is biased downwards if spatial correlation is positive and biased upwards if it is negative. (As we have seen this can only, at best, estimate 'average variance' when the diagonal elements of V are not identical which is the case for some important spatial models.) These issues are discussed in Haining (1988).

If V is known, $\tilde{\mu}$ (4.7) is an unbiased estimator of μ and the asymptotic variance of $\tilde{\mu}$ is (Martin, 1981)

$$\text{Var}(\tilde{\mu}) = (1^T V^{-1} 1)^{-1} \qquad (4.26)$$

Table 4.6. *Standard error of $\hat{\mu}$ (4.4) and $\tilde{\mu}$ (4.7)*

(a) SAR model

Lattice size	Parameter value (ρ)						
	0.0	0.075		0.15		0.225	
	$\hat{\mu}$	$\hat{\mu}$	$\tilde{\mu}$	$\hat{\mu}$	$\tilde{\mu}$	$\hat{\mu}$	$\tilde{\mu}$
5×5	0.200	0.264	0.262	0.398	0.377	0.850	0.624
7×7	0.142	0.193	0.191	0.304	0.288	0.777	0.529
9×9	0.111	0.152	0.151	0.245	0.233	0.699	0.463
11×11	0.090	0.125	0.124	0.205	0.196	0.627	0.414

(b) CAR model

Lattice size	Parameter value (τ)						
	0.0	0.075		0.15		0.225	
	$\hat{\mu}$	$\hat{\mu}$	$\tilde{\mu}$	$\hat{\mu}$	$\tilde{\mu}$	$\hat{\mu}$	$\tilde{\mu}$
5×5	0.200	0.229	0.229	0.281	0.277	0.406	0.377
7×7	0.142	0.165	0.165	0.207	0.204	0.326	0.298
9×9	0.111	0.129	0.129	0.164	0.162	0.273	0.248
11×11	0,090	0.106	0.106	0.136	0.134	0.234	0.213

(c) MA model

Lattice size	Parameter value (v)						
	0.0	0.075		0.15		0.225	
	$\hat{\mu}$	$\hat{\mu}$	$\tilde{\mu}$	$\hat{\mu}$	$\tilde{\mu}$	$\hat{\mu}$	$\tilde{\mu}$
5×5	0.200	0.248	0.247	0.296	0.292	0.345	0.337
7×7	0.142	0.179	0.179	0.216	0.214	0.253	0.249
9×9	0.111	0.140	0.140	0.170	0.169	0.200	0.197
11×11	0.090	0.115	0.115	0.140	0.139	0.165	0.163

so that the relative efficiency of $\hat{\mu}$ with respect to $\bar{\mu}$ is

$$(\mathbf{1}^T\mathbf{V}\mathbf{1})(\mathbf{1}^T\mathbf{V}^{-1}\mathbf{1})/n^2$$

Table 4.6 shows the standard error for $\hat{\mu}$ and $\bar{\mu}$ for different sized lattices and for different first order interaction models.[7] The first column for each table is the independent case (spatial parameter set to zero) and is the 'classical' estimate of the standard error. Figure 4.11 shows the plot of the standard error of $\bar{\mu}$ for increasing lattice sizes for different parameter values of the first order CAR model.

In Chapter 2 it was pointed out that spatial dependency implies a loss of information in the estimation of the mean. One way to quantify this loss is to examine Fishers information measure, and in particular identify the information loss that arises when observations are dependent compared to the situation when observations are independent. If I_μ denotes Fishers information about μ contained in an observation on Y then for $\bar{\mu}$

$$I_\mu = (\mathbf{1}^T\mathbf{V}^{-1}\mathbf{1})$$

(Note that asymptotically $I_\mu^{-1} = \mathrm{Var}(\bar{\mu})$). For the first order CAR model the ratio of the information loss for the case $\tau > 0$ to I_μ for $\tau = 0.0$ is

$$\tau[4 - (2/t) - (2/s)] \qquad (4.27)$$

where t and $s(t=s)$ are the lattice dimensions. For large lattices $(s,t \to \infty)$ then $4\tau(|\tau| < 0.25)$ is a measure of the information loss in the estimation of

Figure 4.11. Plot of the standard error of μ for different lattice sizes and different parameter values of the single parameter CAR model. After Haining (1988).

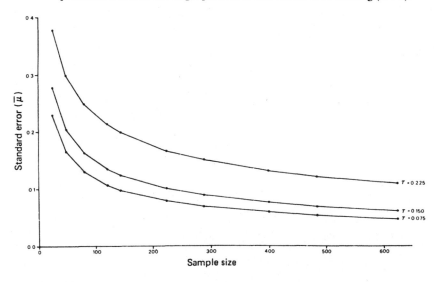

μ for given τ and this measure seems appropriate for any rectangular lattice ($s \neq t$) providing *both* s and t increase to ∞. Figure 4.12 shows a graph of (4.27) as a percentage figure. Table 4.7 provides the same measure of information loss for the directional parameters of a linear trend surface. Further details and results for other models are given in Haining (1988).[8]

A practical consequence of these results in the constant mean case is that if means are being compared (either in the same area at two or more different times or in two or more areas at the same time) then the information loss arising from positive spatial correlation may lead to incorrectly accepting the hypothesis of significant differences between the means if (4.4) is used with its classical variance estimate σ^2/n. If $\hat{\mu}$ is used with $\mathrm{Var}(\hat{\mu})$ as given by (4.25) this is an inefficient estimator for μ.

These difficulties carry over to the trend surface and regression models (see for example Cliff and Ord (1981) and for a detailed discussion in the

Figure 4.12. Plot of the % information loss for the single parameter CAR model for different parameter values and different lattice sizes. After Haining (1988).

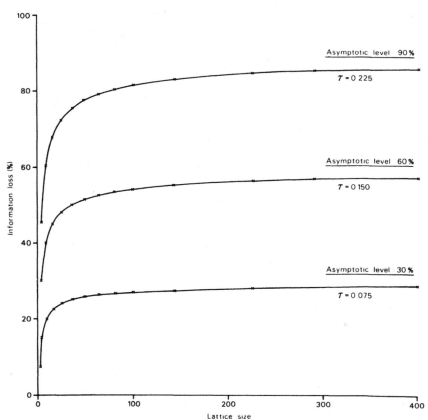

Table 4.7. *Information loss as a percentage of total information (independent case) for each directional parameter of a first order trend surface*

(a) *SAR model*

Lattice size	Spatial parameter (ρ)		
	0.075	0.15	0.225
5 × 5	40.23	69.30	87.19
7 × 7	43.25	73.58	90.99
9 × 9	44.94	75.93	92.96
11 × 11	46.03	77.42	94.16

(b) *CAR model*

Lattice size	Spatial parameter (τ)		
	0.075	0.15	0.225
5 × 5	22.90	45.81	68.72
7 × 7	24.85	49.71	74.57
9 × 9	25.96	51.92	77.89
11 × 11	26.67	53.35	80.03

(c) *MA model*

Lattice size	Spatial parameter (v)		
	0.075	0.15	0.225
5 × 5	33.28	51.74	63.03
7 × 7	35.44	54.46	65.87
9 × 9	36.64	55.94	67.38
11 × 11	37.40	56.87	68.32

time series case with examples see Johnston (1972, Chapter 8)). If classical estimators (4.5) and (4.6) are used in the case of trend surface analysis there is a risk of fitting a surface of too high order and in a regression analysis there is a risk of wrongly retaining non significant variables. These are consequences of underestimating the true sampling variability of (4.5) and (4.6) when errors are correlated. Further, the downward bias in the estimation of σ^2 means that the R^2 goodness of fit measure will be inflated so that the model will appear to be doing better in describing the data than it is. In the case of (4.6) and (4.9):

$$\text{Var}(\hat{\beta}) = (\mathbf{X}^{\mathsf{T}}\mathbf{X})^{-1}\mathbf{X}^{\mathsf{T}}\mathbf{V}\mathbf{X}(\mathbf{X}^{\mathsf{T}}\mathbf{X})^{-1} \text{ and } \text{Var}(\tilde{\beta}) = (\mathbf{X}^{\mathsf{T}}\mathbf{V}^{-1}\mathbf{X})^{-1}$$

respectively.

Kramer and Donninger (1987) have examined the efficiency of (4.6) relative to (4.9) when $\mathbf{V} = \sigma^2[(\mathbf{I} - \rho\mathbf{W})^{\mathsf{T}}(\mathbf{I} - \rho\mathbf{W})]^{-1}$. They quantify the loss in efficiency of (4.6) for certain special data sets and note that the loss can be substantial particularly for intermediate values of ρ. However they also demonstrate that in certain cases the use of (4.6) results in no loss in efficiency for limiting values of ρ for some definitions of $\mathbf{W}$ and $\mathbf{X}$.

The foregoing results assume that $\mathbf{V}$ is known. If as is usually the case $\mathbf{V}$ must be estimated, the loss in efficiency for (4.6) may not be so marked particularly if spatial correlation is weak. Indeed, simulation evidence in Kramer and Donninger (1987) suggests that (4.6) may even outperform (4.9) when $\mathbf{V}$ has to be estimated and if ρ is small. Their results point to the need to consider the form of the regression model and the severity of correlation in the data before embarking on the estimation of spatial parameters.

NOTES

1. Cliff and Ord (1981, pp. 134–9) suggest the use of the partial spatial autocorrelation function (PAF) as an aid to model specification. For example, if the autocorrelation function dies away to zero rapidly but the PAF does not this suggests a moving average model. If the PAF does die rapidly this is suggestive of some form of autoregressive scheme. They also discuss the use of the PAF for determining the order of a CAR scheme which avoids the need to estimate the next higher order scheme.
2. Mardia and Marshall (1984, p. 137) for example, comment that general maximum likelihood procedures for spatial analysis require a computational time proportional to n^3 unless $\mathbf{V}$ can be structured in some simple way.
3. There is no formal requirement that prohibits a first order SAR model with four spatial parameters (one for each compass direction). However these parameters are not estimable unless the east and west parameters are the

same and the north and south parameters are the same. The problem is
that swapping parameters around in the unrestricted model generates
identical probability structures.

4. Computational methods are discussed in, amongst other sources, Mardia
and Marshall (1984), Kitanidis (1983), Hepple (1976) and Ord (1975).
Ripley (1988) shows that the likelihood function may not be unimodal,
advises against using Newton methods of optimisation and suggests that
grid search methods may represent the safest approach.

5. Journel and Huijbregts (1978, pp. 193–4) discuss testing the significance
of semi-variogram models. They obtain fluctuation variances and note
that in view of the magnitude of these variances no goodness of fit test for
the fitting of a theoretical model to an experimental semi-variogram would
ever invalidate the fit. Estimation variances, however, vanish as sample
size increases so that tests can be performed if $\gamma(h)$ is interpreted as the
'local' semi-variogram (see section 4.1).

6. There are difficulties in devising a measure of fit. The coefficient of
determination, R^2 is given by

$$1 - [(\hat{\mathbf{e}}^T\hat{\mathbf{e}})/((\mathbf{y}-\bar{\mathbf{y}})^T(\mathbf{y}-\bar{\mathbf{y}}))]$$

where $\hat{\mathbf{e}}$ is the vector of prediction errors and $\bar{\mathbf{y}}$ is the sample mean. But
quantifying model fit using this measure can be unreliable and unduly
influenced by one or two values which are difficult to predict. In the case of
fitting PGCF models, for example, Starks and Fang (1982) note that since
z_i^2 is approximately χ^2, R^2 could suggest a poor fit when in fact this is due
only to a few values.

7. For the CAR model on a rectangular $n = s \times t$ lattice (4.26) is:

$$\text{Var}(\bar{\mu}) = \sigma^2[st - \tau(4st - 2(s+t))]^{-1} \qquad s,t > 1$$

and for the SAR model

$$\text{Var}(\bar{\mu}) = \sigma^2[st - (2\rho - \rho^2)(4st - 2(s+t)) + \rho^2(12(st-s-t)+8)]^{-1} \qquad s,t > 2$$

8. For a given value of $\tau > 0.0$ and a given number of observations, how
many additional observations would be required to bring the standard
error to the same value as for the independent ($\tau = 0.0$) case? If n is the
number of observations in the independent case, let the lattice size (for
$\tau > 0.0$) be $k \times k$. For the CAR model this means solving for k in the
quadratic expression:

$$(1 - 4\tau)k^2 + 4\tau k - n = 0$$

The correct root is

$$k = (-2\tau + (4\tau^2 + n - 4n\tau)^{\frac{1}{2}})/(1 - 4\tau)$$

Values of k^2 are given in Table A1 for different n and τ. Note that as n
increases, $k^2 \rightarrow n/(1 - 4\tau)$ so that again for the case of a square lattice 4τ is a
measure of information loss (Haining, 1988).

Statistical analysis of spatial populations

Table A1. *Numbers of observations (k^2) required for equivalent standard error between $\tau = 0.0$ case and $\tau > 0.0$ cases (CAR model)*

	Number of observations	
τ	$\tau = 0.0$	$\tau > 0.0$
0.075	10	13
	50	68
	100	138
0.15	10	19
	50	110
	100	228
0.225	10	42
	50	336
	100	754

Spatial data collection and preliminary analysis

5

Sampling spatial populations

5.1   Introduction

This chapter examines the problems that arise in sampling a surface for
purposes of estimating its properties. We shall assume that the surface is
continuous or very nearly so. Often data are made available on a predeter-
mined grid or framework such as a network of established weather stations,
pixels on a remotely sensed image, areas or tracts in a census survey. In this
chapter, however, we consider the situation where the analyst can choose
the framework for data collection. We consider three categories of spatial
sampling problem.

> *Category I:*   Problems concerned with estimating some non-
> spatial characteristic of a spatial population, for example the
> frequency distribution of areal values, or an areal mean, or a total,
> proportion or intensity value. So, we might wish to estimate the
> mean or total level of precipitation in a basin, average levels of
> household income, the proportion of infected individuals in an area
> during an epidemic, or the proportion of an area with high levels of
> pollution.
>
> *Category II:*   Problems where the spatial variation of some vari-
> able is specifically required, either in the form of a map or in the
> form of a summary measure (such as a variogram or correlogram)
> to highlight important scales of variation. Included here are
> situations where the objective is to ensure efficient spatial interpol-
> ation (in geology, geomorphology or meteorology for example) for
> purposes of converting scattered point data to map form.
>
> *Category III:*   Obtain observations that are independent or nearly
> independent so that classical statistical procedures can be used to
> classify data or establish relationships between variables. An
> example of this arises in the development of classification schemes

171

for remotely sensed images in which pixel signatures are matched to known ground characteristics so that pixel values can be used to classify other areas for which ground conditions are not known.

Underlying all spatial sampling problems should be a concern for surface variation and how this influences the method of sampling and the conclusions to be drawn from sampling. In the second category of problems, standard approaches to characterising spatial variation and interpolation are based on classification procedures. To predict the value of some property (Y) at a site (i) the class membership (j) of the site must be specified. Then if y_{ij} is the value of the property Y at place i in class j

$$y_{ij} = \mu_j + e_{ij}$$

where μ_j is the mean of class j and e_{ij} is an independent random variable with mean 0 and variance σ^2. The mean of class j is estimated by

$$\hat{\mu}_j = (1/n(j))\Sigma_{k=1,\ldots,n(j)} y_{k,j}$$

where $\{y_{k,j}\}$ is a set of observations known to fall in class j and $n(j)$ is the number in the set. So for any unobserved site $\hat{y}_{ij} = \hat{\mu}_j$ and the estimation variance is $\sigma^2[1 + n(j)^{-1}]$ consisting of the within class variance (σ^2) and the estimation variance of the class mean ($\sigma^2/n(j)$). As Burgess, Webster and McBratney (1981) note this procedure focuses attention on developing good classifications (minimising σ^2). In fact the issues of sample size and sample design are of little importance. The configuration of points is irrelevant and increasing sample size has, on the whole, little effect on the estimation variance – the crucial term is σ^2 and it is the reduction of this quantity by refining the classification procedure that has the greatest impact on reducing estimation variance.

The underlying spatial model for the standard approach to interpolation and prediction is discontinuous spatial variation in the sense of contiguous areas of more or less uniform Y values (subject only to spatially uncorrelated random variation (e_{ij}) – called the nugget effect) with changes at the boundary from one class to another. In this chapter we shall assume an underlying model in which spatial variation is continuous (or nearly continuous) with neighbouring values spatially correlated. A key issue then becomes the identification of a model to characterise surface variation and provide a basis for interpolation.

Alternatively, suppose the aim is to estimate the mean value of some variable (a category I problem). The standard approach is represented by the model

$$y_i = \mu + e_i$$

172

where y_i denotes the value of Y at site i, μ is the mean and e_i is independent with mean 0 and variance σ^2. (The e term can refer to measurement error rather than a property of the surface.) The estimator for μ is $\bar{y} = (1/n)\Sigma y_i$ and the estimation variance is σ^2/n. In this case sample size (n) is important and n is chosen so that $n = t_\alpha^2 \sigma^2/c$ where t_α is students t at the chosen probability level α and c is the desired limits of estimation of μ. However, the spatial configuration of sample points is still irrelevant. But suppose $\{e_i\}$ is an attribute of the surface and displays continuous and spatially correlated variation. Each observation now carries information about its neighbourhood. As a consequence we might expect to require fewer observations than indicated by classical theory in order to estimate the mean of the surface to a given level of precision. Furthermore with spatial variation now in the model we might expect configuration of sample sites to assume greater importance in the sample design (section 5.3.3).

Spatial sampling ranges from relatively inexpensive site visits to the installation of permanent or semi-permanent manned or unmanned equipment (weather stations, geological bore holes, etc.). This indicates a further consideration. In some situations the underlying spatial patterns are stable and sampling experiments can be designed as one-off exercises to record an essentially static system. Soil and geological surveys frequently fall into this category. By contrast, spatial variation of meteorological events is subject to considerable variation (through time). Thus even if the objective is to describe the variation in precipitation, or estimate totals at one point in time, a sampling grid that might be adequate for a cyclonic system might not be adequate to pick up the spatial variation associated with convection rainfall for example. Sampling schemes may need to be designed to accommodate the range of patterns that may arise over the life of the monitoring system.

Samples are taken from some region (A) and the elementary sampling units are either point sites or quadrats. These are the two forms of sampling to be examined here. The full specification of a sampling experiment requires, apart from the specification of the characteristics to be measured:

(i)   selection of a sample size;
(ii)   selection of a sample design;
(iii)   selection of an estimator for the population characteristic;
(iv)   estimation of sampling variance in order to compute confidence intervals.

The presence of spatial dependency has implications for all these four stages. In fact, herein lies quite an important practical problem. In order to answer many of the above questions about optimal sampling schemes it is necessary to know the nature of surface variation. On the other hand this usually cannot be known without sample data. Partial solutions are to take a pilot survey (transects are often proposed) or to draw on experience from earlier surveys. Fortunately theoretical results in this area indicate that little *specific* knowledge about the form of surface variation is required.

At a theoretical level different models for spatial variation are proposed and results given as to how best to sample each type of surface. (A significant practical problem is then to decide which model best describes the underlying (real) situation.) Spatial properties for the variable (*Y*) are usually of two types:

(i)   first order properties of *Y* (the mean of the surface) to reflect the presence of trend or periodicity;

(ii)   second order properties of *Y* (fluctuations about the mean) to reflect the presence of spatial correlation.

We need results to cover both forms of variation occurring separately or jointly. Surfaces with only mean variation are fixed and inferences relate to properties of the population of values in A. Surfaces with second order variation may be treated as probabilistic, implying that the set of values in A are but one possible realisation. In this case inferences from the sample data could be used either to specify properties of the realised values in *A* or properties of the 'superpopulation' (the set of all possible realisations). Usually in sampling problems it is the 'realised' surface that is of interest for obvious reasons but the 'superpopulation' model remains the basis for the theoretical evaluation of the relative merits of different sampling designs.

Good estimators of population characteristics are, under repeated sampling, unbiased and minimum variance. If $\hat{\theta}$ is an unbiased estimator of a parameter θ (for a fixed population of values in A) then for random, stratified random and systematic sampling with random start points, the sampling variance is

$$\mathrm{Var}(\hat{\theta}-\theta) = E(\hat{\theta}-\theta)^2 = E(\hat{\theta}-E(\hat{\theta}))^2$$

where expectations are taken with respect to repeated sampling. Estimators and sampling designs can be compared in terms of the size of their sampling variations. For biased estimators of θ,

$$E(\hat{\theta}-\theta)^2 = E(\hat{\theta}-E(\hat{\theta}))^2 + (E(\hat{\theta})-\theta)^2$$

which is termed the mean square error (MSE). The MSE is a measure of the *accuracy* of an estimator while the sampling variance is a measure of its *precision*. In the superpopulation model these quantities are evaluated for a fixed realisation and then averaged over all possible realisations.

In category I problems, sites too close together will tend to duplicate information so such a sampling design would be wasteful, but too far apart

174

will give rise to large sampling variances and so be inefficient. Random sampling designs which often result in clusters of sites in some parts of the area and no observations in others would seem unsuited to spatial sampling. We are led to anticipate that some form of systematic sampling will probably be needed which will keep sites at some optimal distance apart while providing a full areal coverage.

Category II problems may require a relatively dense network of sites if the variable nature of the spatial surface is to be characterised. Regular sampling grids provide a good basis for estimating covariances and the semi-variogram. Again, therefore, systematic sampling at an adequate intensity would seem appropriate and prior information will be needed to select the sampling interval.

Problems in the third category require the selection of samples at distances exceeding the range of the variable or the distance at which the covariances are close to zero. A problem may arise if the sample size for subsequent statistical analysis implies a sampling interval which is less than this distance. Problems of statistical analysis with spatially dependent data are treated in later chapters. We shall not comment further on the general nature of this type of sampling problem except to note that some pilot survey or prior information will usually be required in order to specify a 'safe' sampling interval. The need to sample at intervals suggests that a systematic sampling design will be appropriate.

In the next section we summarise spatial sampling designs and in section 5.3 present results on different sampling designs for estimating means. Section 5.4 examines results on sampling for purposes of interpolation. The final section is a discussion of some practical issues and the relationship between the 'ideal' results of sections 5.3 and 5.4 and different types of problems and spatial surfaces.

5.2   Spatial Sampling designs

5.2.1   Point sampling

Sampling designs have been described in detail in Quenouille (1949) and more recently by Ripley (1981) amongst other sources. The following description is relevant to point sampling from a continuous surface or point sites from a lattice. The second case differs from the first in that sampling is from a finite set of elementary units.

There are three basic forms of point sampling in a region A: uniform or random sampling where *n* points are chosen uniformly and independently in A; stratified random sampling where *m* points are chosen uniformly and independently in each of k strata that partition A$(n = k \times m)$ and systematic sampling. One common form of systematic sampling is to partition A into square strata, then locate one point per strata in the centre of each square. This is called an aligned centric systematic sample. Another common form is to randomise the location of the site in the first strata, allocating points in the remaining strata to the same relative position. In the case of homogeneous surfaces, stratification, either for systematic or stratified random sampling, is often systematic in terms of squares but Ripley (1981, p. 19) notes that other forms could be used.

Aligned versions of all three basic schemes can be constructed, and sample designs which are aligned in both directions form a grid. If aligned in only one direction the sample points line up in irregularly spaced columns (rows) with the sample points irregularly spaced on the columns (rows).

Various 'hybrid' sampling designs can be constructed in which sampling differs between the two axes, say, random in one direction and systematic in the other. Still more hybrid forms can be designed by aligning either on both axes, on only one, or on neither, and by varying strata shape. Figure 5.1 gives some examples of various sampling designs for a continuous surface. Most results on the precision of estimators for different sampling designs are for non-hybrid forms with or without alignment. Nonetheless hybrid forms may be useful in practice where surfaces show anisotropy or other forms of directionality.

Figure 5.1. Sampling designs in two dimensions. × = location random in (x,y) co-ordinates, ○ = location fixed in (x,y) co-ordinates, ● = location fixed in x co-ordinate, random in y co-ordinate, + = location fixed in y co-ordinate, random in x co-ordinate. *Sampling designs*: (*a*) Random sampling, (*b*) Stratified random (un-aligned), (*c*) Aligned centric systematic, (*d*) Stratified random (aligned on both axes), (*e*) Systematic unaligned (x axis), stratified unaligned (y axis), (*f*) Systematic unaligned on both axis.

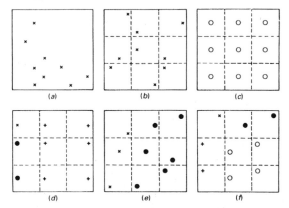

5.2.2 Quadrat and area sampling

One form of quadrat sampling is to sample from a systematic and exhaustive grid that covers the study area. The various forms of point sampling described in section 5.2.1 can be used to select quadrats. Remotely sensed images, made up of large numbers of pixels can be sampled according to these rules.

In ecology, quadrat sampling takes the form of either imposing a systematic and exhaustive grid across the study area or laying down quadrats at random across an area. These sampling methods are discussed in Greig-Smith (1964) and Kershaw (1973). In the case of exhaustive sampling, inferences apply either to a conceptual superpopulation or to a larger region if the exhaustively sampled area is itself one of a number of sampled areas in a larger region. In principle forms of (non-exhaustive) sampling other than random sampling could be used. For example, quadrats could be laid down according to systematic and stratified random designs, with or without alignment.

In geostatistics, area sampling underlies the methods of 'block' kriging (Journel & Huijbregts, 1978). Choice of sites can again be made using the designs of the previous section.

5.3 Sampling spatial surfaces: estimating the mean

Suppose the true mean of a surface can be denoted $\bar{Y}$:

$$\bar{Y} = (1/N)\Sigma_{i=1,\ldots,N}\, y_i$$

where N is the population size and y_i is the value at site i. For a continuous surface

$$\bar{Y} = (1/|A|)\int_A y(\mathbf{x})d\mathbf{x}$$

where $|A|$ is the area of the region A and $y(\mathbf{x})$ is the value at location $\mathbf{x}$. An unbiased estimator for $\bar{Y}$ in either case is given by the sample mean, $\bar{y} = (1/n)\Sigma y_i$, where n is the size of the sample. For any sampling design what is the sampling variance of $\bar{y}$ and which sampling design minimises this sampling variance? Of course, better estimators of $\bar{Y}$ than $\bar{y}$ may exist, for example the ML estimator described in Chapter 4. However, this estimator requires detailed information on the correlation structure which is usually not available. After sampling, and once spatial correlation has been estimated, this estimator could be used. In the following, attention focuses on $\bar{y}$.

We discuss population models that comprise trend, or periodicity (or both), and models where the set of values comprise a spatially correlated process with and without additional first order variation.

5.3.1 Fixed populations with trend or periodicity

If a population consists solely of linear trend then the sampling variances of $\bar{y}$ for systematic and stratified random sampling will both be less than for random sampling. Systematic sampling will give larger variances than stratified random sampling for if the systematic sample is too low in one stratum it will be too low in all the strata, whereas with stratified random sampling there is the opportunity for within stratum errors to cancel (see Figure 5.2(*a*)). Cochran (1963, p. 216) examines the effects mathematically. The variance of the systematic sample can be improved either by using a centrally located sample or by using border corrections (Yates, 1948). Corrections for two-dimensional series are given by Quenouille (1949).

If a population consists of a periodic trend then the effectiveness of systematic sampling depends on the relationship between the size of the strata and the frequency of the periodicity. Two cases are shown in Figure 5.2(*b*). In case A the number of strata equals the periodicity of the curve and the estimate will be poor and no better than a single (random) observation. In case B every systematic sample has a mean equal to the true mean ($\bar{Y}$) since deviations above and below $\bar{Y}$ cancel out. In case B the number of strata is an odd multiple of the half period of the curve. Between these two extremes systematic sampling shows different degrees of precision.

Strict periodicity is rare in natural spatial populations but weak or quasi-periodicity (consistent with a weakly stationary correlated process for example) may be evident and these results indicate the dangers inherent in systematic sampling in such situations. Even the more commonly encountered situation of trend in a spatial population can cause problems for systematic sample designs. In both these cases a stratified random sample may be a safer bet.

5.3.2 Populations with second order variation

We consider the case of second order stationary populations. Results in this case are obtained by taking expectations of the estimators both with respect to the set of possible sampling sites and with respect to the set of all other possible realisations of the surface. This is the so-called 'superpopulation' view. So if $\hat{y}_*$ is an estimator of the mean of the realised surface ($\bar{Y}$) with respect to sampling design $*$, then the sampling variance denotes the effects of repeated sampling on the same surface according to sampling design ∗ and is given by

$$\text{Var}(\hat{y}_* - E(\hat{y}_*)) = \text{Var}[\hat{y}_* - \bar{Y}] = E(\hat{y}_* - \bar{Y})^2$$

if $\hat{y}_*$ is an unbiased estimator of $\bar{Y}$. However the following results are for

$$\mathscr{E}(\text{Var}(\hat{y}_* - \hat{Y}))$$

Where $\mathscr{E}$ is the average or expected value of the sampling variance across all possible realisations of the surface. This is called the average sampling variance. As Cochran (1963, p. 219) notes, the comparison of sampling designs for any single realisation is difficult because of the inherent irregularities in the covariance functions of single realisations so that taking expectations ($\mathscr{E}$) is used to average out the irregularities of each realisation. As previously, the estimator of $\bar{Y}$ is $\bar{y}$.

Figure 5.2. Sampling on surfaces with (*a*) trend, where × = location of systematic sample points in sample region A_1–A_2, $\bigcirc$ = location of stratified random sample points in sample region A_1–A_2. (*b*) periodicity where × = location of systematic sample A. ✳ = location of systematic sample B.

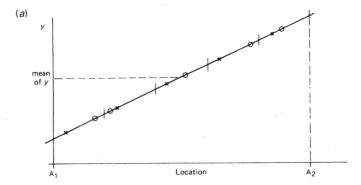

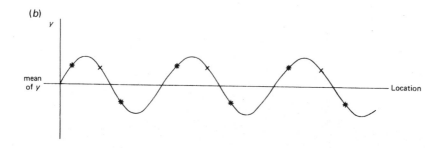

5.3.2 (a) Results for one-dimensional series

Cochran (1963, p. 220) considers the case of a large finite population of size N where

$$\mathcal{E}(y_i) = \mu \text{ and } \mathcal{E}(y_i - \mu)(y_{i+j} - \mu) = \rho_j \sigma^2 \qquad i = 1, \ldots, N$$

where ρ_j is the correlation between two observations j units apart, where $\rho_j \geqslant \rho_k \geqslant 0$ whenever $j < k$. For this class of populations stratified random sampling is always preferable to random sampling. The status of systematic sampling is not clear cut. For correlation functions that are concave upwards, that is

$$\rho_{j+1} + \rho_{j-1} - 2\rho_j \geqslant 0$$

then (Cochran, 1963, Theorem 8.6) systematic sampling is preferable to stratified random sampling. This is because with such functions the

Figure 5.3. Correlation functions in one dimension and the most efficient sampling design for each. (a)(b) Systematic, (c)(d) Stratified random.

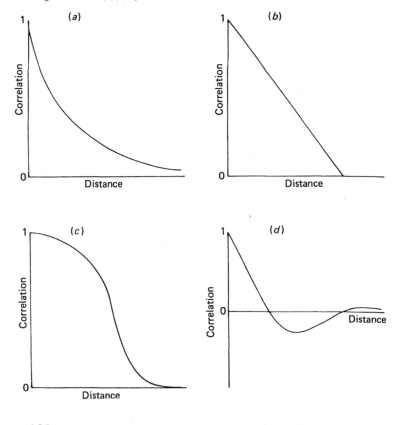

stratified random sample loses more in precision when the distance between sample points is less than the fixed distance (d) between sample points in the systematic sample than it gains in precision when the distance between sample points is larger than d. Quenouille (1949) has generalised this result to the case

$$\mathscr{E}(y_i) = \mu_i \text{ and } \mathscr{E}(y_i - \mu_i)^2 = \sigma_i^2 \qquad i = 1, \ldots, N$$

with μ_i distributed about μ with variance σ^2.

In view of the frequency with which these types of correlation functions seem to arise in practice (the negative exponential is concave upwards) this result underlines the importance of systematic sampling. Quenouille (1949), however, discusses other correlation functions, and Figure 5.3 shows four cases (including the above case) and indicates the most efficient sampling scheme for each. Stratified random sampling is preferable where there is quasi-periodicity (case d) and where there is strong low order correlation (case c) which might also be observed if there was a quasi linear trend in the population. When systematic sampling is the most efficient there is the choice of starting point within the first stratum (which then determines the position of each sample point within the remaining strata). Madow (1953) shows that for monotonic decreasing correlation functions the most efficient systematic samples are centrally located, rather than random start designs.

5.3.2 (b) Results for two-dimensional surfaces

Quenouille (1949) extended the results for one dimensional finite populations to the case of two dimensions. A sample of size $n_1 \times n_2$ is drawn from the elements $y_{ij}(i = 1, \ldots, N_1, j = 1, \ldots, N_2)$ forming a single finite population drawn from a $(N_1 \times N_2)$ superpopulation in which

$$\mathscr{E}(y_{ij}) = \mu \text{ and } \mathscr{E}((y_{i,j} - \mu)(y_{i+u,j+v} - \mu)) = \rho_{u,v}\sigma^2$$

Results are given for the average sampling variance of $\bar{y}$ under different sampling designs with detailed treatment of linear by linear processes in general (where $\rho_{uv} = \rho_u\rho_v$) and the doubly geometric process in particular (where $\rho_u = \rho^u$ and $\rho_v = \rho^v$). Martin (1979) has summarised results on the sampling of linear by linear processes. For this class of models the mean squared error of $\bar{y}$ can be found in terms of the results for one-dimensional series whenever the planar sampling is aligned. So, the efficiencies of different types of aligned sampling designs can be compared using only comparisons made on the line. For the case of linear by linear processes with monotonically decreasing correlation functions the most efficient aligned sampling design is the centrally located aligned systematic sample. Whether aligned sampling is better in general than unaligned sampling is not clear although the weight of evidence seems to favour unaligned sampling (see for example Bellhouse, 1977). Martin (1979, pp. 214–15) assesses the evidence. On the other hand the practical simplicity of taking aligned samples may outweigh the (small) losses in efficiency.

Quenouille (1949) also considers the case of spatial populations with continuous isotropic negative exponential correlations. Again, systematic sampling appears preferable to either stratified random or random sampling in the situations considered. Bellhouse (1977) provides further evidence for the superiority of systematic sampling in two-dimensional situations. Ripley (1981, pp. 19–25) considers continuous spatial processes with a constant mean and takes a superpopulation view of sampling. For general covariance functions, C, stratified random sampling is more efficient than random sampling (see also Rodriguez-Iturbe and Mejia, 1974). For comparisons with systematic sampling, C is assumed stationary. In this case stratified and systematic sampling are seen to have smaller average sampling variances than random sampling, particularly if there is strong local positive correlation. The gains from stratification are greatest when the correlation function is large, up to the diameter of the strata, becoming negligible thereafter. So for monotonically decreasing correlation functions, small strata should be used with, consequently, few sample points per stratum (ideally $k = 1$, $m = n$). Ripley further suggests that systematic sampling will generally outperform stratified random sampling unless the surface contains strong periodicity. Summarising empirical evidence from a number of studies Cochran (1963, pp. 221–4) draws the same conclusion, and results in Matérn (1960, p. 83) also support this view for negative exponential and Bessel type correlation functions. From the earlier discussion the presence of a trend in the population may also favour the use of stratified random sampling over systematic designs. Of the two possible problems the second is usually considered the more serious. Milne (1959), for example, considered the danger to centric systematic sampling of unsuspected periodic variation to be of no consequence.

What types of strata shape are optimal for either systematic or stratified sampling? For isotropic stationary processes the strata should be compact (Ripley, 1981). Squares or hexagons could be used and hence square or triangular sampling grids. Evidence reported in Martin (1979) and Ripley (1981) indicate that triangular sampling grids are optimal for monotonic isotropic covariances. The gains over square sampling grids are generally small (for detailed numerical examples and discussion see Matérn (1960, pp. 72–8)). In the case of anisotropic surfaces the optimal strategy is to elongate the strata in the direction of strongest correlation.

Finally, if trend or periodicity is strong in one direction only, a 'hybrid' sampling scheme may be preferable. In the case of a surface with linear trend plus isotropic second order variation a stratified random design could be used in the direction of the trend with an unaligned systematic sample in the direction orthogonal to the slope.

5.3.3 Standard errors for confidence intervals and selecting sample size

An estimate of the standard error (the square root of the sampling variance) is needed to specify confidence intervals for $\bar{y}$ and, prior to sampling, some idea of its value is essential in order to achieve desired levels of precision. (The sampling distribution of $\bar{y}$ is also needed and we comment on this later.) Standard errors are less (for any given n) than for situations where observations are independent. The size of the reduction depends on the sampling design and the covariance structure of the surface.

In random sampling within a region A

$$\text{Var}(\bar{y} - \bar{Y}) = [\sigma^2 - E(C(\mathbf{x},\mathbf{y}))]/n$$

where $\mathbf{x}$ and $\mathbf{y}$ are two randomly selected points in A and $E(C(\mathbf{x},\mathbf{y}))$ denotes the expected covariance between $\mathbf{x}$ and $\mathbf{y}$ (Ripley, 1981). If A is large and the covariance function decays rapidly to zero (relative to the size of A) then σ^2/n provides a good guide to the sampling variance where σ^2 is the variance of Y_i (or $Y(\mathbf{x})$ in the continuous case). So with a provisional estimate of σ^2, n can be specified to achieve the desired precision by standard methods. Once the data have been collected, an unbiased estimate of the sampling variance is obtained estimating σ^2 by s^2 where

$$s^2 = (1/n-1)\Sigma_{i=1,\ldots,n}(y_i - \bar{y})^2 \tag{5.1}$$

(Ripley, 1981).

For small areas or where the covariance function does not decay rapidly it is necessary to include $E(C(\mathbf{x},\mathbf{y}))$ in the estimation. Now the mean distance between two randomly selected points in a plane convex region is called the 'characteristic correlation distance' (Rodriguez-Iturbe and Mejia, 1974). Results on the distribution of distances between two randomly selected points are given in Ghosh (1951). Matérn (1960) used these results to give distances in regions of different shape and unit area. These are: circle $= 0.5108$; hexagon $= 0.5126$; square $= 0.5214$; equilateral triangle $= 0.5544$; rectangle $(\alpha = 2) = 0.5691$; rectangle $(\alpha = 4) = 0.7137$; rectangle $(\alpha = 16) = 1.3426$, where α denotes the ratio of the longer to the shorter side. For other sized areas adjustment is made by taking a proportionality coefficient made up of the ratio of two corresponding distances in the figure of area A and the figure of unit area. For example, in the case of a rectangle of area $30\,000$ square kilometres with $\alpha = 2$ the diagonal distance is 268 kilometres. For the rectangle of unit area this distance is 1.58 so the characteristic correlation distance for the larger area is $0.5691 \times (268/1.58) = 97$ kilometres. (For further examples see Rodriguez-Iturbe and Mejia, 1974.) Now given the correlation function of the model $(R(s,t))$, $E(C(\mathbf{x},\mathbf{y}))$ can be determined by computing $\sigma^2 R(|\,\mathbf{x}-\mathbf{y}\,|)$. Figure 5.4 shows the variance reduction arising from different sized random samples with Bessel type and negative exponential correlation functions.

However, we have seen that stratified random and systematic sampling give lower average sampling variances so we would normally prefer to

Figure 5.4. Variance reduction factor due to spatial sampling with random design used in the estimation of the areal mean of a rainfall event with (a) Bessel type and (b) negative exponential correlation function. After Rodriguez-Iturbe and Mejia (1974).

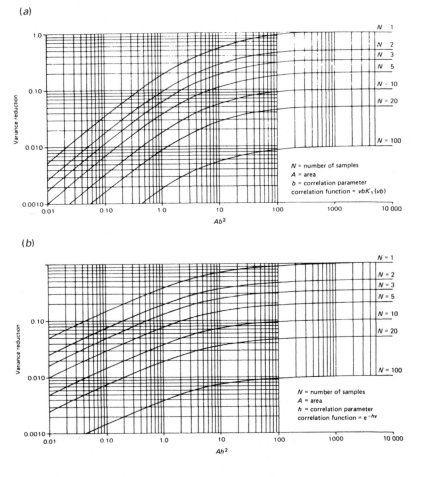

sample with these designs. In these cases an approximate estimate of n, for sampling purposes, could be obtained from σ^2/n, where again σ^2 must be given a prior value. On the other hand this will clearly be an overestimate and if sampling is expensive, will be wasteful. With stratified random sampling (Ripley, 1981)

$$\mathrm{Var}(\bar{y} - \bar{Y}) = [\sigma^2 - \overline{E(C(\mathbf{x}_i, \mathbf{y}_i))}]/n \qquad (5.2)$$

where $\overline{E(C(\mathbf{x}_i, \mathbf{y}_i))}$ is the average expected covariance between two points randomly selected in stratum i (where the average is over all strata). Since the strata will be generally small relative to A the reduction factor will usually be much larger than in the case of random sampling (see Figure 5.5). The

Figure 5.5. Variance reduction factor due to spatial sampling with stratified design used in the estimation of the areal mean of a rainfall event with (a) Bessel type and (b) negative exponential correlation function. After Rodriguez-Iturbe and Mejia (1974).

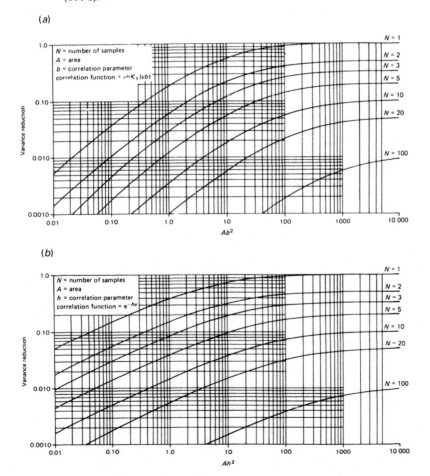

(a)

(b)

precision of the estimate can be identified by the same method used above. But adjusting sampling size adjusts the size of the strata and hence the characteristic correlation distance so this has to be allowed for if an n is to be determined to achieve a desired level of precision.

For stratified sampling the empirical evidence in Cochran (1963) suggests that relative sampling variances between stratified random $(k = n, m = 1)$ and systematic sampling can vary between 1.0 and 3.0 with the higher figure arising on variable surfaces, the lower on uniform surfaces. These results could be used to adjust figures for n for systematic sampling. More exact procedures do not seem to be available.

Quenouille (1949) and Ripley (1981) discuss methods for estimating the sampling variance from the sample data. In the case of stratified random sampling with $m = 2$, (5.1) is estimated for each stratum, the average taken over all strata and the estimate of the sampling variance obtained by dividing through by n. This will be an unbiased estimator. From the previous section the sample design $k = n$, $m = 1$ has lower sampling variance in which case (5.1) cannot be used on each stratum. The same problem confronts systematic sampling where again there is only one sample per stratum. One approach is to regard σ^2/n as the error variance. Quenouille (1949, p. 372) discusses three possibilities. One approach is to construct new, larger strata (with two sites per stratum) and then use the method for stratified random sampling with $m = 2$. The use of σ^2/n, the easiest, does not seem to come out badly in the studies cited by Ripley (1981). Note that in all these cases the standard error is estimated without reference to the underlying correlation structure. Thus, apart from helping to specify n and choosing a sampling design detailed estimates of spatial correlation are not needed.

In order to compute confidence intervals and estimate suitable sample sizes it is necessary to know the sampling distribution of $\bar{y}$. A central limit theorem (on $\bar{y}$) for spatial populations, (Bolthausen, 1982) shows that for large n, the normal distribution is suitable for situations where the process is stationary and mixing (i.e. shows weak dependence at large distances).

5.4   Sampling spatial surfaces: second order variation

5.4.1   *Kriging*

Issues of sampling design and sample size arise when the objective is to interpolate in an optimal fashion and construct contour maps for a variable within a region. The problem of interpolation is also examined in Chapter 7, here we note that if $y(\mathbf{x}_1), \ldots, y(\mathbf{x}_n)$ are n observations on a variable y at n

locations $x_1, \ldots, x_n$ in a region then if x_0 is an unobserved site, then the estimator for $y(x_0)$ is given by

$$\hat{y}(x_0) = \Sigma_{i=1,\ldots,n} \lambda_i y(x_i)$$

where $\{\lambda_i\}$ are weights that sum to 1. This is the 'kriging' estimator of $y(x_0)$ which, for any fixed sampling design, minimises the estimation variance given by:

$$\sigma_y^2 = E[(y(x_0) - \hat{y}(x_0))^2]$$

which depends only on the semi-variogram (or covariance function) and the configuration of observation points, not on the observed values themselves (see Chapter 7). If we now allow the sampling design to vary, we can ask what sampling design minimises σ_y^2 and what sample size ensures a specified upper bound to this minimum.

Burgess, Webster and McBratney (1981) consider this problem in terms of both punctual (point site) and block (area) kriging using semi-variogram methods. In the case of monotonically decreasing spatial correlation, σ_y^2 increases for any point (x_0) the further away it is from the observation set. This implies that systematic sampling will be best and that triangular grids will be better than square or rectangular grids (Figure 5.6). For practical reasons rectangular grids are usually preferred and of these centric aligned systematic grids are best (Webster and Burgess, 1984). Now σ_y^2 is at a maximum for such grids when interpolating a site in the centre of a group of any four sites (since at that position it is as far away as possible from the observation sites). So a plot of this maximum value of the minimised estimation variance against sampling interval, or sample size, can be used to select sample size to achieve a required level of precision (Figure 5.7). (Derivation of the 'classical' estimates in Figure 5.7 can be found in, for example, Burgess, Webster and McBratney, 1981.)

In the case of block kriging, the maximum estimation variance occurs when blocks are central either at the centres of grid cells or on observation points, depending on the size of the blocks in relation to grid mesh. Modification to the sampling design is also required in the case of anisotropic schemes. Both alignment and grid specification should be adjusted to the structure of anisotropy. Note also that to detect such anisotropy, transect pilot surveys should examine correlation patterns in three or more directions (McBratney, Webster and Burgess, 1981).

Hughes and Lettenmaier (1981) also found that uniform sampling produced minimum kriging variances in the cases that they considered. Departures from uniformity could lead to severe losses of information but the degree of severity depended on the structure of covariance. It is possible, for example, that the gains from uniform sampling are less where spatial variation is discontinuous.

The sampling methods and conclusions of this section are also relevant to 'kriging' or interpolation based solutions to the problem of estimating $\bar{Y}$. (See Chapter 7 and McBratney and Webster, 1983a.)

Goodchild (1984) discusses automated methods of constructing maps based on optical scanning of an image (such as an aerial photograph) in which a dense regular grid of points are sampled and the data passed to a contouring algorithm. One of the critical issues here, he notes, is the trade off between increased sample density and map accuracy on the one hand, and sample size and processing time and cost on the other. There appears to have been little work done on these aspects of automated sampling for mapping purposes where the highly variable and complex nature of the surfaces may make it difficult to lay down very general guidelines.

Figure 5.6. Comparison of maximum estimation variances for punctual kriging from square (—) and triangular (– – –) grids at the same sampling densities. The semi-variograms are $\gamma(\mathbf{h}) = C_0 + \mathbf{h}$ with $C_0 = 0$, 1, 2. After Burgess, Webster and McBratney (1981).

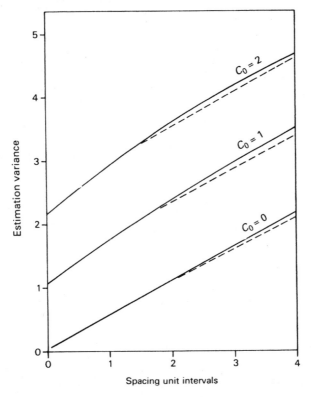

5.4.2 Scales of variation

Spatial variation may be conceptualised as comprising a small number of more or less discrete possibly independent components at several different scales or spatial hierarchies. For example, Burrough (1983) suggests that soil variation arises from independent soil forming processes operating at distinct spatial scales. In social and economic systems distinctive scale effects might be associated with different administrative tiers. In these cases it is important to identify the contribution to total variation that arises from each of the different scales. One approach is to divide the area into large blocks which are subdivided into (an equal number of) smaller blocks which are subdivided in turn and so on. This gives rise to different hierarchical levels each one equivalent to a specific geographical scale. In a social system these areas might be irregularly shaped following a hierarchical structure from township to county to state to region and so on. Suppose Y_{ijk} denotes the variable for county k in state j in region i then, for example, we could write

$$Y_{ijk} = \mu + \alpha_i + \beta_{ij} + \zeta_{ijk}$$

Figure 5.7. Plot of maximum value of the minimised estimation variance against sample size. The semi-variograms are $\gamma(\mathbf{h}) = C_0 + 0.839\mathbf{h}$ (*a*) $C_0 = 34.76$, (*b*) $C_0 = 0$. — kriged estimation error, – – – classical estimation error. After Webster and Burgess (1984).

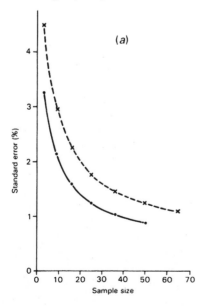

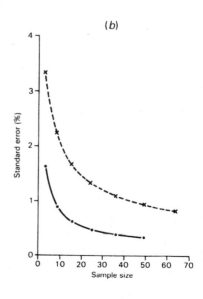

where μ is the grand mean for all counties, α_i is the effect of region i, β_{ij} is the effect of the state ij and ζ_{ijk} is the effect of county ijk. The total sum of squares can be partitioned into these three scales of variation. The approach is discussed in detail in Ripley (1981) with references.

It is evident that for a large region the volume of data required to perform such an analysis could be enormous. The data volume required increases dramatically if small scales of variation are felt to be important necessitating a fine partitioning of the area. One solution is to adopt a nested sampling framework. After subdividing at the first level a number of these units are chosen at random and subdivided. These subdivisions are in turn randomly selected and subdivided and so on. An example is shown in Figure 5.8. This dramatically reduces the data requirements whilst preserving the hierarchical structure of the underlying model. In particular it enables very small scales to be examined without the need for enormous amounts of data. In a situation where there is uncertainty as to what are the significant scales of variation this has considerable appeal (see Krumbein and Slack, 1956).

The lack of independence of spatial data undermines standard analysis of variance of these scale components (using F tests) but another problem is that the sampling design does not control for distance between the sample points. So, if scales of variation are continuously varying and associated

Figure 5.8. Stages in the development of a five-level nested design with random selection of units.

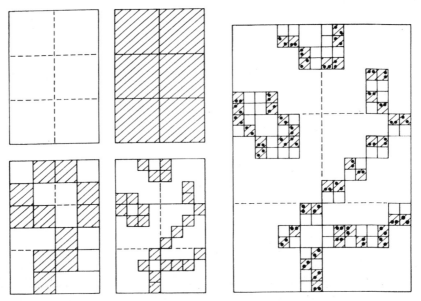

with specific distance bands (rather than partitioned into discrete hierarchical levels) than this method of sampling may obscure the scale contributions. A way round this is to develop a hierarchical sampling scheme which fixes distances between sample points (Youden and Mehlich, 1937). Randomisation may be allowed in the starting point and in the direction taken from each site, but the distance between sample sites is fixed at those intervals the investigator wishes to study closely. Primary sample points are fixed at a distance apart of x_1 metres. At each of these a further sample point is sited at a fixed distance from the primary sample point of x_2 metres. From all these sites further sites are selected a distance x_3 metres away and so on. This is the so-called balanced design and the total variance can be partitioned so that

$$\sigma^2 = \sigma_1{}^2 + \sigma_2{}^2 + \ldots + \sigma_m{}^2$$

where σ^2 is the total variance and $\sigma_i{}^2$ is the variance associated with distance interval i. Since numbers of sample points increase dramatically for small distances an unbalanced design may be preferred where, once sufficient degrees of freedom have been attained, only some of the sample sites at higher levels spawn 'off shoots'. An example is given in Figure 5.9. From the sample data, covariances and semi-variograms can be computed to describe spatial variation. In fact in a nested scheme cumulative variances form a nested variogram. With m levels separated by distances d_1, $d_2, \ldots, d_m$ then (Miesch, 1975):

$$\sigma_m{}^2 = \gamma(d_m)$$
$$\sigma_{m-1}{}^2 + \sigma_m{}^2 = \gamma(d_{m-1})$$
$$\vdots$$

and so on, where $\gamma(d_m)$ is the semi-variogram at distance d_m.

5.5   Sampling applications

Standard sampling theory depicts space as a set of regions. There is a mean value associated with each region although observing this mean value introduces additional independent random error into the data values. Adjacent regions are demarcated by a discontinuity at the boundary. Spatial sampling theory as described here depicts space as a continuously varying surface (reflecting smooth variation in the mean with additional smoothly correlated variation about the mean). If discontinuities are present in the surface these produce sub areas which are treated as separate strata.

The spatial model is an attractive one for sampling many environmental variables which show continuous or semi-continuous spatial variation. Its superiority over the classical model may not, however, seem so strong in the social sciences. Two reasons may be cited. First, the spatial variations of many social and economic variables are perceived as mosaics (such as the social areas of a city) – independent units (households) aggregated into homogeneous or semi-homogeneous areas. Second, even where interdependency is present between household units in terms of social and economic variables the spatial aspect of that variation is often felt to be incidental, the patterns of interaction being highly complex and not constrained by simple distance considerations. In much survey work in the social sciences, sampling involves areal stratification (controlling for important social and economic characteristics) but the choice of units within the strata is made at random. The map both of the individual units and the strata is considered of minor importance in the sampling effort except inasmuch as the distribution of the population might raise the costs of conventional survey work so that for national surveys, for example, a

Figure 5.9. Fixed interval point sampling. After Oliver and Webster (1986).

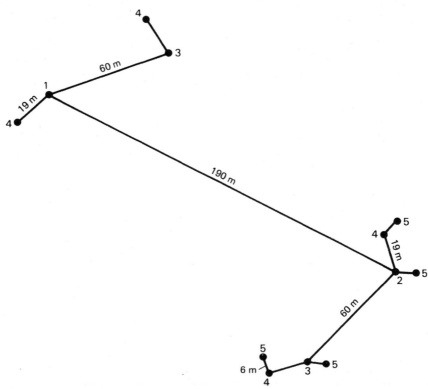

spatially clustered distribution of strata is favoured. It is considered more important to adjust for the effects of a stratified or clustered sampling design than to consider problems of spatial dependency. A modern review of social survey design is given in Rossi, Wright and Anderson (1983).

On the first issue it is debateable whether the spatial discontinuities present in social and economic data are any more pronounced than those found in environmental data (such as land use or soil maps for example). On the second issue the importance of distance effects in social analysis must evidently depend on the problem. Surveys that try to assess awareness of and reactions to some place-specific event, such as noise or atmospheric pollution emanating from a small area or point source, involve spatial effects that ought to be allowed for in the sample design. Measuring adoption (of a new product), awareness (of a rumour) or infection levels (of a disease) merit a spatialised sampling framework because in each case the underlying processes are often dependent on contact and proximity, and may well reflect the spatial structure of social networks and cliques. The spatial model underlying this chapter assumes spatial continuity at the intra-stratum scale and does not simply imply using location as another element of stratification in which intra-distance band variation is merely nugget variation.

Studies of soil and land use variation have paid particular attention to the development of efficient sampling schemes. The monograph by Matérn (1960) includes a comparison of different sampling designs for the estimation of the percentage of land under water and forest in Uppland and provides evidence of the superiority of systematic sampling. Berry and Baker (1968) review different sampling designs for estimating land use proportions and advocate an unaligned systematic sampling procedure. Keyes *et al.* (1976) review the literature and provide further empirical evidence in support of unaligned systematic sampling in land use studies where the structure of spatial correlation is often unknown and periodicity may affect the efficiency of centric systematic sampling.

In a series of papers Burgess, McBratney, Webster and others have applied regionalised variable theory to various aspects of soil mapping. Amongst other problems, they have considered sampling designs in relation to efficient interpolation and the construction of contour maps of soil variability (Burgess, Webster and McBratney, 1981), the estimation of regional means (McBratney and Webster, 1983), and boundary delimitation (Burgess and Webster, 1984). McBratney and Webster (1983) consider the problem of choosing a sampling design where several variables are to be measured with different spatial properties. A summary of this work is given in Webster (1985). Oliver and Webster (1986) apply fixed distance nested sampling methods to the problem of identifying scale components and estimating semi-variograms in soil data. Haggett (1965, p. 266) reports the use of ordinary nested sampling methods (not based on fixed distance sampling) for identifying scale components in land use maps.

Another area where the design of efficient spatial sampling schemes has attracted research interest is in monitoring precipitation levels particularly for flood control. Two aspects of this are the estimation of long term mean areal precipitation (LTMAR) and areal mean rainfall (AMR). The latter quantity is usually estimated for specific rainfall events. In both problems the spatial correlation function is critical and in the case of LTMAR estimation the temporal correlation function as well (Rodriguez-Iturbe and Mejia, 1974). Monitoring networks are required that give good estimates of these quantities at reasonable cost given the range of rainfall events that can occur in the area. An important question is to decide the number and distribution of recording stations.

The papers of Rodriguez-Iturbe and Mejia (1974) and Bras and Rodriguez-Iturbe (1976) deal with the problem of designing efficient networks. Their models are based on covariance function specifications of the spatial properties of rainfall events and they are principally concerned with building up a sampling network from an initially sparse coverage. Interestingly they note that systematic sampling is an impractical and expensive method of sampling, presumably because of the problems of siting and access, although using methods discussed in section 5.3 they note that stratification is necessary to reduce sampling variances. One of the problems here is that different rainfall events have different correlation structures.

Kriging can be used to assess the information content of existing stations by determining to what extent values can be predicted from other (neighbouring) stations. This would seem useful where, as in the case of certain parts of the U.K. the problem is not to extend but rather reduce and rationalise a very intensive monitoring system – although kriging can also be used to indicate areas where additional stations could be sited to advantage, these being areas where prediction errors are large (see for example McCullagh, 1974; Jones *et al.*, 1979; and O'Connell *et al.*, 1979). Figure 5.10 shows an example of this type of analysis. Hughes and Lettenmaier (1981) discuss the design of sampling schemes for monitoring water quality. Switzer (1977) discusses sampling methods for the estimation of air pollution levels.

Evans (1973) argues the superiority of fixed grid systematic sampling for geomorphic interpolation problems although Boehm (1967) notes that variable grid structures are often better in practice, to allow for the variation in surface properties from area to area. Increasingly the construction of these types of maps is becoming automated using optical scanning from aerial photographs where the key factor is processing costs rather than sampling costs.

5.6   Concluding comments

Although in certain cases systematic sampling may prove impractical or too costly, the theoretical evidence stresses the superiority of systematic

Figure 5.10. Rationalising a water grid: Wessex Water Authority region (*a*) shaded regions on maps show the areas for which the root mean square error of optimal interpolation is greater than 1.5 mm for days with widespread rainfall of over 1 mm for two design networks (i) 75 gauges (ii) 220 gauges. (*b*)(i) Percentage of Wessex Water Authority having a root mean square error greater than 1.5 mm (for point interpolation on days with more than 1 mm rainfall) for different networks. (ii) Average root mean square error of interpolation over areas of the Wessex Water Authority for days with widespread rainfall of over 1 mm for different networks. After O'Connell *et al.* (1979).

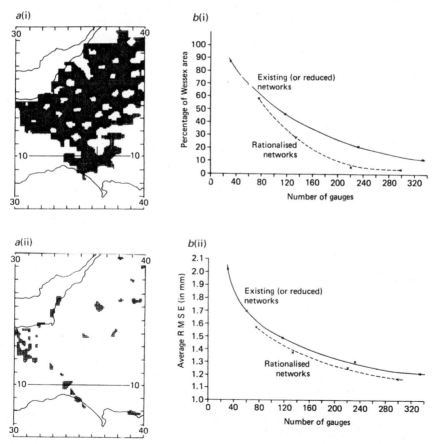

sampling in a variety of spatial situations. The reservations arise where the underlying spatial variation is discontinuous or where there is trend or periodicity in the surface. In these cases it may be preferable to adopt a stratified random scheme and where there are large regions with distinctive spatial variation then different sampling designs may be needed for each.

We have not, so far, mentioned the effects of borders on the methods and results of this chapter and a few comments are in order. Interpolation is affected by borders since prediction errors for sites close to the border are larger than for those sites in the interior. This is simply because there are usually fewer sites nearby. (Figure 7.26 shows an example of this.) Where possible the study area should be encircled by a buffer zone with sample sites that will be used in the interpolation procedures.

Matérn (1960) briefly considers border effects in the estimation of $\bar{Y}$. His analytical results given in an earlier section of this chapter ignore border effects. Matérn shows that sampling variances of $\bar{y}$ depend on the geometry of the study area and if the number of sample points is small this geometry can create a complicated border effect. Ripley (1982) suggests that Matérn's results probably overestimate the real efficiency gains under stratified and systematic sampling.

6

Preliminary analysis of spatial data

This chapter deals with preliminary forms of spatial data analysis, the purpose of which is to make an initial identification of data properties. The emphasis is on data description which should also be useful in developing hypotheses and shaping subsequent statistical analysis. The methods may also be useful for model assessment.

Section 6.1 describes methods for identifying distributional properties of observations. Such summaries may be of substantive interest but also help to indicate whether special data transformations will be required for subsequent statistical analysis. Ways of identifying trend and other forms of spatial arrangement or structure are also considered. This section draws on the methods of EDA (exploratory data analysis: Tukey, 1977; Hoaglin *et al.*, 1983, 1985) adapted to the needs of spatial data analysis. We assume familiarity with standard methods of EDA and the interested reader should refer to these basic sources for more information. Wetherill *et al.* (1986; Chapter 2) provide a good introduction and many of the methods described below can be implemented using MINITAB. Section 6.2 overviews methods for testing for pattern in spatial data (spatial autocorrelation tests). This is an area already well covered in the literature and the section places these methods in context and discusses when different tests are appropriate. This part of the book draws on the work of Cliff and Ord (1981) and that of Hubert, Golledge and colleagues in the use of the generalised cross product statistic for spatial data analysis. Section 6.3 deals with robust estimation of covariances and semi-variograms, extending the procedures in section 4.1. This section draws on developments in geostatistics, many interesting papers appearing in the NATO ASI volume edited by Verly *et al.* (1984).

6.1   Preliminary data analysis: distributional properties and spatial arrangement

EDA is concerned with resistant identification of data properties. This has a number of aspects:

(i)   Identification of 'atypical' values. Such values, also called 'outliers', may be errors arising from measurement or recording practices for example. Automated procedures for detecting and dealing with possible errors may be particularly important when handling large data sets. Often outliers are not errors but rather values that are different enough from the rest of the data to give cause for concern so that the analyst might wonder about the extent to which results will be influenced by these extreme values. Results here may suggest the need for data editing, resistant methods of analysis if the outliers are isolated or robust methods if the outliers are a more continuous string indicating a heavy tailed distribution.

(ii)   Identification of distributional properties of the data such as location, spread, skewness and tail properties. Many statistical procedures depend on distributional assumptions (such as symmetry or normality) and methods are needed to summarise data properties, test assumptions and assess the severity of departures from the assumptions. Results at this stage of analysis may indicate the need for a data transformation, robust methods or the development of a different model based on other distributional assumptions.

(iii)   Identification of the arrangement properties of spatial data including the detection of trends, local spatial relationships, spatial outliers, and spatial properties of multivariate data sets.

(iv)   Identification of properties in multivariate data sets. This aspect of EDA is concerned with the identification of relationships between pairs of groups of variables and between observational units.

6.1.1   Univariate data analysis

The following considers some general applications of EDA. The treatment is not systematic but provides examples of the use of EDA for identifying outliers, distributional and arrangement properties, as well as preliminary examination of hypotheses. All these data sets will be examined in more detail subsequently.

198

Table 6.1. Glasgow Community Medicine Area data: standardised mortality rates. (For locations of CMAs see Figure 6.2)

CMA	All deaths	Accidents	Respira- tory	Cancer	Ischaemic heart disease	Cerebro- vascular	Popula- tion $\times 10^{-3}$
1	114.085	129.065	122.225	108.443	113.819	112.894	7.411
2	111.644	159.104	120.523	95.071	118.121	78.740	6.233
3	94.943	210.182	97.568	76.951	96.200	101.862	10.444
4	98.847	107.072	122.055	97.257	92.726	101.963	7.862
5	98.153	107.465	100.522	96.814	109.606	92.919	11.557
6	74.161	99.641	85.479	85.533	6.480	85.601	7.953
7	87.542	78.844	88.396	86.463	95.130	85.640	9.075
8	101.126	104.942	132.365	79.686	91.031	103.198	7.196
9	98.317	93.839	120.616	97.378	99.872	98.366	11.370
10	92.276	45.969	91.810	101.540	100.667	88.427	16.535
11	106.359	119.758	110.993	101.187	108.577	94.067	17.580
12	119.605	127.171	162.207	89.572	137.545	116.638	13.380
13	120.223	167.436	111.259	97.417	127.372	133.226	13.210
14	99.794	111.524	75.916	95.324	111.691	106.610	21.531
15	116.991	112.177	118.460	98.327	123.784	117.872	4.820
16	88.001	115.375	96.051	73.222	108.849	85.150	7.791
17	90.863	113.781	104.828	78.775	91.742	96.058	17.910
18	70.620	46.687	58.075	82.029	74.121	70.187	12.560
19	72.096	117.692	45.348	83.923	74.967	89.706	14.337
20	79.579	71.174	49.814	91.773	90.320	96.079	12.176
21	109.758	126.395	129.227	104.071	111.458	89.869	18.580
22	116.441	111.115	181.745	109.086	95.878	82.082	3.083
23	125.330	210.496	168.573	123.652	116.616	70.684	3.401
24	118.123	167.646	124.306	101.169	115.486	87.924	7.333
25	134.123	192.021	189.270	128.096	122.325	118.024	13.515
26	100.563	49.142	98.253	103.705	96.315	115.328	18.341
27	119.393	118.470	155.149	127.659	121.011	105.672	10.856
28	109.671	99.840	95.591	117.403	123.033	88.197	19.206
29	102.749	109.484	94.803	111.948	98.417	118.541	4.651
30	87.137	31.907	107.208	72.362	95.743	94.292	9.483
31	79.752	75.843	82.527	86.398	85.090	55.788	6.181
32	90.014	43.883	119.958	86.576	94.520	91.166	6.481
33	71.257	172.677	61.282	78.998	58.820	62.134	11.458
34	88.498	118.985	91.861	105.982	·90.424	60.181	11.856
35	92.127	71.600	73.501	79.578	100.321	98.951	6.119
36	114.284	112.996	104.273	94.385	131.505	119.488	7.585
37	93.388	140.073	80.654	88.527	88.157	93.827	5.357
38	117.701	100.833	148.480	106.200	153.059·	116.056	6.556
39	86.646	76.505	77.470	82.980	80.348	113.726	8.175
40	121.128	298.178	146.156	100.121	107.525	115.961	3.629
41	111.887	138.195	117.003	114.191	112.797	101.317	8.974
42	127.232	198.336	126.890	146.401	86.695	118.804	8.431
43	129.100	155.788	105.582	140.519	136.184	105.697	3.845
44	131.399	102.561	147.285	117.230	124.470	116.197	10.573
45	141.336	260.690	190.628	145.028	111.744	135.961	8.607
46	105.539	97.900	107.303	102.550	102.735	104.940	15.996
47	92.099	52.383	77.834	95.897	101.570	92.830	8.062
48	118.743	193.428	154.079	119.363	113.469	86.066	2.862
49	116.921	99.267	124.584	118.599	114.614	105.353	8.178

Table 6.1 (*cont.*)

CMA	All deaths	Accidents	Respira-tory	Cancer	Ischaemic heart disease	Cerebro-vascular	Popula-tion $\times 10^{-3}$
50	111.683	150.419	133.475	101.650	96.695	123.630	9.352
51	101.643	97.805	100.059	96.913	103.641	103.127	10.903
52	86.493	141.746	90.470	74.337	94.202	99.481	8.129
53	109.376	76.730	130.024	107.088	104.664	122.406	16.848
54	98.471	41.872	84.503	95.905	112.177	98.361	12.496
55	125.110	72.116	179.470	115.673	111.338	126.680	19.554
56	125.377	90.817	161.727	122.531	103.394	91.900	16.394
57	92.328	78.871	88.181	93.809	100.434	99.032	14.634
58	125.857	125.616	146.364	130.448	103.995	116.503	12.488
59	98.659	81.158	92.422	103.382	84.177	107.335	22.805
60	77.747	40.116	63.323	86.860	81.749	81.259	21.423
61	97.925	75.369	71.468	96.044	106.511	84.559	18.379
62	99.890	67.087	70.405	97.090	91.951	129.768	13.834
63	97.987	99.953	81.949	89.003	100.415	106.074	21.193
64	88.585	117.396	76.800	96.967	90.355	86.947	22.369
65	108.839	98.066	111.912	111.477	109.194	109.712	15.042
66	106.423	86.894	120.708	111.485	100.203	94.443	13.809
67	85.019	70.340	62.866	84.785	93.291	71.474	9.938
68	77.662	59.878	62.269	77.086	83.768	75.205	8.733
69	79.751	94.648	65.736	77.245	92.085	86.952	7.591
70	70.003	86.280	46.452	76.000	80.390	66.101	7.417
71	74.284	31.396	75.892	75.269	75.398	119.805	4.793
72	81.826	109.278	50.669	93.768	70.644	83.600	7.867
73	61.088	00.000	43.727	65.359	58.510	69.031	7.997
74	74.271	58.815	63.743	84.797	81.256	75.620	9.857
75	81.383	98.589	106.859	76.098	64.783	101.805	7.811
76	95.893	45.261	95.941	102.234	100.074	106.536	10.158
77	94.391	82.398	78.956	93.972	104.597	107.269	11.057
78	101.745	72.069	86.037	118.235	102.294	104.903	10.939
79	85.160	47.671	78.335	82.826	88.311	91.778	13.088
80	115.335	99.126	135.515	122.038	103.179	128.892	13.905
81	107.173	36.923	79.968	123.275	108.241	124.506	7.009
82	103.522	92.446	94.313	108.630	90.819	117.860	15.981
83	91.421	119.665	85.980	84.286	91.587	103.441	6.648
84	122.548	158.655	113.591	118.660	113.910	132.510	16.489
85	108.146	133.566	82.698	119.862	115.920	96.177	17.781
86	126.539	91.690	166.835	123.551	109.832	120.578	15.623
87	105.595	104.522	93.290	92.429	120.534	124.854	7.247

6.1.1 (a) General distributional properties

Table 6.1 records (indirect) standardised mortality rates (SMRs) for 87 community medicine areas (CMAs) in Glasgow (1980–2). The CMAs were constructed by aggregating enumeration districts so as to preserve social and economic homogeneity.

SMRs are mortality rates adjusted for the age–sex composition of the

200

population. The SMR for the ith CMA is equal to the observed number of deaths due to a given disease (O_i) divided by the expected number of deaths (E_i) given the age–sex composition of the ith area. This value is then multiplied by 100. Age and sex specific death rates for the whole of Glasgow are used. Figure 6.1(a) shows the map for the cancer data. This type of standardisation is a useful way of representing data for a set of areas where areas differ markedly in size (absolute values and rates are sensitive to population size) and where it is necessary to allow for differences in population characteristics between areas. An important problem is to decide the set of population characteristics and the geographical extent of the reference group that will determine the value of E_i. The chi-square statistic $(O_i - E_i)^2/E_i$ is also resistant to the problem of variable population size. Visvalingam (1983) in the context of other social indicators suggests signing the chi-square value depending on whether $O_i > E_i$ or $O_i < E_i$ (negative). But all these statistics are very sensitive when used on rare events.

Figure 6.1(b) shows box plots for the 6 SMRs. Most of the distributions are positively skewed (the median is closer to the lower quartile, F_L, than the upper quartile, F_U). Figure 6.1(c) shows stem and leaf and rankit plots (SMR values plotted against their normal scores) for the 'all causes' data. These data summarising techniques show evidence of a symmetric distribution. The other five SMR data sets have similar distributional features. Figure 6.1(b) shows the presence of outliers in the case of three of the SMRs. An observation (y_i) is classified as an outlier if

$$y_i > F_U + 1.5(F_U - F_L) \text{ or } y_i < F_L - 1.5(F_U - F_L) \tag{6.1}$$

where $(F_U - F_L)$ is called the interquartile range. The CMAs with outlier values are shown in Figure 6.2. Accident deaths are rare, and counts low, compared to the other causes of death.

To what extent do neighbouring CMAs have similar mortality rates? Figure 6.3 shows plots of y_i against $(\mathbf{W}y)_i$ where $(\mathbf{W}y)_i$ is the average of the SMRs in those areas that share a common boundary with area i. So $\mathbf{W}$ is an 87×87 connectivity matrix with row sums standardised to unity and $(\mathbf{W}y)_i$ is the ith entry in the resultant vector. The plots for all causes of death and cancer rates are shown and indicate a relationship between neighbouring values in both cases. The relationship is quantified by fitting a regression line using the ordinary least squares estimator and computing the r^2 goodness of fit measure. The t ratio for the slope coefficient is given in brackets underneath the estimate. In both cases there is evidence of a relationship.

As a second example Table 6.2 records petrol prices (for four star petrol) at each of the service stations open in S.W. Sheffield observed on a given day

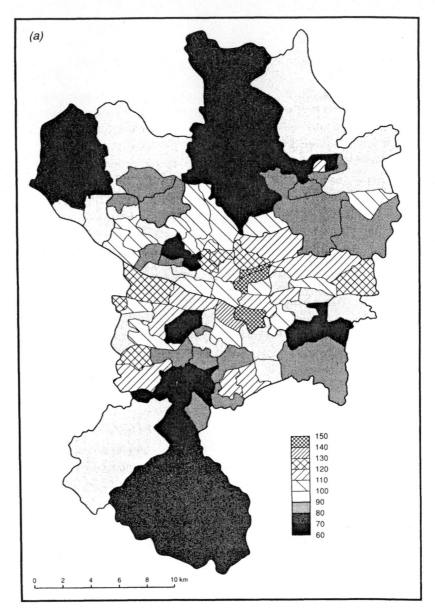

Figure 6.1. (*a*) Standardised mortality rates for the Glasgow cancer data, (*b*) Box plots for the Glasgow health data. (*c*) Stem and leaf and rankit plots for the all-causes data.

(*b*)

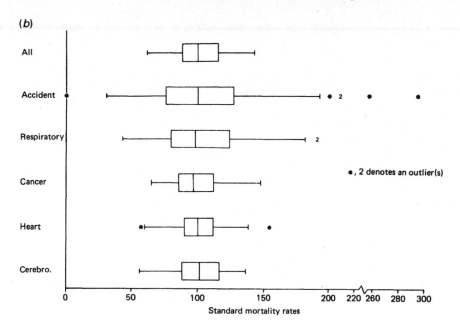

Standard mortality rates

*, 2 denotes an outlier(s)

(*c*)

Leaf unit = 1.0   *N* = 87

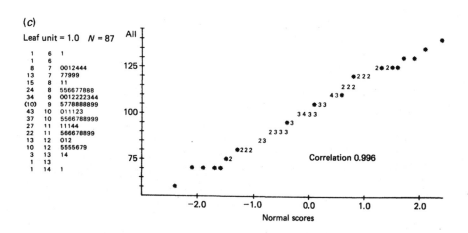

```
 1    6    1
 1    6
 8    7    0012444
13    7    77999
15    8    11
24    8    556677888
34    9    0012222344
(10)  9    5778888899
43   10    011123
37   10    5566788999
27   11    11144
22   11    566678899
13   12    012
10   12    5555679
 3   13    14
 1   13
 1   14    1
```

Correlation 0.996

Normal scores

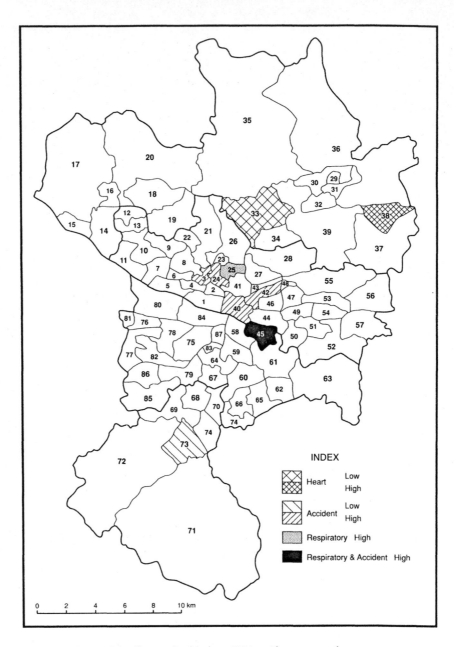

Figure 6.2. Glasgow health data: CMAs with extreme values.

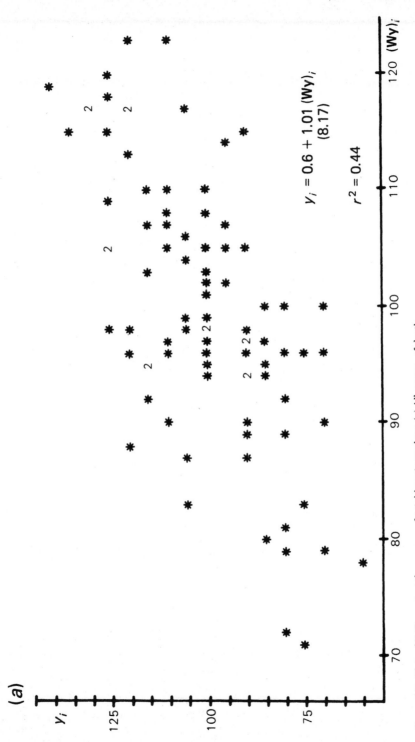

$$y_i = 0.6 + 1.01 \, (Wy)_i$$
$$(8.17)$$

$$r^2 = 0.44$$

Figure 6.3. Plots of SMRs against the average of neighbouring values. (a) All causes of death.

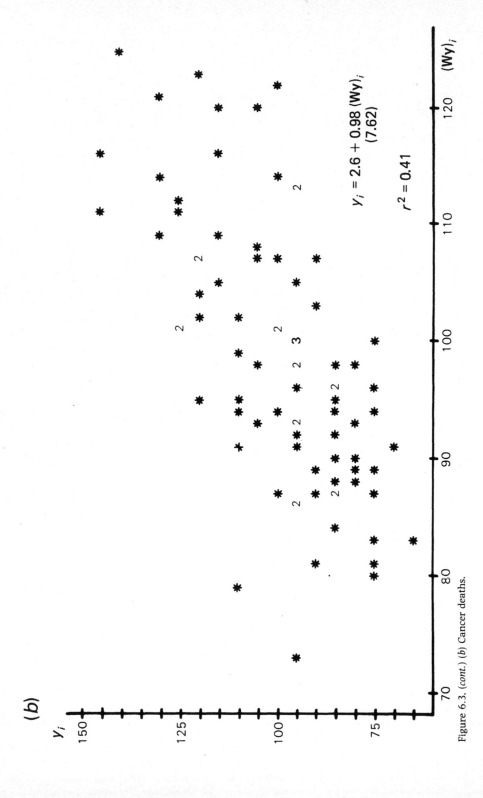

Figure 6.3. (*cont.*) (*b*) Cancer deaths.

in each month from January 1982 to January 1983. The period January to March was a period of rapidly falling prices while August to January was a period of slowly rising and then falling prices (Haining, 1983a, 1986). The graphical data summaries in Figure 6.4 show that unlike the SMR data, the price data, particularly during the period of falling prices is strongly skewed and clearly asymmetric. The 'comet-like' form of the box plots over these first three months is marked by a lengthening rear whisker and a shortening forward whisker as prices are first strung out by the price fall (January) eventually bunching within a narrow competitive band by March. By contrast the period of more stable prices in November is marked by a more symmetric box plot and no outliers.

Figure 6.5 is a map of all 88 petrol stations. The map distinguishes those stations that are on main roads from those on side streets and also identifies groups of stations that lie along different principal radial routeways leading into the city centre. The lines joining sites reflect an attempt to order the sites to reflect possible patterns of competition. Two rules have been used in constructing the linkages: proximity along the principal radial routeways and proximity at important road intersections. Many of the smaller stations off the main roads combine petrol retailing with other functions and are often not in competition with the large petrol-only stations on the main roads. Other orderings could be explored.

Are there important price differences between groups of stations and to what extent are changes from month to month in the distribution of prices the result of shifts among specific groups of stations? Figure 6.6 shows box plots of prices for the first three months for each of the different locational groups of Figure 6.5. The retailers off the main roads seem to be particularly important in accounting for the overall price dispersion while the individual routes differ in terms of their median and spread price characteristics. On the other hand the change in spread characteristics between January and March, noted above, does not seem to be due either to a convergence in median price levels between routes or reductions in spread. An alternative explanation might be that there has been a rather more general increase in local price competition with price differences narrowing between neighbouring stations. Figure 6.7 plots the lower price against the higher price for all pairs of stations joined by a link as shown in Figure 6.5, for the first three months. There is some evidence, comparing the graphs for January and March, to suggest that this is what has happened with a substantial increase in the number of site pairs charging identical or nearly identical prices.

Table 6.2. *Petrol price data: petrol stations in southwest Sheffield*

	\multicolumn{8}{c}{1982}	1983
	Jan.	Feb.	March	Aug.	Sept.	Oct.	Nov.	Dec.	Jan.
1	154.6	151.0	141.9	157.8	169.6	170.0	173.7	174.6	162.8
2	154.6	151.0	141.9	157.8	170.5	171.0	174.6	175.0	164.6
3	164.0	164.0	149.0	169.0	170.0	—	—	—	—
4	155.9	151.9	141.9	157.8	170.5	170.0	174.6	175.0	165.5
5	169.9	151.0	145.0	161.0	171.0	172.0	177.0	175.0	165.0
6	154.6	151.0	144.6	174.1	172.3	170.0	174.6	173.2	165.5
7	156.0	151.9	141.9	157.8	170.5	171.9	174.6	175.0	165.5
8	156.0	151.9	143.7	159.6	—	—	—	—	—
9	154.5	149.5	145.9	159.6	170.0	171.8	176.0	174.1	164.1
10	155.0	150.0	142.0	158.0	170.0	172.0	176.0	175.0	163.0
11	155.0	151.4	141.9	157.8	170.5	171.9	176.9	175.0	162.8
12	170.0	—	142.0	157.8	169.6	171.8	174.0	173.0	163.7
13	155.5	150.5	145.5	174.1	172.3	172.8	174.6	174.6	165.5
14	154.6	150.5	141.9	163.2	169.6	171.9	174.6	173.7	163.7
15	159.6	151.9	146.9	163.2	—	—	—	—	—
16	168.0	168.0	150.0	168.0	172.0	175.0	178.0	178.0	178.0
17	157.8	151.9	142.3	159.6	170.5	172.8	176.9	174.6	163.7
18	157.0	154.0	149.0	162.9	172.9	173.0	178.0	177.0	164.0
19	155.0	152.0	142.0	158.0	170.0	172.0	176.0	175.0	164.6
20	159.0	152.0	146.0	—	—	—	—	—	—
21	159.0	152.0	146.0	160.0	173.0	175.0	178.0	182.0	182.0
22	155.0	150.0	142.0	158.0	170.0	172.0	176.0	175.0	166.0
23	155.0	150.0	142.0	158.0	170.0	172.0	176.0	175.0	164.0
24	156.9	150.9	143.9	161.9	169.6	171.9	175.5	175.0	163.7
25	154.8	150.8	143.8	161.9	169.6	—	—	—	—
26	154.6	149.5	141.9	158.2	169.6	170.5	174.6	173.7	162.8
27	155.0	150.0	142.0	158.0	170.0	172.0	176.0	175.0	166.0
28	157.0	157.0	150.0	158.0	170.0	175.0	179.0	178.0	169.0
29	155.0	150.0	142.0	158.0	170.0	172.0	176.0	175.0	165.0
30	157.8	154.6	146.4	160.5	170.0	172.8	177.8	175.0	165.0
31	166.4	162.8	162.8	172.3	170.0	171.9	174.9	173.7	164.1
32	158.7	151.0	146.0	161.9	170.5	172.8	175.5	175.0	166.9
33	156.0	150.0	144.1	159.6	170.5	172.8	176.0	176.0	163.7
34	156.9	149.6	141.9	157.8	170.5	171.9	175.5	175.5	164.1
35	156.0	149.6	141.8	159.6	170.5	171.9	175.5	175.0	163.7
36	157.0	152.0	141.0	162.0	170.0	174.0	178.0	175.0	166.0
37	154.0	—	—	—	—	—	—	—	—
38	155.9	150.9	141.9	159.5	170.5	171.9	176.5	174.1	164.1
39	158.0	150.0	141.9	172.3	172.3	175.5	177.8	176.9	165.5
40	158.0	152.0	141.9	161.0	171.0	171.5	175.0	175.0	164.5
41	158.0	150.0	142.0	172.0	171.0	172.0	176.0	175.0	167.0
42	156.0	149.6	141.9	172.3	170.5	171.9	175.0	174.6	163.7
43	156.0	149.6	141.9	157.9	170.5	171.9	175.0	174.6	163.7
44	156.0	149.5	141.9	172.3	170.5	171.9	175.0	174.6	163.7

Table 6.2. (*cont.*)

	1982								1983
	Jan.	Feb.	March	Aug.	Sept.	Oct.	Nov.	Dec.	Jan.
45	160.0	152.0	145.0	161.0	170.5	171.5	175.5	175.0	163.5
46	156.9	149.6	144.6	—	172.3	175.5	175.5	176.4	165.5
47	159.0	152.0	146.0	159.6	171.8	173.7	178.0	176.4	165.5
48	163.0	154.0	146.0	165.5	172.0	174.0	176.0	176.0	166.5
49	154.0	—	—	—	—	—	—	—	—
50	156.9	150.0	145.9	161.9	172.3	172.8	176.9	175.0	165.5
51	154.6	149.6	144.6	160.5	170.5	171.9	175.5	175.0	164.1
52	166.0	154.0	150.0	159.0	—	—	—	—	—
53	157.8	155.0	141.9	162.8	172.3	171.9	177.8	175.5	166.0
54	155.0	151.9	141.9	158.2	—	—	—	—	—
55	159.1	151.9	144.1	162.8	172.8	173.7	178.2	178.7	166.4
56	156.9	150.5	144.6	161.9	170.5	171.9	176.4	175.9	164.1
57	164.0	159.0	149.0	166.0	172.0	175.0	177.0	177.0	169.0
58	155.0	149.6	141.9	163.2	169.6	170.5	174.6	173.7	162.8
59	155.0	150.0	142.0	—	—	—	—	—	—
60	170.0	170.0	—	172.0	172.0	172.0	—	—	—
61	153.9	148.9	141.9	157.9	169.9	171.9	174.9	173.9	163.9
62	154.4	149.8	141.8	157.8	169.8	170.5	174.6	173.7	162.8
63	154.5	150.0	141.9	157.9	170.5	171.5	175.5	173.5	162.9
64	156.0	152.0	142.0	159.0	170.0	173.0	178.0	178.0	164.0
65	169.0	162.0	149.0	165.0	172.0	172.0	—	—	—
66	155.5	150.5	141.9	157.8	169.6	171.9	174.6	175.0	163.7
67	157.8	155.0	146.0	157.8	170.5	172.8	175.5	176.4	163.7
68	158.0	153.0	142.0	157.8	170.5	171.9	174.6	176.5	164.1
69	155.0	150.5	142.0	159.0	169.5	171.5	174.0	174.5	163.5
70	155.5	150.5	142.0	157.8	170.5	172.8	175.5	176.9	164.1
71	156.9	153.9	145.9	161.9	170.0	171.9	175.5	175.5	163.7
72	154.6	149.6	141.9	172.3	170.5	171.9	175.5	175.0	164.1
73	155.5	152.3	144.1	172.3	170.5	172.8	177.8	176.4	166.9
74	154.6	148.7	141.9	172.3	170.5	171.9	174.6	175.0	164.1
75	159.6	152.3	144.1	172.8	170.5	173.7	178.2	177.3	165.5
76	164.0	160.0	149.0	183.0	172.0	174.0	178.0	178.0	167.0
77	155.5	150.9	143.9	172.3	169.6	171.9	175.5	175.0	164.1
78	155.5	152.2	—	—	—	171.9	175.5	175.0	164.1
79	155.0	151.0	148.0	172.0	172.0	—	—	—	—
80	170.0	170.0	—	—	—	—	—	—	—
81	172.0	172.0	172.0	—	—	—	—	—	—
82	159.0	159.0	152.0	168.0	172.0	175.5	178.0	178.0	168.0
83	161.0	156.0	148.0	171.0	172.0	174.0	177.0	174.0	169.0
84	160.0	160.0	160.0	160.0	170.0	174.0	174.0	174.0	170.0
85	157.0	155.0	142.0	160.0	172.0	175.0	178.0	176.0	166.0
86	—	—	—	—	170.5	175.0	174.6	175.0	166.0
87	—	—	—	—	—	—	178.0	175.0	165.0
88	—	—	—	—	—	—	—	—	163.8

(*a*)

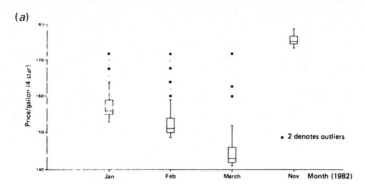

(*b*)

(*c*)

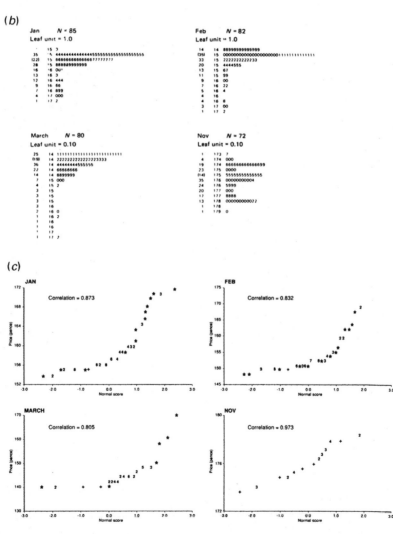

Figure 6.5. Map of petrol stations, neighbours and sub areas: Sheffield. Stations circled are not on a main road.

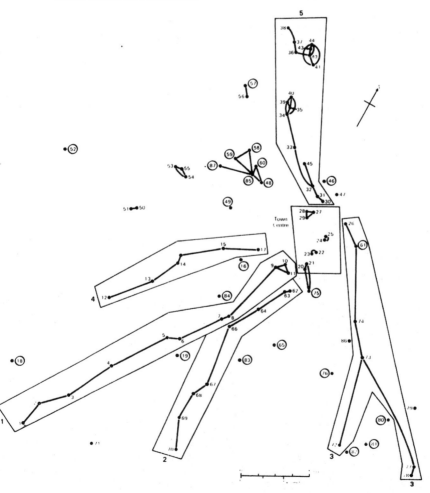

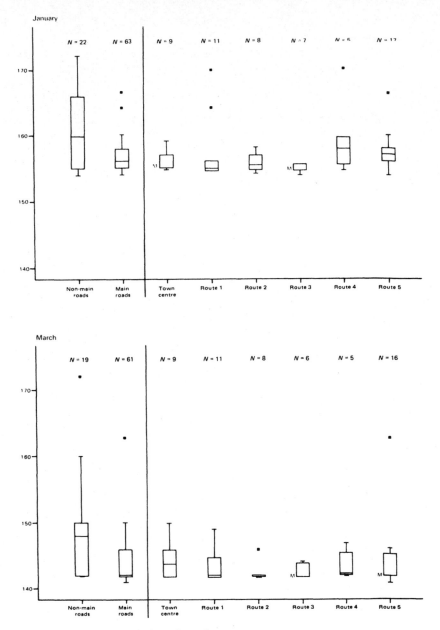

Figure 6.6. Box plots of petrol prices.

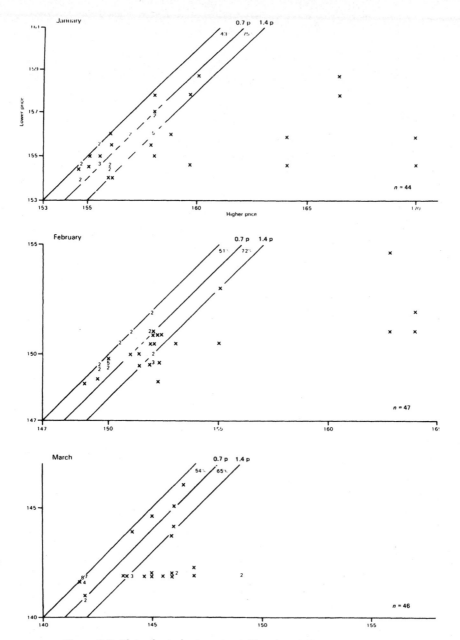

Figure 6.7. Plots of petrol prices at neighbouring stations.

6.1.1 (b) Spatial outliers

In addition to the types of distributional outliers discussed above, spatial data may contain another type of outlier, a data value which is unusual with respect to its *neighbouring* values. Such spatial outliers could have a serious distorting effect on the values of non-resistant statistical estimators of spatial properties. An indication that there may be a spatial outlier problem can be obtained by plotting graphs of pairs of data values. For example if $y(s_1,s_2)$ is the value observed at site (s_1,s_2) of a rectangular lattice then plots of $\{y(s_1,s_2),y(s_1+1,s_2)\}$ and $\{y(s_1,s_2),y(s_1,s_2+1)\}$ may help show up local outliers as will a plot of $y(s_1,s_2)$ against the (average of) four (or eight) nearest neighbours. For non-lattice data the plot of y_i against $(Wy)_i$ where the row sums of W are standardised to unity may provide similar guidance on the presence of outliers. Figure 6.8 shows plots of y_i against $(Wy)_i$ for the SMR data relating to accidents in the 87 CMAs in Glasgow. The least squares regression line is fitted to the data. Observations with large standardised residuals ($\geqslant 2.0$) are circled and the standardised value given in brackets. The standardised residual of 3.45 is probably a cause for concern in any subsequent analysis, possibly also the observation with a value of 2.72. Note that the CMA with the value of 3.45 is the central area and also has one of the lowest population levels. A note of caution, however, in the interpretation of these statistics. The observations are not independent and ordinary least squares does not provide the best estimator of the relationship between y and (Wy) and in the case of positive correlation tends to overestimate the slope parameter and underestimate the residual standard error. This has been discussed in Chapter 4. So this

Figure 6.8. Detection of spatial outliers: accident SMR data.

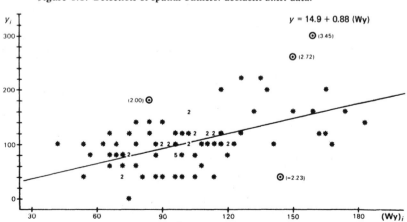

approach (rather than that described in Chapter 4) may overstate the outlier problem. It is important to emphasise the preliminary nature of this analysis.

Local outliers on lattice data can also be detected by treating each row and column as separate data sets and using the criterion (6.1) on each. The presence of a possible outlier is indicated by a large (positive or negative) value on either a row or a column. Cressie (1984) suggests testing for the presence of outliers by computing for each row and column

$$D = n^{\frac{1}{2}}(\bar{y} - \text{med}(y))/(\sigma(0.5708)^{\frac{1}{2}})$$

where $\bar{y}$ is the sample mean, $\text{med}(y)$ the sample median, n is the sample size and $\sigma = $ (interquartile range)$/1.349$. The value of D is then assessed as a standard normal deviate. Cressie (1984) gives an example of the use of this procedure. Data values do not need to be recorded at every lattice point to operate these tests. These tests can also be used on non-lattice data (providing the pattern of sites is not too irregular) by assigning sites or areas to the closest point of some regular lattice overlay although results in this case will be dependent on the assignment procedure and grid spacing. The D statistic assumes independence so, in the presence of positive spatial correlation, will tend to overstate the problem of outliers. Using these tests on each row and column may produce small sample sizes so that both tests may be rather sensitive to the estimate of the interquartile range.

6.1.1   (c) *Spatial trends*

Distributional characteristics, the presence of outliers and arrangement properties such as spatial correlation may be linked to trends or spatially heterogeneous structures in the data. If these are detected and removed the residuals may be better behaved. Trend surface analysis (see Chapters 3 and 7) represents a fairly sophisticated approach to the detection of non-stationarity in the mean. A simpler EDA approach is through mean or median polishing. These methods are best suited to lattice data though they can be adapted to non-lattice data using the regular lattice overlay method described earlier (Cressie and Read, 1989). Of the two forms of polishing, median polishing will be resistant to outliers.

The principle behind median polishing is to treat lattice data as if it were a two-way table. In the case of a two-way analysis of variance, the underlying model depicts an additive decomposition of each data value of the form:

$$\text{data value} = \underbrace{\begin{array}{ccc} \text{common} \\ \text{effect} \end{array} + \begin{array}{c} \text{row} \\ \text{effect} \end{array} + \begin{array}{c} \text{column} \\ \text{effect} \end{array}}_{\text{the fit}} + \text{residual} \qquad (6.2)$$

If $y(s_1, s_2)$ is an observation then the four elements give:

$$y(s_1, s_2) = y(.\,,.) + [y(s_1, .) - y(.\,,.)] + [y(.\,, s_2) - y(.\,,.)]$$
$$+ [y(s_1, s_2) - y(s_1, .) - y(.\,, s_2) + y(.\,,.)]$$

where '.' denotes averaging over the axis. In the case of median polishing, (6.2) is obtained by using medians rather than means. The technique is described in Tukey (1977, Chapters 10 and 11), Emerson and Hoaglin (1983b) and Emerson and Wong (1985) and an application discussed in Cressie (1984). To check for row and column interactions an extended or 'plus one fit' (Tukey, 1977, Chapter 10) can be tried in which (6.2) is replaced by the model

$$\text{data value} = \frac{\text{common}}{\text{effect}} + \frac{\text{row}}{\text{effect}} + \frac{\text{column}}{\text{effect}} \qquad (6.3)$$
$$+ k \left\{ \frac{\text{row} \times \text{column}}{\text{common}} \right\} + \text{residual}$$

and k estimated by plotting the residual in (6.2) against the 'comparison' value given by the expression in brackets in (6.3) and fitting a resistant line through the scatter of points or estimating by eye (see Tukey, Chapter 10). If the slope of the line is nearly zero ($k \approx 0$) then the purely additive model (6.2) is deemed acceptable. Extended fit median polishing is discussed in Emerson and Wong (1985).

Table 6.3 shows the application of full median polishing (using 6.2) to a set of remotely sensed data measuring reflectance values from an area of coastal water (denoted area 1) where sewage disposal has taken place (Haining, 1987c). Higher values indicate higher levels of pollution. Table 6.3(a) gives the original data, 6.3(b) the fit and 6.3(c) the residual values together with the row and column effects together with the common effect in the bottom right corner. Figure 6.9 plots the set of residuals against their comparison values. A resistant fit gives a slope of -1.72. There appears to be some row–column interaction as we would expect. The row and column effects values in Table 6.3(c) point to the presence of strong and quite complicated, high order trends in both directions. For comparison, and because these data sets will be analysed again in Chapter 7, we show the results of a median polish on two other areas giving the original data and residual values together with row, column and common effects (Tables 6.4 and 6.5). Area 3 shows only weak evidence of trend while area 2 shows a more pronounced (linear) trend but less severe than area 1. The purely additive model is fit in both cases. These results may be useful in providing a preliminary indication of the order of trend surface model that may be required in order to describe the surface.

Table 6.3. *Application of median polish to remotely sensed data: area 1*

(a) *Data*

32	35	36	37	38	47	34	35	31
38	39	43	41	55	42	38	34	37
50	62	46	39	55	37	40	32	28
45	50	43	33	24	38	44	42	39
40	36	16	18	31	37	52	30	24
37	14	10	21	26	30	35	41	19
10	12	5	12	17	18	20	24	23
50	62	19	6	14	17	17	5	6
46	35	0	4	5	5	6	0	0

(b) *Fit*

45	41	36	31	36	37	37	35	31
50	46	41	36	41	42	42	40	36
50	46	41	36	41	42	42	40	36
52	48	43	38	43	44	44	42	38
40	36	31	26	31	32	32	30	26
35	31	26	21	26	27	27	25	21
26	22	17	12	17	18	18	16	12
25	21	16	11	16	17	17	15	11
14	10	5	0	5	6	6	4	0

(c) *Residuals*

									Row effects
−13	−16	0	6	2	10	−3	0	0	5
−12	−7	2	5	14	0	−4	−6	1	10
0	16	5	3	14	−5	−2	−8	−8	10
−7	2	0	−5	[(−19)]	−6	0	0	1	12
0	0	−15	−8	0	5	(20)	0	−2	0
2	−17	−16	0	0	3	8	16	−2	−5
−16	−10	−12	0	0	0	2	8	11	−14
[25]	[41]	3	−5	−2	0	0	−10	−5	−15
[32]	[25]	−5	4	0	−1	0	−4	0	−26
Column effects 9	5	0	−5	0	1	1	−1	−5	31 Common effect

[ ] = row outlier values, ( ) = column outlier values.

Median polishing provides a fit that is close to being optimal in a least absolute residuals sense. Properties of the method are discussed in Emerson and Hoaglin (1983b) where other polishing methods are also discussed. Breakdown bounds identify the number of data values that can be outliers without affecting row or column effect estimates. In the worst case all such values would be in one column (or row) in which case the estimate for that column would be unaffected providing the number of 'bad' values was less than or equal to

$$[1/(2\max(I,J))] - [2 - d\{\min(I,J)\}]/2IJ$$

where $d\{n\}$ equals 0 if n is even and 1 if n is odd, I and J are the number of rows and columns of data. This provides the most pessimistic case. Results for the best situation where the observations are scattered through the table are given in Emerson and Hoaglin (1983b, pp. 193–7). If the number of outliers is excessive the analyst should consider modifying the outliers in order to assess their effect on the fit. An important practical feature of median polishing is that it can be applied where there are gaps in the data, although the effect of this is sometimes greatly to increase the number of iterations required for the polish and substantially alter the breakdown bound. A more disconcerting feature, however, is that median polishing can give different results if polishing starts on the rows first or the columns first and this underlines its role as a *preliminary* tool for data analysis. Results of fitting also depend on the number of iterations performed.

If trend is evident in a set of data after the application of a median polish it is natural to apply the earlier data analytical methods to the residuals. Cressie (1984, 1986) suggests, in a kriging context, using median polish to remove data trends and from then on working with the residuals for purposes of estimating the variogram. So the crucial issue is now how

Figure 6.9. Residual versus comparison plot (area 1).

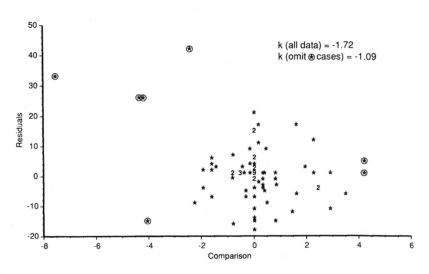

218

Table 6.4. *Application of median polish to remotely sensed data: area 2*

(a) *Data*

16	16	23	22	31	28	22	23	22
27	23	23	23	26	25	24	24	25
29	20	27	33	27	37	32	30	27
15	21	32	28	30	26	18	29	20
12	19	21	23	22	23	24	17	25
17	16	17	15	22	21	25	21	19
16	26	19	15	24	16	20	18	24
15	19	19	20	15	16	22	23	21
17	16	16	19	15	24	21	16	8

(b) *Residuals*

									Row effects	
−3	−3	2	−1	7	5	−2	0	0	0	
6	2	0	−2	0	0	−2	−1	1	2	
3	−6	−1	3	−4	7	1	0	−2	7	
−7	−1	8	2	3	0	−9	3	−5	3	
−7	0	0	0	−2	0	0	−6	3	0	
0	−1	−2	−6	0	0	3	0	−1	−2	
0	10	1	−5	3	−4	−1	−2	5	−3	
−1	3	1	0	−6	−4	1	3	2	−3	
2	1	−1	0	−5	1	−3	−10	−4		
Column effects: −4	−4	−2	0	1	0	1	0	−1	23 Common effect	

the *residuals*, not the original data, are distributed and whether there is evidence that outliers are present. In the case of the three sets of pollution data, only the residuals from area 1 show evidence of a marked asymmetry with outliers. The D statistics for the rows and columns of area 1 are shown in Table 6.6 (starting in the top left).

Large values are evident in rows 4, 8 and 9 and columns 5 and 7 and the suspect values have been circled in Table 6.3(*c*). Figure 6.10 plots residual values against neighbours in the next row above. This plot highlights possible outliers: 25 (column 1 row 8), 41 and 19. Figure 6.11 plots residual values against neighbours in the next column to the right and highlights: 25 (column 2 row 9), 25 (column 1 row 8) and 41. These plots are better at detecting outliers than Figure 6.12 which plots residual values against the mean of the neighbours (denoted $z(s_1,s_2)$) and so tends to smooth differences. Further, all three plots indicate a tendency for neighbouring observations to be correlated.

Table 6.5. *Application of median polish to remotely sensed data: area 3*

(a) Data

3	6	9	11	3	4	8	9	9
3	5	9	7	8	6	6	8	7
7	10	10	8	12	7	8	11	10
13	18	13	14	12	15	10	12	4
16	14	11	12	12	10	11	11	8
10	14	16	10	7	9	11	8	5
7	11	13	9	6	4	2	5	5
6	9	7	11	7	9	3	6	4
1	3	11	9	4	9	14	10	10

(b) Residuals

									Row effects
-4	-4	0	2	-5	-3	1	1	2	-1
-3	-4	1	-1	1	0	0	1	1	-2
-1	-1	0	-2	3	-1	0	2	2	0
2	4	0	1	0	4	-1	0	-7	3
6	1	-1	0	1	0	1	0	-2	2
1	2	5	-1	-3	0	2	-2	-4	1
2	3	6	2	0	-1	-3	-1	0	-3
0	0	-1	3	0	3	-3	-1	-2	-2
-8	-9	0	-2	-6	0	5	0	1	1
Column effect -1	2	1	1	0	-1	-1	0	-1	9 Common effect

Table 6.6. *Row and column D statistics for the residual values for area 1 (Table 6.3c)*

Row	1	2	3	4	5	6	7	8	9
D	-0.47	-0.52	0.89	-3.77	0.0	-0.71	-0.84	3.49	6.07
Column	1	2	3	4	5	6	7	8	9
D	0.46	1.13	-1.61	0.0	2.67	0.89	3.12	-0.39	-0.79

When the distribution of sites or areas is very irregular (as in the case of the Glasgow CMAs) applying methods that depend on reassigning values to a lattice may be too subjective. In the case of the CMA data we might be interested in seeing if, for any SMR, rates follow a trend with respect to distance from the city centre. Figure 6.13 shows a series of box plots for three SMRs (retaining the individual values in the display) for groups of CMAs that are grouped in terms of 'lagged' distance from the city centre (CMA 40). The CMAs that share a common boundary with CMA 40 are at distance band 1, all the CMAs that have common boundaries with the CMAs at distance band 1 are at distance band 2 and so on. The plots in figure 6.13 do not allow for direction; moreover the numbers of CMAs falling into each group are quite variable. The groups could be sub-divided if some directional hypothesis was to be explored (for example, do rates differ between the north and south sides of the city) and McGill *et al.* (1978) suggest making box widths a function of sample size when numbers of observations differ between box plots that are to be compared. The all causes and cancer SMRs show evidence of a decline with distance from the central area and increasing spread in the case of the all causes rates. The rates for respiratory diseases as a cause of death show some evidence of trend and considerable spread at all distance bands. This case also reveals outliers at distance bands 1 and 3 which indicates values that are extreme given their distance from the city centre. The spatial correlation in the all causes SMRs revealed by Figure 6.3 appears now to be due to this trend. After removing the distance dependent median values from the rates and plotting these values against the local averages (Figure 6.14) the local pattern is considerably weaker and the regression fit confirms this.

Figure 6.10. Plot of observed values against North–South neighbours (area 1). ⊗ = possible spatial outlier.

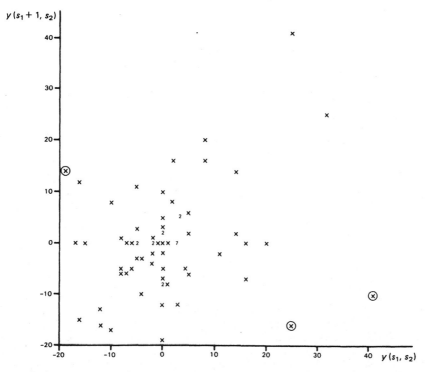

6.1.1 (d) Second order non-stationarity

For many techniques it is necessary to check whether there are non-stationarities in the covariance (C) or correlation (R) structure of the data. Although sometimes mentioned this is a condition that can only, at best, be partially checked. Given earlier comments on the behaviour of estimators for C, R or the semi-variogram in Chapter 4, it will be difficult to say with any certainty whether observed differences in estimated values (in different parts of the map) are real differences or merely sampling variation. One approach is to partition the region into disjoint parts and then compute separate estimates for each of the parts and compare. This is a diagnostic check that is only useful for large lattice data sets. In the case of non-lattice data, site pattern effects (as a part of overall sampling variation) may be too severe to permit any meaningful comparison. For area data, where areas differ greatly in size, non-constant variance should be checked for (see page 49). Cressie and Read (1989) provide an example, partitioning and then plotting data values against area sizes.

Figure 6.11. Plot of observed values against East–West neighbours (area 1). $\otimes$ = possible spatial outlier.

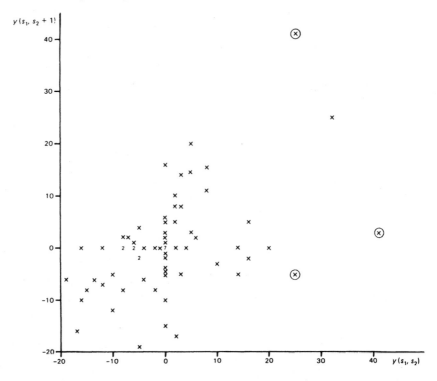

6.1.1 *(e) Regional subdivisions*

Techniques of univariate classification applied to the set of observations but including contiguity constraints may be a useful way of detecting and defining regional subdivisions and can be extended to multivariate data sets. For reviews of spatial classification methods see for example Semple and Green (1984) and Balling (1984). It is likely that such techniques will be particularly important with large data sets or sets that cover extensive areas or areas with considerable physical or social and economic diversity. After a model has been fitted regional patches or mosaics may be indicated by the presence of large spatial clusters of residuals of identical sign and similar size.

6.1.2 *Multivariate data analysis*

The simultaneous exploration of several variables in their spatial context raises special problems when the number of variables exceeds two or three. There are two aspects: to develop methods that present multivariate data

Figure 6.12. Plot of observed values against the average of all neighbours.

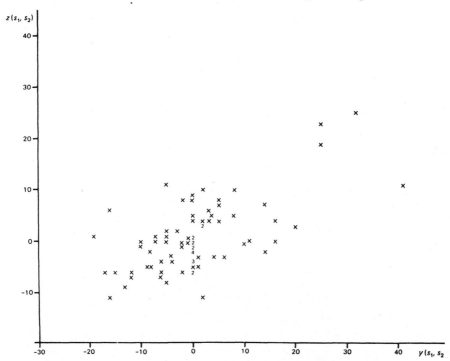

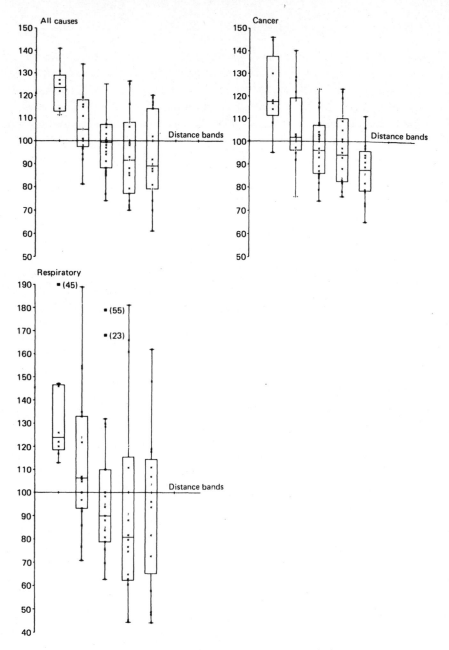

Figure 6.13. Box plots with distance from the centre of Glasgow: SMR data.

sets in a way that assists exploratory analysis, and to display the observational units in their spatial context, again in a way that assists data exploration.

When the numbers of variables are relatively few (less than four or five) bar diagrams, wind roses and glyphs are traditional devices for presenting multivariate data over a group of areas. When there are a larger number of variables some data reduction technique such as principal components might be used to eliminate variables or construct new hybrids which still account for a large proportion of the variability in the original data. Scatterplots are used to examine variable relationships: to check for linearity and whether the scatter is uniform over the range of the explanatory variable. The use of different graphical methods is discussed in Chambers *et al.* (1983) and Cleveland (1985).

In EDA it is useful to retain contact with the original data for as long as possible before resorting to techniques that transform or simplify the data. When there are large numbers of variables, finding useful ways to present the data is not easy. One interesting idea proposed by Chernoff was the use of faces, each variate in the data determining a particular characteristic of the face. Up to 18 variables can be plotted this way. The human eye is used to comparing faces although we may tend to be more sensitive to some facial features (such as the eyes and mouth) than others. Huff and Black (1978) provide an application and discussion of Chernoff faces to multivariate urban data consisting of 16 variables measured for 60 American cities. Mortality rates determined nose length, climatic data the shape of the head,

Figure 6.14. Plot of SMR values against the average of all neighbours after removal of distance related medians: All causes SMR data.

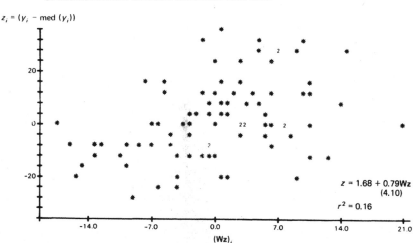

socio-economic data the position and shape of the eyes, pupils and brows and pollution data the position and shape of the mouth. Long noses mean high mortality rates. Wide sad mouths signify high levels of pollution. However detection of spatial attributes require the data to be mapped which raises the second of the two issues. Figure 6.15 shows the Chernoff faces computed by Huff and Black on a base map of the U.S.A. This does help in the detection of regional clusters (New England, California) and to identify anomalous cities (the pollution problem in Boston, the atypical socio-economic characteristics in New York and York Pennsylvania). Sub-sequently the data can be input to a clustering algorithm (with distance or contiguity as a restraining characteristic in the classification procedure to ensure contiguous regions). Whilst mapping the data would seem to be essential if the analyst wishes to examine spatial properties of the data, there is a wide choice of metrics for representing inter-point distances or area size – an issue which applies equally to the spatial representation of univariate data sets. Inter-point distances may be measured in terms of such quantities as time distance, cost distance, interaction or movement distance (Gatrell, 1983). Points or area size may be transformed to reflect variate character-istics. On the other hand, the use of these transformations can make it difficult to display the area and retain other important characteristics, especially relative location. (For an interesting early discussion of map transformations in geography see Bunge, 1966.) It may be preferable therefore to subsume important area characteristics within the carto-

Figure 6.15. Chernoff faces: 60 U.S. cities.

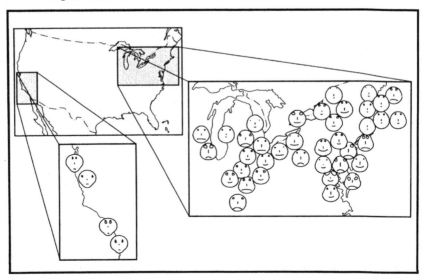

graphic device used to represent the other variables. For example, a multivariate health map of Glasgow could be constructed using Chernoff faces (one for each CMA) with head size representing population size. A range of possibilities should be explored (particularly matching variables to different facial features) – the only limitation is the analyst's imagination, and the available technology, for nearly all the methods described here require plotting equipment and software which is not yet widely available, if they are to be useful for EDA (as opposed to providing one-off data displays). EDA is essentially interactive in nature, for which efficient flexible software and access to good graphics facilities for obtaining rapid terminal displays and hard copy print outs are needed (Cleveland and McGill, 1988). Moellering (1984) provides an overview of developments in computerised cartography and map transformations.

6.1.3 Data transformations

For many techniques of spatial data analysis the data are assumed to be normal. In some situations the requirement is that $y = (y_1, \ldots, y_n)$ is a (single) drawing from a multivariate normal distribution (Chapter 3). With only one drawing available this assumption cannot be tested. Moreover, a histogram of the set $\{y_i\}$ need not necessarily look normal even if y is from a MVN distribution. Current tests that a set of observations are drawn from a normal distribution such as those due to Shapiro and Wilk and Shapiro and Francia (see Wetherill, 1986) and the robust method in Hoaglin *et al.* (1983, p. 425) assume independence. Departures from the independence assumption in the case of regression residuals can, apparently, have a serious affect on test levels (Wetherill *et al.*, 1986, p. 196), although the author is not aware of any systematic examination of this in the case of more general correlated data sets. Using independent subsets of the data or obtaining uncorrelated residuals (providing an appropriate model for the data is known or can be identified) might be tried, though again the author is not aware of any systematic exploration of these options. Once independent observations have been obtained there may be the problem of transforming data that is suspected of being non-normal. A widely used class of data transformations that assume independence are the class of Box–Cox transformations where a scaler λ is sought such that:

$$Y^{(\lambda)} \begin{cases} = (Y^{\lambda} - 1)/\lambda & \lambda \neq 0 \\ = \log_e Y & \lambda = 0 \end{cases}$$

and $Y^{(\lambda)}$ is normally distributed (see Wetherill, 1986, pp. 166–8). Hinkley (1977) describes a robust method for estimating λ.

In regression, transformations of the response variable may be needed to normalise residuals and transformations of the response and/or the explanatory variables may be needed to linearise relationships and stabilise variances. No one transformation may accomplish all these objectives.

6.2 Preliminary data analysis: detecting spatial pattern, testing for spatial autocorrelation

In this section we examine tests for spatial autocorrelation. Data $\{y_i\}$ are said to be autocorrelated if neighbouring values (as defined in section 3.1) are more alike than those further apart. Cliff and Ord (1981, p. 8) provide a rather more formal definition which is discussed in Haining (1980) and Upton and Fingleton (1985). The main emphasis of the work in this field is in testing a hypothesis (H_1) that observations are autocorrelated in space against a null hypothesis (H_0) that the observations are not autocorrelated. (Generally speaking these are not tests for independence for which the procedures of Chapter 4 are required.) A number of statistics have been developed to test for this property and here we merely overview the available methods and offer a summary of practical guidelines. For a detailed and extensive treatment of the problem the interested reader is referred to Cliff and Ord (1981). There is also an extended treatment in Upton and Fingleton (1985) with many examples, and also Goodchild (1986), Griffith (1987) and Odland (1988).

We discuss available test statistics, construction of spatial autocorrelation tests (that is, identifying the distributional behaviour of test statistics and moment properties) and interpretation of results. We conclude with an assessment of available methods. Testing regression residuals for spatial autocorrelation has been treated in Chapter 4.

6.2.1 Available test statistics

The simplest nominal scale data are represented by a binary classification – presence/absence or black/white (BW). If similar values cluster together on a map (positive spatial autocorrelation) then the number of joins where contiguous counties are both black (BB), and the number of joins where both counties are white (WW) would be 'large' relative to the number of joins where one county was black and the other white (BW). If colours alternated (negative spatial autocorrelation) by the same reasoning BB and WW counts would be 'small' relative to the BW count. A random pattern, with no autocorrelation would lie somewhere in between in terms of these counts (see Figure 6.16).

Detecting spatial pattern, testing for spatial autocorrelation

The associated test is called the 'join-count' test. Formally let

$y_i = 1$ if the ith county is B
$\quad\ 0$ if the ith county is W

and let

$\delta_{ij} = 1$ if i and j are contiguous $(i \neq j)$
$\quad\ 0$ otherwise

The observed BB count of joins is

$$BB = \tfrac{1}{2}\Sigma_{(2)}\delta_{ij}y_iy_j$$
$$BW = \tfrac{1}{2}\Sigma_{(2)}\delta_{ij}(y_i - y_j)^2$$
$$WW = A - (BB + BW)$$

where $A(=\tfrac{1}{2}\Sigma_{(2)}\delta_{ij})$ is the total number of joins in the county system and

$$\Sigma_{(2)} = \Sigma_i\Sigma_{j(i \neq j)}$$

An important generalisation of this test is the k colour statistic $(k>2)$, constructed along identical lines for the case of k (unordered) colours. These tests $(k \geqslant 2)$ can be further modified for generalised fixed proximity matrices, replacing δ_{ij} with w_{ij} where $\{w_{ij}\} = W$ (see Chapter 3). Such generalisation permits differential weighting between contiguous neighbours, that is

$w_{ij} \geqslant 0$ if i and j are contiguous $(i \neq j)$
$\quad\ = 0$ otherwise

where non-zero values of w_{ij} are specified and fixed in advance of testing (section 3.1). Also positive values of w_{ij} can be assigned to non-contiguous pairs at different distances apart so that tests for spatial autocorrelation can be conducted at different scales.[1]

Figure 6.16. Clustered, random and alternating map patterns.

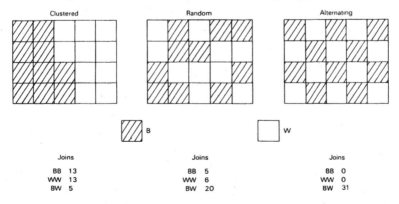

229

Now consider data $\{y_i\}$ measured at the ordinal and interval scale. If the map pattern was clustered we would expect that if counties i and j were contiguous, on average $(y_i - y_j)^2$ would be 'small'. So for such a pattern the sum of such squared differences over all contiguous pairs would be 'small' relative to a random pattern. For an alternating map pattern we would expect the same sum to be 'large' relative to a random pattern. A test, based on this principle, due to Geary is given by:[2]

$$c = [(n-1)/4A](\Sigma_{(2)}\delta_{ij}(y_i - y_j)^2 / \Sigma_{i=1,\ldots,n} z_i^2)$$

where $\bar{y}$ is the mean of the n observations $\{y_i\}$ and $z_i = y_i - \bar{y}$. On the other hand suppose cross products $z_i z_j$ are computed. In the case of a clustered pattern these products will tend to be positive while in the case of an alternating pattern these products will tend to be negative. A test based on this principle, due to Moran, is given by

$$I = [n/2A](\Sigma_{(2)}\delta_{ij}z_i z_j / \Sigma_{i=1,\ldots,n} z_i^2)$$

As with the join-count tests, both the I and c tests can be generalised by replacing $\{\delta_{ij}\}$ by $\{w_{ij}\}$ in order to accommodate different forms of intercounty connectivity and orders of contiguity. The form chosen reflects the structure of interdependence that is assumed in the alternative hypothesis.

These statistics are all special cases of the general cross product statistic

$$\Gamma = \Sigma_i \Sigma_j G_{ij} C_{ij} \tag{6.4}$$

where G_{ij} is a measure of the spatial proximity of locations i and j and C_{ij} is a measure of the proximity of i and j in terms of variate values. The raw index is then usually subject to some standardisation to obtain a final descriptive measure. This generalisation shows that the problem of measuring spatial autocorrelation can be seen in terms of comparing two matrices $\mathbf{G} = \{G_{ij}\}$ and $\mathbf{C} = \{C_{ij}\}$ using (6.4). For example, the Moran coefficient is obtained by setting $G_{ij} = w_{ij}$ and $C_{ij} = (y_i - \bar{y})(y_j - \bar{y})$, while for the Geary coefficient $C_{ij} = (y_i - y_j)^2$. The links between (6.4) and different measures of spatial autocorrelation as well as the general utility of (6.4) for examining spatial autocorrelation are discussed in Hubert *et al.* (1981).

Since there is often arbitrariness in the choice of values for G, a test based on order relationships between the $\{G_{ij}\}$ may reflect more closely the true extent of our knowledge about the spatial system under study and reduce the risk of arriving at decisions based on arbitrary assessments of spatial relationships. Hubert *et al.* (1981) discuss a spatial autocorrelation test, derived from (6.4) in which it is only necessary to specify, for any site (i) whether $G_{ij} > G_{ik}$ or $G_{ij} < G_{ik}$. For example we can write formally:

$a(i,j,k) = 1$ if i,j,k are distinct and $G_{ij} < G_{ik}$
$\qquad = 0$ otherwise
$b(i,j,k) = 1$ if i,j,k are distinct and $C_{ij} > C_{ik}$
$\qquad = 0$ otherwise
$$\Gamma = \Sigma_i \Sigma_j \Sigma_k a(i,j,k) b(i,j,k) \qquad\qquad (6.5)$$

In this definition large G values denote spatial proximity but small C values denote variable similarity (as in the Geary coefficient). Γ in (6.5) counts the number of consistent relationships, that is the number of times (for any row of the matrix, i.e. county) $G_{ij} < G_{ik}$ and $C_{ij} > C_{ik}$. All comparisons are based on single rows so only entries within a single row of G need to be commensurable. Hubert *et al.* (1981) give a worked example using $C_{ij} = |y_i - y_j|$. If comparisons *between* rows are also permissible the statistic is easily adapted to count consistent relationships between *all* pairs of entries in the two matrices (Hubert *et al.*, 1981).

The statistic (6.4) can be used to formulate a variety of tests including tests for unidimensional series (Hubert *et al.*, 1985a), space–time interaction (Mantel, 1967) and tests for directionality in spatial flow data (Hubert *et al.*, 1983) as well as more general forms for C_{ij} 'tailored specifically to the aims of a given research project and data set' (Hubert *et al.*, 1985a). Multivariate tests, for spatial association between two or more variables, can also be constructed and these will be discussed in Chapter 8.

6.2.2 Constructing a test

We now consider the problem of testing the null hypothesis (H_0) of no spatial autocorrelation against the two tailed alternative (H_1) that the data are spatially autocorrelated. In some cases a one tailed test is appropriate with H_1 either specifying that the data are positively autocorrelated or negatively autocorrelated. In the case of a one tailed test, testing is carried out on one of the ends of the distribution.

The distributional behaviour and moments of statistics, under H_0, are evaluated under one of two assumptions; either the normality assumption or the randomisation assumption. Under the normality assumption for H_0 the observed map, consisting of n observations, is the result of n independent drawings from a normal population. The observed map is then one possible realisation of an underlying normal probability model. The reference distribution for a statistic is obtained by considering the set of all possible realisations and the associated set of realised values of the statistic. Under the randomisation assumption for H_0 the observed map is one possible arrangement of the set of n values. The reference distribution for the statistic

231

is obtained by considering the $n!$ possible permutations of the n values (assuming that the analyst does not wish to introduce any constraints into the set of possible arrangements). Unlike the normality assumption, the second assumption deals only with the given set of values and inferences are conditional on this set. In the case of the join-count statistics the equivalent assumptions are free and non-free sampling respectively. Under free sampling each individual county is coded B or W independently with identical probabilities p and $1-p$ respectively. Under non-free sampling, each county has the same probability, a priori, of being B or W but coding is subject to the overall constraint that there are n_1 counties coloured B, n_2 coloured W with $n_1 + n_2 = n$ (Cliff and Ord, 1981, Chapter 1). In all these cases the test procedure is to see whether the observed statistic is an extreme member of its sampling distribution or reference set.

Cliff and Ord (1981, pp. 46–53) demonstrate the asymptotic normality of the I, c and join-count statistics under both sets of assumptions for H_0. An exception to this is the case of the BB join count statistic where for p small, under non-free sampling, a Poisson approximation should be used. The asymptotic normality of the I statistic may be preserved even if the data are non-normal (for details see Sen, 1976). The most serious threat to asymptotic normality appears to arise from testing under sparse W matrices where only a small number of lattice points figure in most of the joins. In such cases asymptotic normality will not hold. Cliff and Ord (1981, p. 50) give the example of the star lattice and Ripley (1981, p. 99) gives a similar example.

The moment properties for these tests are available in the cited literature. Display 6.1 reproduces the first two moments for Γ and Moran's I. Upton and Fingleton (1985, Chapter 3) develop the moments of all these tests, under the randomisation assumption, from the moments of Γ. For any test statistic r, $(r - E(r))/(\mathrm{Var}(r))^{\frac{1}{2}}$ can in many situations be treated as a $N(0,1)$ deviate, where $\mathrm{Var}(r) = E(r^2) - (E(r))^2$. Ord (1980) shows that, for the case of I, $E_R[I^2]$ is an unbiased estimator for the expected value of I^2 for *any* distribution so can be used both when assuming randomisation under H_0 or when assuming independent drawings from any non-normal distribution.

Of considerable practical importance is the validity of the normal approximation in the case of small and moderate sized lattices. Cliff and Ord (1981, pp. 53–63) report the results of extensive simulation work on the distribution of the I, c and join-count tests under the randomisation assumption. They consider sample sizes less than 50, varying the structure of weights, the distribution of the data and the average number of joins per county. They indicate, particularly in the case of I and c, that the normal distribution provides a satisfactory approximation to the tails of the

Display 6.1. *First two moments of autocorrelation statistics.*

(a) First two moments of the general cross product statistic (Γ) under the randomisation assumption.

$$E_R(\Gamma) = S_0 T_0 / n(n-1)$$
$$E_R(\Gamma^2) = [S_1 T_1 / 2n^{(2)}] + [(S_2 - 2S_1)(T_2 - 2T_1)/4n^{(3)}]$$
$$+ [(S_0^2 + S_1 - S_2)(T_0^2 + T_1 - T_2/n^{(4)}]$$

(b) First two moments of Moran's I statistic for general **W** under randomisation assumption (Cliff and Ord, 1981, p. 46)

$$E_R(I) = -1/(n-1)$$
$$E_R(I^2) = \{n[(n^2 - 3n + 3)S_1 - nS_2 + 3S_0^2]$$
$$- b_2[(n^2 - n)S_1 - 2nS_2 + 6S_0^2]\}/(n-1^{(3)}S_0^2$$

(c) First two moments of Moran's I statistic for general **W** under the normality assumption (Cliff and Ord, 1981, p. 44)

$$E_N(I) = -1/(n-1)$$
$$E_N(I^2) = (n^2 S_1 - nS_2 + 3S_0^2)/[(n^2 - 1)S_0^2]$$

Notation:

$S_0 = \Sigma_{(2)} w_{ij}; \quad S_1 = \frac{1}{2}\Sigma_{(2)}(w_{ij} + w_{ji})^2; \quad S_2 = \Sigma_i(w_i + w_{\cdot i})^2; \quad w_i = \Sigma_j w_{ij}; \quad w_{\cdot i} = \Sigma_j w_{ji};$
$n^{(b)} = n(n-1)(n-2) \ldots (n-b+1); \quad b_2 = m_4/m_2^2$ where $m_k = n^{-1}\Sigma_i(y_i - \bar{y})^k$.
T_k as for S_k with w_{ij} replaced by c_{ij}.
n is the number of observations.

sampling distributions in a wide variety of situations so that inferences 'will not be seriously in error'. They propose corrections to improve the fit. It seems clear however that the normal approximation should not be used without care particularly in the case of small lattices under the randomisation hypothesis. Evidence for this has come partly from empirical and theoretical work on the general cross product statistic where, in the case of moderate n, the sampling distributions may show considerable skewness (see for example Mielke, 1978; Costanzo et al., 1983). A number of approaches to the problem of obtaining the reference set of a statistic have been proposed. These have included using the exact moments to provide crude Chebychev bounds (which may be adequate if the test statistic is large) and various curve fitting procedures to approximate the complete distribution based on the first four moments (Siemiatycki, 1978). Costanzo et al., (1983) examine the use of a Pearson Type III function (a function with an exact skewness parameter that can be estimated) as an alternative to the normal approximation.

A more widely used method is to approximate the distribution by randomly sampling from the $n!$ permutations (if n is small a complete enumeration may be possible). If n_s permutations are obtained, the significance level of the observed value of the statistic is assessed by computing the number of *generated* values of the statistic that are more extreme. The topic is discussed in detail by Besag and Diggle (1977), Edgington (1969) and Cliff and Ord (1981). A value of $n_s = 99$ is often deemed adequate but as Costanzo *et al.* (1983) point out, the distribution may be lumpy and sampling bias can occur so that it may be beneficial to check results by using another method such as fitting the Pearson Type III distribution. All these modifications may add considerably to the effort required in testing for spatial autocorrelation.

6.2.3 Interpretation

Under the randomisation (non-free sampling) assumption the statistics test for the presence of pattern, or organisation, in the distribution of observed values relative to the set of all possible patterns in the given data set. Rejection of H_0 implies the presence of pattern as defined by the structure of **W**. When the row sums of **W** are scaled to unity these statistics may also be interpreted as providing a test for the 'predictability' of a given map pattern, that is, the extent to which it is possible to predict observations on the basis of their neighbours as specified in **W** where the predicted value is

$$\hat{y}_i = \Sigma_j w_{ij} y_j \qquad i \neq j$$

(Gatrell, 1977).

Under the normality (free sampling) assumption the statistics test whether the data are autocorrelated or not, providing test assumptions are met. The normality assumption assumes independent drawings in each county from the *same* distribution (constant mean and variance) so that if a surface contains trend, even if the variables are independent around the trend, H_0 may be rejected. The free and non-free sampling assumptions assume that p, the probability that a county is B, is *identical* for all counties. Although I has the general form of a spatial correlation coefficient and c the form of a semi-variogram coefficient neither should be used as model specification devices for which more specialist procedures are needed (see Chapter 4).

In their original forms the autocorrelation statistics do not provide an index or measure of autocorrelation that can be interpreted in a similar way to, say, an ordinary correlation coefficient. This is because, in the case of the I coefficient for example:

$$|I| \leqslant n/S_0 [\Sigma_i (\Sigma_j w_{ij} (y_j - \bar{y}))^2 / \Sigma_i (y_i - \bar{y})^2]^{\frac{1}{2}} \qquad (6.6)$$

where $S_0 = \Sigma_{(2)} w_{ij}$ (Cliff and Ord, 1981, p. 21). The value in (6.6) is generally less than 1 and depends on the $\{w_{ij}\}$. The statistic I could be used as an index of spatial autocorrelation after scaling by the right hand side of (6.6), but when appropriate its presentation as a standard normal deviate seems more informative.[3]

More detail on the underlying structure of spatial autocorrelation is obtained by carrying out a sequence of tests with alternative W matrices. Haggett (1976) gives an example using data on the spread of measles in S.W. England where different W matrices reflect different assumptions about the way the disease might have spread. In this study, analysis seeks to detect different scales and types of pattern and also give some indication of the role of different spatial mechanisms at different periods of the epidemic. A problem here, though, is that several alternative forms may yield significant results and patterns cannot be tested against one another (the null hypothesis is always an absence of pattern). As has often been pointed out there is usually an overabundance of choices for W and the difficulty is in selecting a form which provides a decisive alternative, Not only may there be arbitrariness in the choice of weighting values but also in the choice of elements of W to make non-zero. Kooijman (1976) has suggested finding the W matrix that maximises the spatial autocorrelation statistic, as one approach to these types of issues, but this is likely to prove computationally demanding.

Carrying out tests with alternative W matrices provides a more detailed description and test of pattern and also avoids the risk of failing to detect pattern which may only be present at certain scales. By defining W to identify pairs of sites separated by a specific distance or lag on a network and repeating this over successive distance bands or lags, pattern properties can be examined at different scales. The plot of the value of the statistic against distance is usually termed the spatial correlogram and can be computed using any of the previous statistics although the Moran and Geary statistics are often favoured because of their links with covariance and semi-variogram estimators. (Details for the Moran case are in Cliff and Ord, 1981, p. 119.) Indeed there appears to be a convergence of method as well as objective in computing spatial correlograms on the one hand or estimating covariances or semi-variograms on the other (see section 6.3). However, spatial correlograms are limited in use to the examination and testing of spatial pattern and do not provide a rigorous basis for developing models for the data in the sense of Chapters 3 and 4. They are most often used in areas of research where such model building is deemed inappropriate because of the severe forms of non-stationarity that are believed to underlie the data. This is often the case, for example, when data are in aggregated form over an irregular partition where regions differ greatly in size. Upton and Fingleton (1985, Chapter 3.5) provide numerous examples of the use of correlograms and how they may be evaluated in a program of research. Since bounds differ at different lags (see (6.6)) they suggest scaling any plot by dividing each estimate by its upper bound. One advantage claimed for computing the spatial correlogram is that it may be possible to distinguish patterns due to data trends from patterns arising from patchiness and second order variation. (Sokal and Oden, 1978; Sokal, 1979).

Neighbouring values of the correlogram are highly correlated so its usefulness is restricted to detecting broad structure rather than detail. Oden (1984) develops a test for the significance of an entire correlogram. If I^* is a k-dimensional vector of Moran correlogram estimates then under the null hypothesis of no correlation (at any lag)

$$Q = (I^* - E(I^*))\Sigma^{-1}(I^* - E(I^*))$$

is approximately χ_k^2 where Σ^{-1} contains the theoretical variances and covariances of I^* under the null hypothesis. Rejection of the null hypothesis still leaves the problem of examining correlogram estimates whilst controlling for effects at other distances. A possibility is to compute the partial correlogram. If r_j denotes the Moran coefficient calculated at distance class j and scaled (see 6.6) then the interdependence of the estimates is defined by the Yule–Walker equations

$$r_j = \Sigma_{k=1,\ldots,d}\, p_{d,k} r_{j-k}$$

where $p_{d,d}$ is the partial correlogram at distance class d. Writing

$$p = (p_{d,1}, \ldots, p_{d,d})^{\mathsf{T}};\ q = (r_1, \ldots, r_d)^{\mathsf{T}}$$

$$R = \begin{bmatrix} 1 & r_1 & r_2 & \cdots\cdots\cdots & r_{d-1} \\ r_1 & 1 & r_1 & \cdots\cdots\cdots & r_{d-2} \\ . & . & . & \cdots\cdots\cdots & . \\ . & . & . & \cdots\cdots\cdots & . \\ r_{d-1} & r_{d-2} & r_{d-3} & \cdots\cdots\cdots & 1 \end{bmatrix}$$

then $p_{d,d}$ is obtained by solving

$$p = R^{-1}q$$

and this must be repeated for each d. The partial correlogram is the plot of $p_{d,d}$ against d. Upton and Fingleton (1985) give examples and note the problem of constructing formal significance tests for these estimates. There is further discussion in Cliff and Ord (1981, p. 135) and an alternative computational procedure. The function eliminates periodicity effects (where correlogram peaks at large scales are due to effects at smaller scales) and gives a measure of pattern at each distance.

6.2.4 Choosing a test

With so many tests and forms of test to choose from we conclude with some comments that seem important in making a choice. The principal issues dealt with here are: what statistic to choose, what form of W to adopt and whether to test under the normality or randomisation hypotheses.

The comments here on choice of statistic largely refer to the more common tests. Although a great many types of univariate autocorrelation test can be constructed using the generalised cross product statistic, there has been, to date, little comparative analysis. In the case of nominal data the join count tests are widely used with the BW statistic preferred to the BB since both are equally easy to compute but the former considered more informative (Upton and Fingleton, 1985, p. 211). For ordinal and interval scale data the main choice is between the I and c tests. Cliff and Ord (1981, p. 174) report power comparisons between these two statistics when the underlying model is a first order single parameter SAR model. They conclude that generally I performs better than c.

Since c computes squared differences it may be more sensitive to outlier values than I, on the other hand I unlike c requires the estimation of the mean ($\bar{y}$) which is also sensitive to outliers. The generalised cross product statistic offers the possibility of developing robust tests tailored to the nature of the data (evaluating H_0 under the randomisation hypothesis). One such example is a test based on evaluating absolute differences, computing $|y_i - y_j|$, (see Hubert *et al.*, 1981 and Royaltey *et al.*, 1975). The effects of mean variation could be eliminated by a median polish prior to testing, a modification which might prove useful in the application of several of the previous statistics in order to make a provisional separation of first and second order attributes of spatial variation.

The distributional behaviour and moment estimators for both I and c are robust to 'reasonable' departures from normality for large n (Cliff and Ord, 1981, p. 14). However, the empirical results presented by Cliff and Ord (1981, pp. 54–6) suggest that the normal approximation holds (slightly) better for the tails of the distribution of I than c when the sample size is small.

Since these statistics are often used as exploratory devices, a statistic such as (6.5) based on orderings rather than quantitative measures for $\{w_{ij}\}$ may be safer. Alternatively, a binary matrix embracing relatively few non-zero entries may be used. A danger with complex W matrices is that if correlation effects are restricted to specific scales then these may be swamped or cancelled out in an analysis which seeks to incorporate (within a single analysis) a number of possible scale effects through an extensive W matrix. It might seem safer, and more informative, to consider several tests with different simple forms for W. In each case the structure of the W matrix is organised to test for autocorrelation at different distances or to reflect reasonable expectations about the possible existence and nature of autocorrelation in the data. This is particularly important if the analyst hopes to extend work to shed light on possible underlying mechanisms, although a problem with such a program of testing is that results from different tests will not be independent.

The choice between the randomisation and normality (independence) hypotheses is partly a substantive choice. Cliff and Ord (1981, p. 12) use a retailing example. Randomisation is appropriate for a study of a shopping area map where there has been planning to control the mix (but not the locations) of shops. Independence is appropriate if the mix and location of shops is unplanned and uncontrolled. If the normality assumption is used it should be possible to imagine the phenomena recurring as in the case of say an epidemic or a distribution of prices across a set of retail outlets, or even a set of regional crop yields. Randomisation demands no such precursor and is considered, by its advocates, conceptually simple. It is also closer in spirit to the principles and methods of exploratory data analysis for it classifies data and suggests the presence of data structure solely in terms of what is given in the data. In addition, randomisation has certain practical advantages: it enables the analyst to either construct or sample from the exact sampling distribution for the statistic, whilst as indicated above, randomisation moments provide good estimates, in the case of I, of the expected value of I^2 under the independence assumption for *any* distribution. In the case of join count tests, the free sampling assumption requires that p, the probability that a region is coloured B, is either known a priori or can be estimated from prior realisations or occurrences.

Often what an analyst is looking for is a quick, simple but reliable test for the presence of pattern before engaging in a more detailed examination of its structure and the reasons for it. In the case of ordinal or interval scale data that seems to be provided by the I statistic and providing the frequency plot of the data is not seriously non-normal and providing the lattice is well behaved (in the sense that the join structure is not dominated by one or two regions) and providing $n \geqslant 20$ the statistic can be tested as a standard normal deviate using $E_N[I]$ and either $E_R[I^2]$ or $E_N[I^2]$. The summary of Upton and Fingleton (1985, p. 211) provides further discussion of these points.

6.3 Describing spatial variation: robust estimation of spatial variation

The statistics of section 6.2, when evaluated (and tested) over a succession of distances or lags provide a quantitative description of spatial variation at various scales. A description of spatial variation is also provided by estimating spatial covariances $\{C(k)\}$ or the semi-variogram $\{\gamma(k)\}$ across a range of distances or lags. However, a more specific reason for estimating these functions is that they provide a basis for specifying a model for spatial variation (Chapter 3) essential for many areas of univariate and multivariate data analysis, and this should be borne in mind throughout.

Estimators for $\{C(k)\}$ and $\{\gamma(k)\}$ have been discussed in Chapter 4. It was noted there, however, that the usual estimator for $\gamma(k)$, $\hat{\gamma}(k)$, is sensitive to non-normality in the distribution of data values and to the presence of outliers. Not only has this led to the adoption of data analytical methods to detect the presence of these characteristics, it has also led to the development of robust and resistant estimators for $\{\gamma(k)\}$. Since the usual semi-variogram estimator is based on squared differences it is particularly sensitive to outlier values. Although the usual spatial covariance estimator is less likely to be as sensitive to outliers, nonetheless, since it depends on $\bar{y}$ (rather than a resistant estimator of location) resistant versions of the statistic should also be considered (Cressie, 1984).

Non-normality and outliers may be identified using the methods of section 6.1. However, it is particularly useful to identify which covariance estimates or semi-variogram estimates may be most affected by the presence of outliers. Chauvet (1982) proposed the semi-variogram 'cloud' which is a plot of $\{(y_i - y_j)^2, d_{ij}\}$ where d_{ij} is the distance between sites i and j. Cressie (1984) suggests the use of the square root differences cloud instead which is a plot of $\{|y_i - y_j|^{\frac{1}{2}}, d_{ij}\}$. Even if the data are Gaussian, $\{(y_i - y_j)^2\}$ will be highly skewed in distribution since $(Y_i - Y_j)^2$ are χ^2 distributed. It will be difficult then to tell from a plot of Chauvet's cloud whether large values are due to atypical values or the inherent skewness of the underlying distribution. Cressie (1984) gives an example using lattice data (see Figures 6.17 and 6.18), comparing the two plots as box plots, and identifying the lagged distances at which the semi-variogram is most affected by outliers. Kitanidis (1983) also gives examples using both types of plot. Cressie (1984)

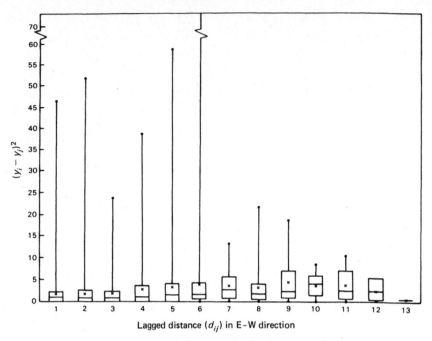

Figure 6.17. Box plots of Chauvet's cloud. After Cressie (1984).

Figure 6.18. Box plots of square root differences cloud. After Cressie (1984).

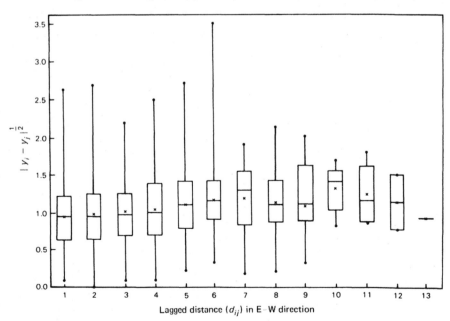

also discusses the use of the pocket plot for identifying the contribution made by each *row* of the data (on a regular grid) to the semi-variogram estimator at different lags. The value of these procedures is that they help identify the extent to which atypical data values might influence the analysis of second order data properties. It is quite possible that outliers detected by the methods described in section 6.1 do not seriously influence estimates of the semi-variogram. Conversely new problem areas of the data might be highlighted. In the case of covariance estimates an equivalent to Chauvet's cloud is the plot $\{(y_i - \bar{y})(y_j - \bar{y}), d_{ij}\}$ and pocket plots can be designed to identify the contribution made by each row of the data to the covariance estimator at different lags.

This discussion indicates that one approach to dealing with these problems is to use conventional estimators after first editing the data to remove outliers and transforming the data to normalise. We now examine an alternative approach based on the theory of robust estimation.

6.3.1 Robust estimators of the semi-variogram

The design of robust estimators for the semi-variogram ($\gamma(\mathbf{h})$) depends on the point of view adopted in estimating

$$2\gamma(\mathbf{h}) = \mathrm{Var}[(Y(\mathbf{u}+\mathbf{h}) - Y(\mathbf{u})] \tag{6.7}$$

$$= E[(Y(\mathbf{u}+\mathbf{h}) - Y(\mathbf{u}))^2] \tag{6.8}$$

(see Chapter 3). The traditional estimator ($\hat{\gamma}(\mathbf{h})$) may be viewed either as an estimator of *scale* for the differences (6.7) or as an estimator of *location* for the squared differences (6.8). Now $\hat{\gamma}(\mathbf{h})$ is a robust estimator of scale in the case of jointly Gaussian variables but is not resistant to outliers. In the non-Gaussian case $\hat{\gamma}(\mathbf{h})$ is neither resistant nor robust. On the other hand $\hat{\gamma}(\mathbf{h})$ is neither a robust nor resistant estimator of location even for the Gaussian case (Dowd, 1984). The design of robust and resistant estimators reflects this difference. (Other approaches are discussed in Cressie and Hawkins, 1980.)

Cressie and Hawkins (1980) examine several robust location estimators for $\{(y(\mathbf{u}_i) - y(\mathbf{u}_j))^2\}$. Included in these are four estimators from the class of M-estimators (Huber, 1978). Cressie and Hawkins examine the Huber, Tukey, Hempel and Andrews M-estimators together with mean and median estimators, trimmed mean estimators and $\hat{\gamma}(\mathbf{h})$ using a bias correction. They generate contaminated normal data laid out along a traverse with an imposed unilateral autoregressive structure (identical to a first order temporal autoregression).[4] Although the M-estimators perform well, the mean estimator, with bias correction performs very well and is easy to compute. (*M*-estimators involve considerable computation particularly if the sample size is large.) The mean estimator is given by:

$$2\gamma_{CH}(\mathbf{h}) = f(n_h)[(1/n_h)\Sigma_{i,j\epsilon D_h}(|y(\mathbf{u}_i) - y(\mathbf{u}_j)|)^{\frac{1}{2}}]^4 \qquad (6.9)$$

where

$$f(n_h) = (0.457 + 0.494n_h^{-1} + 0.045n_h^{-2})^{-1}$$

D_h denotes the set of sites separated by distance **h** and n_h denotes the number of such sites. As n_h increases, the correction factor (for bias), $f(n_h)$, simplifies to $(0.457)^{-1}$. An alternative to (6.9) is obtained by computing the median rather than the mean of the square root of the absolute differences, and the estimator can be written

$$2\gamma_{CH}(\mathbf{h}) = f(n_h)\{\text{median}[|y(\mathbf{u}_i) - y(\mathbf{u}_j)|^{\frac{1}{2}}]\}^4 \qquad (6.10)$$

This estimator is, for large *n*, equivalent to the quantile estimator

$$2\gamma_q(\mathbf{h}) = Q_q[y(\mathbf{u}_i) - y(\mathbf{u}_j)]^2 \qquad (6.11)$$

where Q_q denotes the qth quantile of the squared differences, providing $q = 0.5$ and the correction is made.

Median based estimators are the simplest resistant estimators of location but are not robust estimators for strongly asymmetric distributions which in fact are very likely to arise when squared differences are taken. The Cressie and Hawkins estimator (6.9) possesses similar resistance and robustness properties (Dowd, 1984).

A simple scale estimator of the semi-variogram is the mean absolute difference estimator

$$2\gamma_D(\mathbf{h}) = [(1/n_h)\Sigma_{i,j\epsilon D_h}|y(\mathbf{u}_i) - y(\mathbf{u}_j)|]^2 \qquad (6.12)$$

Another simple estimator is the median absolute difference estimator given by either

$$2\gamma_H(\mathbf{h}) = 2.198\{\text{median}|y(\mathbf{u}_i) - y(\mathbf{u}_j)|\}^2 \qquad (6.13)$$

or, setting $z_{ij}(\mathbf{h}) = (y(\mathbf{u}_i) - y(\mathbf{u}_j))$

$$2\gamma_{H*}(\mathbf{h}) = 2.198\{\text{median}|z_{ij}(\mathbf{h}) - \text{med}(z(\mathbf{h}))|\}^2 \qquad (6.13a)$$

where $\text{med}(z(\mathbf{h}))$ is the median of $\{z_{ij}(\mathbf{h})\}$. Medians in (6.13) and (6.13a) are taken over all pairs separated by distance **h** and 2.198 is a correction factor. Both estimators are resistant to outliers. Other resistant scale estimators have been suggested by Lax (1975) and Armstrong and Delfiner (reported in Dowd, 1984) the latter adapted from Huber's scale estimator.

242

These estimators are designed to cope with distributional and outlier effects but not sampling problems as may arise when an irregular spatial sampling design leads to clusters of sample sites in some areas and sparse coverage in others. Omre (1984) suggests a weighted estimator

$$2\gamma_0(\mathbf{h}) = \Sigma_{i,j\epsilon D_h} \tilde{a}_{ij}(y(\mathbf{u}_i) - y(\mathbf{u}_j))^2 \tag{6.14}$$

where $\tilde{a}_{ij}$ is a non-negative empirically estimated weighting in which observations from sparsely sampled areas receive more weight than those in heavily sampled areas.[5] If sampling is uniform $\tilde{a}_{ij} = (n_h)^{-1}$. The derivation of the weights follows an approach by Switzer (1977).

A major problem is to decide when to use the different estimators. The traditional estimator is best when the data are normal, there are no outliers and when spatial sampling is uniform. If the first two conditions do not hold then one solution is to transform the data and/or eliminate outliers. The Omre estimator (6.14) seems to be the only estimator available to cope with irregular sampling designs.

Theoretical comparison of the different estimators is difficult because spatial observations are not independent and it is possible, for example, that estimators may be sensitive to the form of inter-site correlation in the data. Reported experiments based on real and simulated data suggest the following general observations. The Cressie and Hawkins estimator defined by (6.9) works well on contaminated normal data unless there are severe levels of contamination in which case (6.10) may be preferable. Hawkins and Cressie (1984) discuss further the bias and sampling variability properties of (6.9) and (6.10) and conclude that overall (6.9) is better for large n and small n and (6.10) for intermediate values of n. However, these results are for contaminated normal data and both these estimators are sensitive to distributional deviations from normality. The M-estimators examined in Cressie and Hawkins (1980) did not perform as well and the quartile estimator (6.11) gave limited success in the cases examined by Armstrong (1984). The effect of a biased sampling design on (6.9) was examined by Omre (1984) who found that this estimator performed poorly relative to (6.14) and in some cases relative to the traditional estimator. In a comprehensive examination of many of the estimators discussed above using two sets of real (non-normal) data Dowd (1984) found that the median absolute difference estimator given by (6.13a) was the best performer, although (6.9) performed well, as did the Lax estimator. A problem with some of the M-estimators examined by Cressie and Hawkins and the Huberised scale estimators is that they are much more difficult to compute which also appears to be true of (6.14). Further discussion of issues on the choice of resistant and robust semi-variogram estimators can be found in Armstrong (1984). In many cases the range of distributional assumptions under which different estimators are optimal is an area still requiring further research. The performance of estimators, including the traditional estimator, is often improved by ensuring that the data have been carefully detrended.

6.3.2   Robust estimation of covariances

Cressie (1984) suggests replacing $C^*(\mathbf{h})$ (see Chapter 4) with

$$C^+(\mathbf{h}) = (1/n_h)\Sigma_{i, j \in D_h}(y(\mathbf{u}_i) - \text{med}(y))(y(\mathbf{u}_j) - \text{med}(y))$$

where $\text{med}(y)$ is the median of the set of observations. This estimator is both resistant to outliers and less biased than $C^*(\mathbf{h})$ when the surface mean is unknown.

Since for stationary processes $\gamma(\mathbf{h}) = C(0) - C(\mathbf{h})$ the estimation problem for the covariance function and the semi-variogram are similar. Robust and resistant methods for $\gamma(\mathbf{h})$ should therefore be appropriate. Chung (1984) compares the performance of $C^*(\mathbf{h})$ with that of two versions of the Andrews M-estimator, one version a direct form of the estimator, the other using the fourth root power transformation also used by Cressie and Hawkins (1980). Tests are carried out on one-dimensional simulated normal data with negative exponential correlation function. Ten percent of the data are corrupted by random noise to reflect the influence of outliers. Chung's results show that $C^*(\mathbf{h})$ performs very poorly in comparison with the Andrew's estimators.

6.4   Concluding remarks

The methods of sections 6.1 and 6.2 are useful for preliminary stages of analysis when the objective is to identify data properties and perhaps stimulate questions from the data. They are also useful at later stages of analysis, after the fitting of a model, to help assess the validity of the model. This will include the identification of residual properties including the identification of areas of the map where the model may be giving a poor fit.

The exploratory nature of the methods should be emphasized and the fact that when applied to spatially correlated data some of the methods of this chapter are of unknown reliability having been developed initially, like so many areas of statistics, for situations where observations are independent.

An important theme of this chapter has been the presentation of methods for detecting problem values. Problem values may have a disproportionate influence on later stages of analysis. It may be necessary to use resistant methods when such data problems are encountered. Some areas of application of resistant methods have been discussed in section 6.3. However, it is easy to stress the need to be alert to data problems, not so easy to devise good methods for detecting them nor for deciding when their effects are serious. In the case of spatial data, values may be extreme either in a distributional sense or spatial sense (or both) and the implications of these two situations in a modelling context may be different and difficult to assess. This issue is taken up in more detail in the next two chapters.

NOTES

1. Dacey (1965) introduced a non-parametric test for two colour maps illustrated for BW joins and since generalised by Cliff, Martin and Ord (1975) to cover BB joins and general k colour maps. The test involves counting the number of counties with $0,1,2,\ldots,R$ joins of the same (BB) colour (or different colours, BW) where R is the maximum number of joins for any county in the region. The expected number of joins under the null hypothesis of independence can be derived thus producing a $2 \times (R+1)$ contingency table of observed and expected counts. The test is reviewed in Cliff and Ord (1981, pp. 194–6). Since entries in the contingency table are not independent the usual χ^2 test cannot be employed and Cliff, Martin and Ord (1975) provide correction factors. Another non-parametric test is given in Sen and Soot (1977).
2. Another test, due to Royaltey, Astrachan and Sokal (1975) uses the absolute difference between y_i and y_j. Their test is for ranked data with binary spatial weights. Hubert (1978) has since generalised the statistic to interval scale data and arbitrary weights using the general cross product statistic (6.4).
3. The maximum also depends on the specific $\{y_i\}$ observed. De Jong *et al.* (1984) obtain extreme values of I and c for all sets of observations $\{y_i\}$

$$\text{For } I: I_{\max}=(n/2A)m_{\max} \qquad I_{\min}=(n/2A)m_{\min}$$
$$\text{For } c: c_{\max}=((n-1)/4A)m_{\max} \qquad c_{\min}=((n-1)/4A)m_{\min}$$

where $A=\frac{1}{2}\Sigma_{(2)}w_{ij}$ and $m_{\max}$ and $m_{\min}$ are the largest and smallest eigenvalues of $\mathbf{PWP}$ where $\mathbf{P}=(\mathbf{I}-\mathbf{C})$, $\mathbf{I}$ is the identity matrix, $\mathbf{C}=\{c_{ij}\}$ and $c_{ij}=1/n$.

De Jong *et al.* (1984) discuss how these limits might be used to indicate whether the normal approximation is likely to be valid in testing the significance of the I statistic. Note that I adjusted by (6.6) is the correlation coefficient on $\{y_i, \Sigma_j w_{ij} y_j\}$.
4. Contaminated normal data $\{U_t\}$ may be generated as a mixture of distributions where

$$U_t \sim \begin{matrix} N(0,\sigma^2) & \text{with probability } 1-\varepsilon \\ H & \text{with probability } \varepsilon \end{matrix}$$

where $\varepsilon \ll 1$ and H is some heavy tailed distribution with mean zero. While the bulk of the distribution comes from the $N(0,\sigma^2)$ distribution a small amount (representing outliers) comes from H. A choice for H might be $N(0,k\sigma^2)$ where $k \gg 1$. A widely used model for contaminated data is to set $\varepsilon = 0.05$ and $k = 9$.

Suppose there is a smooth (no nugget variance) stationary normal process $\{X_t\}$ with variogram $2\gamma_X(\mathbf{h})$ but suppose that it is the process $\{Z_t\}$ that is observed where $Z_t = X_t + U_t$. $\{U_t\}$ is independent of $\{X_t\}$. The variogram for the $\{Z_t\}$ process is

$$2\gamma_z(\mathbf{h}) = 2\{\gamma(\mathbf{h}) + \sigma^2[1 + \varepsilon(k-1)]\}$$

so that contamination introduces a nugget effect and displaces the variogram upward from the uncontaminated case ($\varepsilon = 0$) by a constant amount $\varepsilon\sigma^2(k-1)$. (Hawkins and Cressie, 1984).

5.   This statistic is similar in appearance to the numerator of the Geary coefficient, c, defined for a specific distance band and with arbitrary weights $\{w_{ij}\}$. However, the weightings $\{w_{ij}\}$ in c do not adjust for variable spatial coverage, only for variable distance separation within the specified distance band.

PART D

Modelling spatial data

7

Analysing univariate data sets

Section 7.1 considers the problem of finding a good description of a spatial surface. We begin by considering how to fit a trend surface model in which spatial variation is assumed to contain three scale components: a large scale or regional trend, a local component of continuous (spatially correlated) variation and a site or area level random (or noise) component. Comparisons with geostatistical approaches are made. Models for non-normal and presence/absence data are briefly considered. Good description of spatial variation is important for summarising data and for interpolation. If data are available on the same variable in other areas it may be of interest to compare surfaces. If data are available on the same variable in the same area but over a succession of time periods comparative study might be concerned with identifying how the surface is changing through time and to relate this to changes in conditions. Over a relatively short span of time this might take the form of a 'before and after' study, such as monitoring and assessing the effects of an anti-pollution campaign.

In section 7.2 spatial interpolation problems are discussed. Observations on a continuous surface are made at a number (n) of point sites in an area. These might be soil or vegetation measurements, or data from geological samples. Estimates may be required for sites that have not been visited (together with error estimates); perhaps a map is required. The theory and practice of kriging (almost synonymous with geostatistics) represents an important approach to interpolation widely used in the natural and environmental sciences for characterising and assessing resource availability. The approach is *sequential*: develop a description for the *observed* data first and then use this to estimate values at the unknown sites.

Consider a slightly different version of the same problem. Measurements have been taken at a fixed number (n) of sites (weather stations for example) or perhaps have been collected for n areas into which a region has been subdivided (such as census tracts, or pixels on a remotely sensed image). For one reason or another observations are missing – one or more weather stations may have been temporarily out of commission, census data might have been lost or never reported; monthly employment data on an area basis may be incomplete because of office closure, transfer or industrial dispute; the sensor device might have failed along a scan line or patchy cloud obliterated small sections of a remotely sensed image. The analyst may wish to estimate these 'missing values' in order to complete the record. If analysis is proposed with the incomplete data and with records taken across the same area but at other periods of time, (which may also be incomplete but not necessarily in the same locations) then 'completing' the records may help to retain comparability. If several variables are recorded at the *n* sites, and some multivariate analysis is proposed such as multiple regression, deleting sites with missing observations, or excluding variables which are not recorded at all sites, could be very wasteful of data or give misleading results if important variables have to be discarded. Again it might be important to try to complete the record. There might be another reason for estimating values. The techniques of Chapter 6 may have highlighted some observations which the analyst now feels certain are wrong, or if not wrong may have a disproportionate influence on model fitting. The analyst may wish to carry out a sensitivity analysis by replacing these values with estimates that are consistent (in either a local or global sense) with other values of the same variable. Unlike the kriging problem the *n* sites are treated as the spatial population and the missing values are gaps in the specified data matrix. In this case it might be appropriate to base model selection on the observed data, but to treat the problems of model fitting and missing value estimation *simultaneously*, rather than sequentially.

7.1 Describing spatial variation

The decomposition of surfaces into different components of variation is a difficult exercise. It is far from clear how different components should be isolated, indeed whether the procedures currently available allow them to be isolated satisfactorily. The problems are particularly severe with social and environmental data where surface variation may be very complex.

Filter mapping is perhaps the simplest method used. Filter mapping consists of passing a rectangular grid of a given size across the map and computing averages (or ratios) for each cell of the partition. The set of average values are then mapped. The operation is repeated using a larger grid with the cells of the next lower order grid nesting in 2s or 4s for example within the larger grid squares. Successive levels filter out detail at each scale leaving only larger scale variation. At each level residual maps can be drawn. Different forms of filter mapping are described in, for example, Haggett (1965, p. 269), Tobler (1969), Curry (1970) and Cressie and Read (1989). A limitation with this method is that results vary with the grid size chosen. The methods considered now are less sensitive to such choices and often provide more precise quantitative description of surface properties.

7.1.1 Non-stationary mean, stationary second order variation: trend surface models with correlated errors

A trend surface model for a variable (y) is defined

$$\mathbf{y} = \mathbf{A}\boldsymbol{\theta} + \mathbf{e} \tag{7.1}$$

where $\mathbf{y}$ is an $n \times 1$ vector corresponding to the n sample sites, $\mathbf{A}$ is the matrix of location co-ordinates for the n sites and $\boldsymbol{\theta}$ is the vector of trend surface parameters. Both $\mathbf{A}$ and $\boldsymbol{\theta}$ depend on the order of trend surface. The structure of $\mathbf{A}\boldsymbol{\theta}$ is described in section 3.2. In the usual form of the model, $\mathbf{e}$ is an unobserved vector of independent random variables, where $E[\mathbf{e}] = 0$ and $E[\mathbf{e}\mathbf{e}^\top] = \sigma^2 \mathbf{I}$ usually interpreted as site effects or measurement error.

Apart from providing a description of a surface, (7.1) has also been used to identify the scale components of an areal pattern and give insights into the importance of underlying process elements in those cases where areal variation can be linked to specific spatial processes. It is best at showing the broad features of a surface, often being used to remove those broad features in order to examine local scales of variation. It may also be used as part of an interpolation procedure. The model has been used in a number of areas of the physical sciences particularly geology and geophysics but also in geography, archaeology and meteorology. In the earth sciences it has been widely used in the analysis of magnetic field data, stratigraphic (facies) maps, isopach maps and continues to be used as a primary tool in geological analysis for the identification and separation of surface features, estimating mineral resources and for data interpolation and smoothing. A summary introduction to applications in various fields can be found, in, amongst other sources, Agterberg (1984), Hodder and Orton (1976), Chorley and Haggett (1965), Krumbein and Graybill (1965).

Model (7.1) can be fitted by ordinary least squares as a special case of multiple regression. Special precautions have to be taken since polynomial regression is an ill-conditioned least squares problem – a problem accentuated if data points are clustered. Even for cubic order surfaces $\mathbf{A}^\top \mathbf{A}$ can be nearly singular so that matrix inversion to obtain an estimate of $\boldsymbol{\theta}$ becomes a problem. Scaling the study area to the unit square and the use of orthogonalising multiple regression software packages reduces this problem. Failure to observe these points can result in very unreliable estimates of θ. Details are discussed in Ripley (1981, pp. 30–4) and Upton and Fingleton (1985, p. 324). A further problem is order selection. Residuals from a trend surface tend to be autocorrelated, especially if the order of the fitted model is too low. In that case least squares parameter estimates are inefficient and errors 'carry less information than the (statistical) theory suggests and we would usually be led to fit a surface of too high an order' (Ripley, 1981, p. 34). These issues have been discussed in section 4.5. Other problems such as border and frame shape effects as well as clustered distributions of point sites may also affect results (see for example Upton and Fingleton, 1985, pp. 324–5; Unwin and Wrigley, 1987b).

The main interest in this section is with a generalisation of (7.1) in which

$$y = A\theta + u \tag{7.2}$$

where $E[u] = 0$ and $E[uu^T] = V$ where V need not be diagonal. Permissible forms for V are described in Chapter 3. The non-diagonal form of V introduces an additional component into the variation of y. While $A\theta$ in (7.1) describes large scale variation and e describes local scale 'noise', u in (7.2) introduces a third component with an intermediate scale of variation. A model such as (7.2) might be advocated if it is suspected that there are processes operating at this scale. With remotely sensed data, (7.2) might provide a better structure for describing spatial variation, with V capturing the underlying correlated measurement error associated with each scan line. Model (7.2) might also provide a more parsimonious description of a surface (for purposes of interpolation and data smoothing) if underlying trends are high order polynomials. Even if none of these conditions hold, in view of the problem of residual spatial correlation in fitting (7.1), (7.2) may often be a statistically safer model to fit providing a suitable form for V can be specified. However, given that patterns of spatial variation associated with many models for V may exhibit trend-like behaviour (especially over small regions) it is often safer to think of (7.2) as providing a data description rather than trying to ascribe process significance to the components of variation. It is certainly unwise to try to infer process from such a decomposition. Some of these issues have been considered recently with examples by Agterberg (1984), Hoeksema and Kitanidis (1985) and Haining (1987c).

Fitting procedures for (7.2) have been discussed in sections 4.2 and 4.3 where it is noted that there is the additional problem of specifying a model for V for results may be quite sensitive to its chosen form. We now consider some applications of these methods, which develop these issues.

Example 7.1: A spatial comparative analysis (Haining, 1987c)
Figure 7.1 shows reflectance values for three areas taken from an extensive areal survey monitoring pollution levels due to the pumping of waste material into the English Channel. The sampled areas were at different distances from exhaust points. Areas 1 and 2 are equal distances from the pipe but area 1 is closer to a discharge point on the pipe. Area 3 is further offshore. These data sets were analysed by median polish in Chapter 6. There is evidence of trend in at least two of the data sets. Estimation of this trend would provide a measure of the dispersal of pollutants from the source point, parameter values of the trend surface indicating the dispersal gradient. Allowing for a non-diagonal matrix V is justified on two counts. First, remotely sensed data are correlated because of the nature of the recording device (Chapter 2). Second, at a process level, pollution in any small area (pixel) will be affected by local mixing and local dispersal arising from small scale turbulence and wave action.

Different orders of trend surface were fitted using the iterative maximum likelihood procedure of section 4.2, testing the least squares residuals at the first iteration for residual autocorrelation. When significant residual autocorrelation was found correlations were estimated up to lag 6 in order to help select an appropriate model for V. These tests are performed and correlation estimates obtained each time a new order of trend surface is fitted. Significance tests on model parameters use the estimate of the asymptotic variance covariance matrix of the estimators (see section 4.2).

Three forms for the specification of V were examined: (i) a single parameter SAR model, (ii) empirical estimates of the lag correlations, and (iii) Agterberg's model for lag covariance estimates (see section 4.3). Approaches (ii) and (iii) are consistent with an assumption of local stationarity whereas (i) is non-stationary. Agterberg's function generally underestimated the empirical covariances (as expected), moreover this method and the method based on empirical covariances did not always converge so that no trend surface parameters could be estimated. (This problem could probably have been overcome by setting small spatial covariance estimates to zero or by using a model for the estimated covariances.) Of the three methods the SAR model worked best (it always converged) and in addition guarantees a positive definite form for V. Table 7.1 gives the (least squares) estimates of the parameters of the zero order model together with standard errors and the generalised Moran coefficient as a standard normal deviate. All three data sets have autocorrelated residuals. Table 7.2 gives the lag correlation estimates. The monotonic decay is a common feature and indicates why the SAR model often provides

253

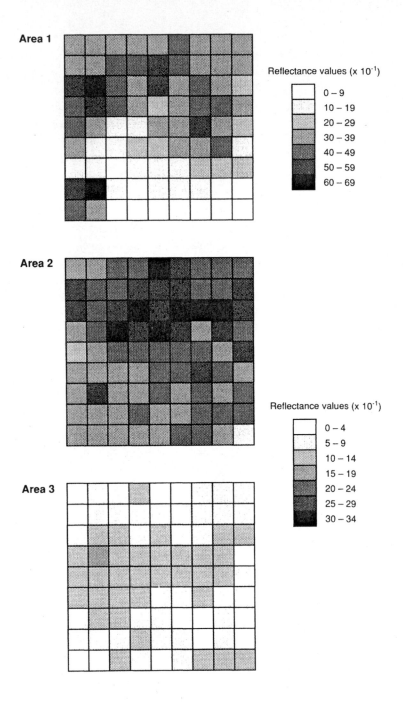

Figure 7.1. Pollution data maps. After Haining (1987c).

Table 7.1. *Ordinary least squares estimates for the zero order trend surface model (7.1): pollution data*

	Area		
Estimates	1	2	3
$\hat{\beta}_{oo}$	2.95	2.17	0.87
$SE(\hat{\beta}_{oo})$	0.17	0.05	0.03
GMC	8.16	5.26	5.79

Table 7.2. *Lag spatial correlations for the least squares residuals from the zero order trend surface model: pollution data*

Area	$R^*(0)$	$R^*(1)$	$R^*(2)$	$R^*(3)$	$R^*(4)$	$R^*(5)$	$R^*(6)$
1	1.0	0.66	0.43	0.26	0.14	0.03	-0.06
2	1.0	0.42	0.31	0.26	0.14	0.06	-0.09
3	1.0	0.46	0.22	0.05	-0.05	-0.13	-0.08

an acceptable fit. Tables 7.3 and 7.4 repeat the information for the fit of a first order (linear) trend surface. Note that the values in 7.4 are smaller than in 7.2 for low order lags indicating that some of the lag correlation in 7.2 is due to order misspecification. Tables 7.5 and 7.6 provide details of the iterative maximum likelihood estimation for zero and first order trend surfaces using the SAR model. For comparison and to indicate the importance of the specification of V, Tables 7.7 and 7.8 report parameter estimates under the three different assumptions for V. Subsequent model fitting assumed the SAR form for V (page 82). Data set 3 is best described by a zero order model with SAR errors. Neither of the higher order coefficients in the first order model are significant. It seems that data set 1 can be described either by a zero order model with SAR errors or a first order model with SAR errors. The second choice seems preferable even though the (adjusted) R^2 is only slightly higher. The estimate of the SAR parameter in the zero order case is close to the stationary maximum and may be unreliable. [1] In the case of the first order model both the higher order coefficients are significant and it is possible therefore that the large value of the SAR parameter estimate in the zero order case is due to the presence of this trend. It is important to check in the case of data sets 1 and 2 whether a second (or higher) order model provides a better fit. Table 7.9 provides a summary of results. The

Table 7.3. *Ordinary least squares estimates for the first order (linear) trend surface model: pollution data*

Estimates	Area		
	1	2	3
$\hat{\beta}_{00}$	5.45	2.49	0.99
$\hat{\beta}_{01}$	−3.38	−0.94	−0.03
	(0.44)	(0.17)	(0.13)
$\hat{\beta}_{10}$	−1.12	0.37	−0.19
	(0.44)	(0.17)	(0.13)
$R^2 \times 100$	43.7	30.3	2.4
GMC	5.74	2.55	5.74

Note: standard errors in brackets.

Table 7.4. *Lag spatial correlations for the least squares residuals from the first order (linear) trend surface model: pollution data*

Area	$R^*(0)$	$R^*(1)$	$R^*(2)$	$R^*(3)$	$R^*(4)$	$R^*(5)$	$R^*(6)$
1	1.0	0.43	0.08	−0.08	−0.11	−0.09	−0.07
2	1.0	0.17	0.02	0.00	−0.05	0.00	−0.05
3	1.0	0.29	0.04	−0.07	−0.09	−0.10	0.00

Table 7.5. *Estimates using an SAR model for* **V** *in (7.2): order 0 model on pollution data*

Area	$\tilde{\beta}_{00}$	$SE(\tilde{\beta}_{00})$	$\tilde{\rho}$	$SE(\tilde{\rho})$	$\tilde{\sigma}^2$	$R^2 \times 100$
1	2.51	0.44	0.23	0.01	0.77	67.1
2	2.09	0.11	0.16	0.02	0.19	30.9
3	0.78	0.08	0.18	0.01	0.07	41.2

Table 7.6. *Estimates using an SAR model for* V *in (7.2): first order model on pollution data*

Area	$\tilde{\beta}_{00}$	$\tilde{\beta}_{01}$	$\tilde{\beta}_{10}$	$\tilde{\rho}$	SE($\tilde{\rho}$)	$\tilde{\sigma}^2$	$R^2 \times 100$
1	5.09	-2.57	-1.71	0.20	0.01	0.79	67.6
	(0.75)	(0.86)	(0.86)				
2	2.45	-0.89	0.35	0.08	0.03	0.17	33.6
	(0.18)	(0.22)	(0.22)				
3	0.68	0.03	0.12	0.19	0.01	0.07	41.5
	(0.21)	(0.25)	(0.25)				

Note: Standard errors of the trend surface coefficients are in brackets.

Table 7.7. *Maximum likelihood estimates for (7.2): zero order model on pollution data*

Area	Models for V		
	SAR model	Empirical cov.	Agterberg model
	$\tilde{\beta}_{00}$	$\tilde{\beta}_{00}$	$\tilde{\beta}_{00}$
1	2.51	2.84	2.81
2	2.09	2.01	NC
3	0.78	0.81	0.87

Table 7.8. *Maximum likelihood estimates for (7.2): first order model on pollution data*

Area	Models for V								
	SAR model			Empirical cov.			Agterberg model		
	$\tilde{\beta}_{00}$	$\tilde{\beta}_{01}$	$\tilde{\beta}_{10}$	$\tilde{\beta}_{00}$	$\tilde{\beta}_{01}$	$\tilde{\beta}_{10}$	$\tilde{\beta}_{00}$	$\tilde{\beta}_{01}$	$\tilde{\beta}_{10}$
1	5.09	-2.57	-1.71	5.34	-3.15	-1.15	NC	NC	NC
2	2.45	-0.89	0.35	2.48	-0.91	0.37	2.51	-0.94	0.42
3	0.68	0.03	0.12	NC	NC	NC	0.93	-0.06	-0.03

NC = no convergence, cov = covariance.

Table 7.9. *Maximum likelihood estimates using an SAR model for* **V** *in (7.2): second order model on pollution data*

	Area 1	Area 2
$\tilde{\beta}_{00}$	3.28	1.60
	(1.21)	(0.34)
$\tilde{\beta}_{01}$	4.81	0.78
	(3.33)	(0.90)
$\tilde{\beta}_{10}$	−1.02	2.77
	(3.33)	(0.90)
$\tilde{\beta}_{02}$	−4.65	−1.29
	(2.70)	(0.72)
$\tilde{\beta}_{20}$	1.51	−1.93
	(2.70)	(0.72)
$\tilde{\beta}_{11}$	−4.17	−0.47
	(2.20)	(0.63)
$\tilde{\sigma}^2$	0.70	0.16
$R^2 \times 100$	68.2	37.7
GMC*	5.92	2.15
$\tilde{\rho}$	0.19	0.05

* on OLS residuals; standard errors are given in brackets.

second order model with SAR errors does not provide a better fit to data set 1 but does to data set 2. Data set 2 was best described by a second order model with SAR errors. 'These surface descriptions appear to be quite reasonable in terms of the dispersal characteristics of water borne pollutants, and the position of the three areas relative to the discharge points' (Haining, 1987c, p. 468). These orders of fit are also suggested by the evidence of the median polish (see Tables 6.1, 6.2 and 6.3).

Finally note that the original maps can be decomposed into their three components of variation. In model (7.2) writing $\mathbf{V} = \sigma^2 \mathbf{LL}^\mathsf{T}$ then:

$$\mathbf{L}^{-1}[\mathbf{y} - \mathbf{A}\boldsymbol{\theta}] = \mathbf{e} \tag{7.3}$$

where $E[\mathbf{e}] = 0$ and $E[\mathbf{ee}^\mathsf{T}] = \sigma^2\mathbf{I}$. The three map components are

trend: $\mathbf{A}\tilde{\boldsymbol{\theta}}$; noise: $\tilde{\mathbf{e}} = \tilde{\mathbf{L}}^{-1}(\mathbf{y} - \mathbf{A}\tilde{\boldsymbol{\theta}})$; signal: $\mathbf{y} - \mathbf{A}\tilde{\boldsymbol{\theta}} - \tilde{\mathbf{e}}$

where ∼ denotes ML estimates. Figure 7.2 shows area 1 decomposed into these three components with $\tilde{\mathbf{L}}^{-1} = (\mathbf{I} - \tilde{\rho}\mathbf{W})$.

Example 7.2: A space-time comparative analysis of trend (Haining, 1978a) Model (7.2), with an autocorrelated error structure may be used to test

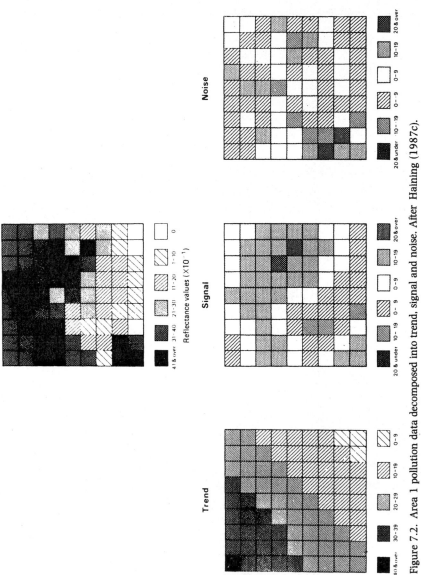

Figure 7.2. Area 1 pollution data decomposed into trend, signal and noise. After Haining (1987c).

259

hypotheses relating to spatio-temporal variation. In a study of crop yield variation in the American High Plains (Haining, 1978a) it was argued that soil and precipitation gradients would tend to encourage a decline from east to west and that temperature gradients would tend to produce a decline from north to south in corn and wheat yields. It was further argued that these gradients would be most striking in the early pioneering years (1879–1909) when there was a lack of technical knowledge, but would steadily disappear after the first world war as methods of dry farming were introduced and developed. The main interest focussed on estimating the broad features of the regional crop yield pattern (the trend component) for each of the years for which data was available. The corn yield data provided the most striking evidence of this but analysis only went as far as fitting linear surfaces. Figure 7.3 shows corn yield data for six of the years analysed and Figure 7.4 shows contour maps of the best fitting estimated trend surface models from a subsequent analysis of the data. V was specified as a first order symmetric SAR scheme (page 82).

Example 7.3: A space-time comparative analysis of the correlation structure (Haining, 1981b)
In some cases spatial correlation properties are of more interest than the trend properties. Haining (1981b) reviewed various theories of rural population density distribution and then examined data from Iowa and Wisconsin over the period 1930–70 in order to identify whether spatial correlation was present in the data and determine whether it could be described by an inverse distance function as predicted. Model (7.1) was used to extract broad regional trends (by least squares estimation) and the residuals were then analysed for the presence of spatial correlation. The procedure is equivalent to the first cycle of the iterative procedure discussed above and was used because of the large volume of data ($n = 219$ for Iowa; $n = 155$ for Wisconsin). The full iterative cycle would have necessitated the inversion of a 219 by 219 matrix, in the case of Iowa. The pattern of spatial correlation was stable over the period for both areas analysed but the model fitting (rather crude relative to the methods described in Chapter 4) did not support the inverse distance model. Figure 7.5 shows the results for Iowa.

Boundary or edge effects create a problem in trend surface analysis, described by Ripley (1981, p. 30) as a 'tendency to wave the edges to fit points in the centre'. Unwin and Wrigley (1987a,b) note that high leverage values are found for sites near edges (especially corners) so that the fit of any model will be particularly sensitive to errors in recording near the boundary.[2] However in (7.2) there is the additional problem of specifying V. Tables 7.7 and 7.8 show that estimates of θ are sensitive to the specification of V. Boundary effects on the estimation of unknown parameters in V are discussed in section 3.2. The recommended procedure of extending the spatial coverage beyond the study area in order to reduce edge effects in fitting (7.1) is probably even more important in the case of (7.2). Where data are not available the creation of an artificial data border along the lines described at the end of section 3.4 may be worth considering.

An uneven spatial distribution of sample sites with clustering in parts of the region and sparse coverage in others will accentuate spatial correlation in the residuals and will lead, in the case of (7.1), to a model which is weighted to that part of the map which is most intensively sampled although the relative weight of each point within any cluster is reduced relative to the case of uniform coverage (Ripley, 1981, p. 35). In sparsely sampled areas, the fit of the model will tend to be excessively dependent on

Figure 7.3. County corn yield (in bushels per acre) data maps for an area of Kansas/Nebraska 1889–1959.

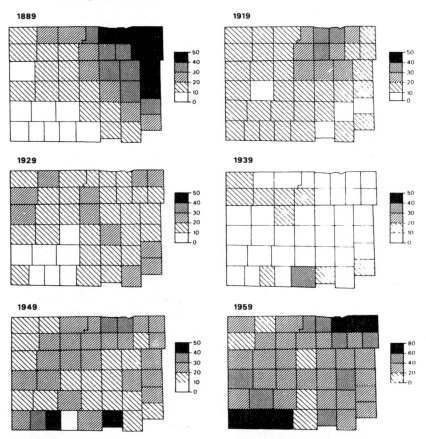

the few isolated data points for these will have high leverage. In these areas the fitted model will be sensitive to measurement error at those sites although the relative influence of different points will depend to some extent on the order of surface fitted. How serious a problem this is, is not clear, and Upton and Fingleton (1985) conclude that mild clustering is probably not a problem. Unwin and Wrigley (1987a,b) examine cases of severe data clustering.

Frame effects, the shape of the study area, may also be influential. Unwin and Wrigley (1987b) report the work of Doveton and Parsley showing how contours of higher order surfaces are distorted so as to lie parallel to the long

Figure 7.4. Trend surfaces for the county corn yield data.

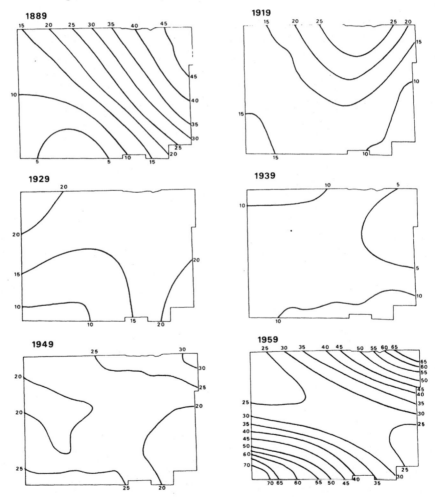

axis of any distribution of data points that are aligned in any particular direction. This problem like the others cited above appear to be more significant when fitting higher order models.

In the case of (7.2), estimating $\boldsymbol{\theta}$ by $\tilde{\boldsymbol{\theta}} = (A^T V^{-1} A)^{-1} (A^T V^{-1} y)$ down-weights the contribution of points in data clusters since these are the points with the strongest correlation and the greatest information redundancy.

Figure 7.5. Correlograms for NW Iowa spatial population density variation.

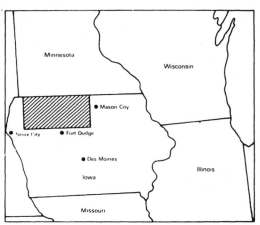

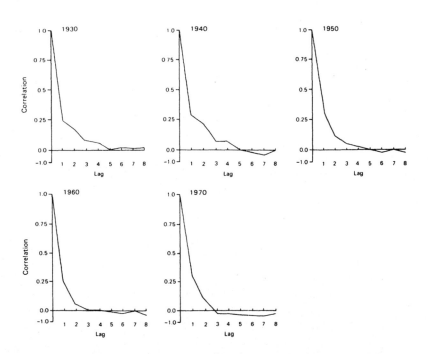

This reduces the dependence of the fitted model on areas that are heavily sampled. On the other hand, if there are sparsely sampled areas, the form of the model will still be heavily dependent on these scattered observations. Indeed the contribution of these points will be relatively greater, worsening leverage problems.

The problem of model sensitivity to outliers (distributional and spatial) was raised in Chapter 6. Two simple options are to estimate model parameters (both trend and second order) by maximum likelihood or ordinary least squares after transforming the data (to control the spread of values) or after editing the data by replacing outliers with, say, their local spatial average. Median polish provides a simple method of resistant trend fitting but is not well suited to irregularly spaced data points nor does it provide parameter estimates. Another resistant approach which yields parameter estimates of the trend surface is to minimise sums of absolute values of deviations (least absolute residuals estimation) rather than sums of squares. An early application of this estimation principle to trend surface analysis is given in Dougherty and Smith (1966) but the method has not been widely used. Li (1985, p. 295) notes that high leverage values (which occur on the boundaries even when regularly spaced data are available) can cause the estimator to break down. Estimates may not be unique and explicit formulae for the parameter estimates cannot be written down. There is further discussion of this type of fitting in Gentle (1977) and Emerson and Hoaglin (1983a,b). Although LAR estimation has good resistance properties it has a related problem which is to overstress small residual values in the fitting procedure. For (7.2) the LAR estimator minimises

$$\Sigma_i \, | \, (\mathbf{L}^{-1}\mathbf{y})_i - (\mathbf{L}^{-1}\mathbf{A})_i \boldsymbol{\theta} \, |$$

with respect to $\boldsymbol{\theta}$. $\mathbf{L}$ may also be estimated by resistant methods (see Chapter 6).

When the model residuals indicate a heavy tailed distribution it may be appropriate to use the robust regression methods described in Chapter 2. For example, the W-estimator for $\boldsymbol{\theta}$ in (7.2) is:

$$\hat{\boldsymbol{\theta}} = (\mathbf{A}^{\mathsf{T}}(\hat{\mathbf{L}}(\mathbf{W}^*)^{-1}\hat{\mathbf{L}}^{\mathsf{T}})^{-1}\mathbf{A})^{-1}(\mathbf{A}^{\mathsf{T}}(\hat{\mathbf{L}}(\mathbf{W}^*)^{-1}\hat{\mathbf{L}}^{\mathsf{T}})^{-1}\mathbf{y})$$

where $\hat{\mathbf{V}} = \sigma^2 \hat{\mathbf{L}}\hat{\mathbf{L}}^{\mathsf{T}}$ and $\mathbf{W}^*$ is a diagonal matrix which downweights the contribution of observations with large residuals (Chapter 2). So the regression parameter $\boldsymbol{\theta}$ is estimated using $(\mathbf{W}^*)^{\frac{1}{2}}\hat{\mathbf{L}}^{-1}\mathbf{y}$ and $(\mathbf{W}^*)^{\frac{1}{2}}\hat{\mathbf{L}}^{-1}\mathbf{A}$. $\mathbf{L}$ may be estimated by resistant methods also (see Chapter 6). At the first stage let $\mathbf{W}^* = \mathbf{I}$ and use the final set of (independent) residuals ($\tilde{\mathbf{e}}$) to specify a form for $\mathbf{W}^*$. Then $\mathbf{L}$ and $\boldsymbol{\theta}$ will need to be re-estimated and the new residuals used to specify a new matrix **W***. The procedure is iterated to convergence as described in Chapter 4. The case where V is an interaction model is considered in the appendix to Chapter 8. The relative merits of all these different approaches needs exploration.

If the surface is to be fit to density data where the sizes of areal units differ there may be a size–variance relationship. Small areal units may have larger variances and we might wish to downweight their influence in the model fit. This will also reduce the dependence of the model fit on those areas of the surface which have a large number of observations based on a large number of small areas. In this case the procedure recommended is weighted least squares where W^+ replaces W^* and is specified once and for all by reference to the areas of the observation units rather than the size of the residuals. The diagonal elements of W^+ are related to the size of the areal units (pages 49–50).

Least squares is fairly resistant to departures from residual normality. However, if evidence indicates a serious departure from normality then appropriate data transformations (Chapter 6) should again be considered.

Example 7.4: Surface fitting to irregular areal data
The methods of Chapter 6 showed that standardised mortality rates for all forms of cancer in Glasgow (Table 6.1) tend to decline with distance from the central CMA. We consider here the problem of identifying significant spatial trends in the data and identifying CMAs that have particularly high or low rates with respect to those trends.

Figure 7.6 is a plot of SMRs for cancer against distance from the central

Figure 7.6. Plots of cancer SMRs against distance from the central CMA (40) of Glasgow.

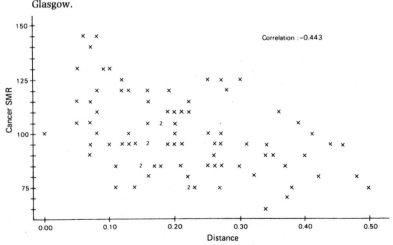

Table 7.10. *Glasgow health data:*

(*a*) contiguous CMAs

CMA	Contiguous CMAs	#
1.	2, 4, 5, 40, 84	5
2.	1, 3, 4, 24, 40, 41	6
3.	2, 4, 6, 8, 21, 24	6
4.	1, 2, 3, 5, 6	5
5.	1, 4, 6, 7, 80	5
6.	3, 4, 5, 7, 8	5
7.	5, 6, 8, 9, 10, 11, 80	7
8.	3, 6, 7, 9, 21, 22	6
9.	7, 8, 10, 19, 22	5
10.	7, 9, 11, 13, 14, 19	6
11.	7, 10, 14, 80	4
12.	13, 14, 16, 18, 19, 20	6
13.	10, 12, 14, 19	4
14.	10, 11, 12, 13, 15, 16, 17	7
15.	14, 17	2
16.	14, 17, 20	3
17.	14, 15, 16, 20	4
18.	12, 19, 20	3
19.	9, 10, 12, 13, 18, 20, 21, 22	8
20.	12, 16, 17, 18, 19, 35	6
21.	3, 8, 19, 22, 26, 35	6
22.	8, 9, 19, 21	4
23.	24, 25, 26	3
24.	2, 3, 23, 25, 26	5
25.	23, 24, 26, 27, 41	5
26.	21, 23, 24, 25, 27, 28, 33, 35	8
27.	25, 26, 28, 41, 43	5
28.	26, 27, 33, 34, 39, 43, 48, 55	8
29.	30, 31	2
30.	29, 31, 35, 36, 39	5
31.	29, 30, 32, 36, 39	5
32.	31, 36, 39	3
33.	26, 28, 34, 35, 39	5
34.	28, 33, 39	3
35.	19, 20, 21, 26, 30, 33, 36, 39	8
36.	30, 31, 32, 35, 38, 39	6
37.	38, 39, 55, 56	4
38.	36, 37, 39	3
39.	28, 30, 31, 32, 33, 34, 35, 36, 37, 38, 55	11
40.	1, 2, 41, 42, 44, 45, 58, 84	8
41.	2, 24, 25, 27, 40, 42, 43	7
42.	40, 41, 43, 44, 46, 48	6
43.	27, 28, 41, 42, 48	5
44.	40, 42, 45, 46, 49, 50, 61	7

Table 7.10 (*cont.*)

(*a*) contiguous CMAs

CMA	Contiguous CMAs	#
45.	40, 44, 58, 61	4
46.	42, 44, 47, 49	4
47.	46, 48, 49, 53, 55	5
48.	28, 42, 47, 55	4
49.	44, 46, 47, 50, 51, 53, 54	7
50.	44, 49, 51, 52, 61	5
51.	49, 50, 52, 54	4
52.	50, 51, 54, 57, 61, 63	6
53.	47, 49, 54, 55, 56, 57	6
54.	49, 51, 52, 53, 57	5
55.	28, 37, 39, 47, 48, 53, 56	7
56.	37, 53, 55, 57	4
57.	52, 53, 54, 56	4
58.	40, 45, 59, 61, 84, 87	6
59.	58, 60, 61, 64, 83, 87	6
60.	59, 61, 62, 64, 65, 66, 67, 70, 74	9
61.	44, 45, 50, 52, 58, 59, 60, 62, 63	9
62.	60, 61, 63, 65	4
63.	52, 61, 62	3
64.	59, 60, 67, 75, 79, 83	6
65.	60, 62, 66, 74	4
66.	60, 65, 74	3
67.	60, 64, 68, 70, 79	5
68.	67, 69, 70, 74, 79	5
69.	68, 72, 73, 74, 79, 85	6
70.	60, 67, 68, 74	4
71.	72, 73	2
72.	69, 73	2
73.	69, 71, 72, 74	4
74.	60, 65, 66, 68, 69, 70, 73	7
75.	64, 78, 79, 82, 83, 84, 87	7
76.	77, 78, 80, 81	4
77.	76, 78, 82, 86	4
78.	75, 76, 77, 80, 82, 84	6
79.	64, 67, 68, 69, 75, 82, 85, 86	8
80.	5, 7, 11, 76, 78, 81, 84	7
81.	76	1
82.	75, 77, 78, 79, 86	5
83.	59, 64, 75, 87	4
84.	1, 40, 58, 75, 78, 80, 87	7
85.	69, 79, 86	3
86.	77, 79, 82, 85	4
87.	58, 59, 75, 83, 84	5

(*b*) digitised co-ordinates for each CMA centre

CMA	x	y
1.	0.41	0.54
2.	0.43	0.56
3.	0.41	0.59
4.	0.40	0.57
5.	0.34	0.57
6.	0.36	0.59
7.	0.30	0.61
8.	0.37	0.63
9.	0.33	0.66
10.	0.28	0.66
11.	0.24	0.63
12.	0.23	0.74
13.	0.25	0.71
14.	0.19	0.69
15.	0.13	0.71
16.	0.21	0.78
17.	0.12	0.83
18.	0.30	0.78
19.	0.35	0.73
20.	0.30	0.87
21.	0.43	0.69
22.	0.38	0.68
23.	0.45	0.63
24.	0.44	0.58
25.	0.49	0.62
26.	0.47	0.68
27.	0.54	0.60
28.	0.60	0.63
29.	0.70	0.81
30.	0.69	0.83
31.	0.73	0.80
32.	0.67	0.76
33.	0.52	0.74
34.	0.57	0.68
35.	0.52	0.94
36.	0.71	0.90
37.	0.81	0.66
38.	0.82	0.73
39.	0.69	0.69
40.	0.49	0.52
41.	0.49	0.57
42.	0.54	0.55
43.	0.54	0.57
44.	0.56	0.50
45.	0.55	0.46
46.	0.56	0.53

(*b*) digitised co-ordinates for each CMA centre

CMA	x	y
47.	0.61	0.55
48.	0.60	0.56
49.	0.62	0.51
50.	0.61	0.46
51.	0.65	0.48
52.	0.69	0.44
53.	0.69	0.54
54.	0.68	0.51
55.	0.70	0.58
56.	0.79	0.55
57.	0.76	0.49
58.	0.49	0.47
59.	0.48	0.47
60.	0.50	0.37
61.	0.57	0.40
62.	0.57	0.34
63.	0.69	0.36
64.	0.43	0.40
65.	0.53	0.32
66.	0.49	0.31
67.	0.43	0.36
68.	0.39	0.32
69.	0.34	0.29
70.	0.43	0.30
71.	0.37	0.03
72.	0.22	0.15
73.	0.36	0.21
74.	0.44	0.25
75.	0.38	0.44
76.	0.29	0.49
77.	0.24	0.45
78.	0.34	0.46
79.	0.39	0.37
80.	0.30	0.54
81.	0.24	0.50
82.	0.29	0.42
83.	0.42	0.43
84.	0.41	0.50
85.	0.29	0.32
86.	0.27	0.37
87.	0.44	0.47

Table 7.11. *Ordinary least squares estimates for the second order trend surface model (7.1): cancer data*

	Constant	(E–W)	(N–S)	(E–W)^2	(N–S)^2	(E–W) (N–S)
Parameter estimates	63.4	−15.0	168.8	−33.7	−196.5	91.4
Standard deviation	22.4	73.7	45.6	60.1	45.1	68.2
t value	2.83	−0.20	3.7	−0.5	−4.3	1.3

$R^2 \times 100 = 22.0$, R^2 (adj) $\times 100 = 17.2$, $\hat{\sigma} = 18.40$.

CMA (#40). The plot shows the decrease in rates with distance from the centre but also shows that it is the absence of higher rates at larger distances from the centre that underlies the apparent decrease.

Table 7.11 summarises the fitting of model (7.1) to the Cancer SMR data after digitising area midpoints which were defined by eye (Table 7.10). A second order (quadratic) model appears to provide the best fit in the sense that the R^2 for this model is 22.0% whereas a linear surface has an R^2 of only 1.9% and a 3rd order cubic surface has an R^2 of 27.3%. If an adjusted R^2 is computed (adjusting for the number of parameters fitted) then the third order model has an R^2 of 18.8%. The evidence of this fit is that there is a north–south component of variation but no significant east–west variation. The peak of the surface, as expected, is near the city centre so there is, apparently, a significant trend in rates away from the centre of Glasgow.

Figure 7.7 summarises important regression diagnostics. Three areas stand out with large (positive) studentised and standardised residuals (CMAs 42, 43 and 45). However CMA 43 has the smallest population level of the 87 areas so the large value may be a consequence of this. The border CMAs in the north and south have high leverages and the DFITS coefficient for observation i, which measures the effect of deleting this observation on the fit of the model, further highlights the two large CMAs (71 and 72) in the south.[3] The importance of these boundary CMAs in influencing the significant north–south variation needs investigation.

Figure 7.8 plots residuals against normal scores. There seems to be no serious departure from normality nor are there any serious outliers. Figure 7.9(*a*) plots each residual value against the average of the residuals of contiguous CMAs. There appears to be a positive relationship which disappears for higher order contiguity (Figure 7.9(*b,c,d,e*)). The generalised

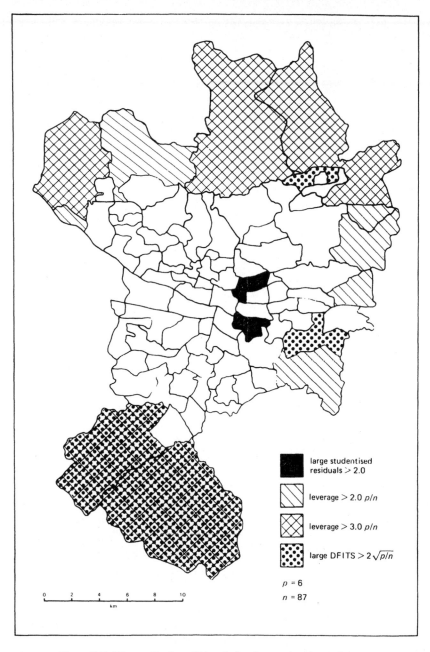

Figure 7.7. Diagnostics from fitting 2nd order trend surface: Glasgow cancer SMR data.

271

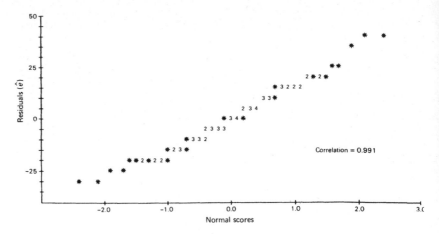

Figure 7.8. Plots of residuals from 2nd order trend surface against normal scores: Glasgow cancer SMR data.

Moran coefficient (see Chapter 6) applied to the OLS residuals gives a value of 1.277 for first order neighbours which is 4.917 as a standard normal deviate so there is significant spatial autocorrelation in the residuals.

Figure 7.10 plots each residual against the population size of the CMA. The plot suggests that as population size increases error variance decreases. We would expect SMRs based on large populations to be more reliable than those based on small populations and so might wish to make an adjustment for this reason. (Note that if error variances are not constant then the test figure for the Moran coefficient may be misleading.)

Separate analyses were run in order to identify the effects of boundary CMAs, non constant variance and residual autocorrelation on model fit.

Variance stabilising data transformations or weighted least squares are commonly used to remedy the problem of heteroscedastic errors. The SMR for the ith CMA is $(100 \times O_i/E_i)$ where O_i and E_i are the observed and expected number of deaths for the ith CMA. Now O_i is approximately Poisson distributed.[4] A common variance stabilising transformation in this case is to compute the square roots of the responses $(y^{\frac{1}{2}})$. If some SMRs are very small or zero the Freeman–Tukey transformation is used: $(y^{\frac{1}{2}} + (y+1)^{\frac{1}{2}})$. These transformations are described in Weisberg (1985, p. 133). The square root transformation was used. Under this transformation of the response variable the N–S axis variables remain significant and CMAs 42, 43 and 45 had standardised residuals > 2.0. Reducing the data set by deleting the boundary CMAs with high leverage scores did not affect the conclusions. Using other data transformations including $\log(y)$ also did not affect the conclusions.

272

The effects of residual spatial correlation were examined by fitting (7.2). Three models

$$V_{SAR} = \sigma^2[(I - \rho W^T)(I - \rho W)]^{-1}$$
$$V_{CAR} = \sigma^2[I - \tau W]^{-1}$$
$$V_{MA} = \sigma^2[(I + v W^T)(I + v W)]$$

were examined for the errors. V_{SAR}, the first order simultaneous autoregressive model gave the best fit in the sense of the highest R^2, with a total R^2 (regression + error model) of 41.1%. The contiguity matrix W consisted of

Figure 7.9. Plots of residuals from 2nd order trend surface: Glasgow cancer SMR data.

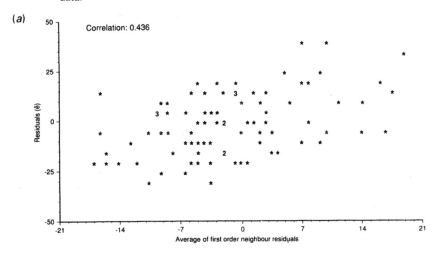

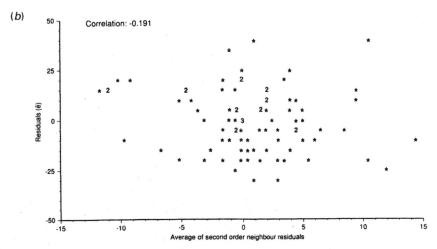

(*c*)

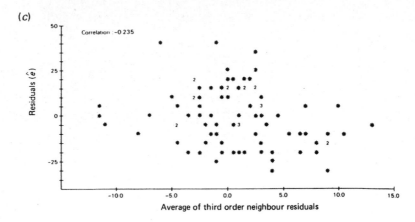

(*d*)

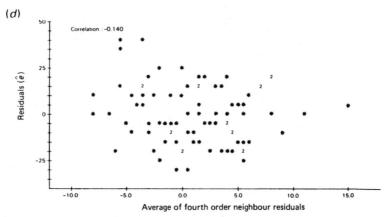

(*e*)

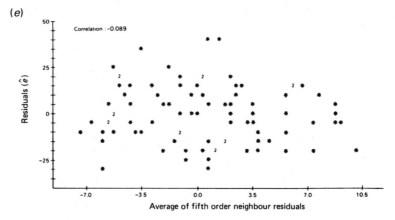

Os and 1s (Table 7.7) which constrained ρ to lie in the interval $(-0.32, 0.16)$. Maximum likelihood estimates are given in Table 7.12. North–South variation is still significant although only in the quadratic term.

Figure 7.11 shows diagnostics of leverages and standardised residuals. CMA 43 no longer has a large residual but 73 has a significantly low (negative) residual. The leverage problem is worse with this model as Table 7.13 shows.

Figure 7.12(a) shows the plot of the independent residuals ($\hat{e}$) against the average of their neighbours under model (7.2) and V_{SAR}. The positive correlation noted when fitting model (7.1) has disappeared. (Note that it is not possible to use the GMC test on generalised least square residuals to provide a formal test.) There is no apparent trend in the residuals with distance from the city centre (Figure 7.12(b)). It appears that model 7.2 has accounted for the main spatial attributes of the data.

Model (7.2) with V_{SAR} assumes that CMAs with a large number of contiguous neighbours will have larger variances. This property of the SAR model was discussed in Chapter 3 and Figure 7.13 shows the plot of the diagonal elements of V_{SAR} against the number of contiguous neighbours of each CMA. The CMAs with most contiguous neighbours are those nearest the centre. Population density is greatest near the centre and so many of these small areas have large populations. We would not necessarily expect SMRs for such areas to have large error variances (see section 2.3.3(h)). One approach would be to re-analyse the data setting row sums of W to unity. Given the evidence of Figure 7.10, an additional matrix W^+ could be

Figure 7.10. Plots of residuals from 2nd order trend surface against CMA populations.

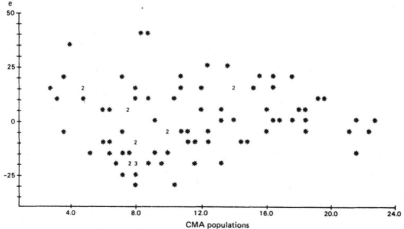

Table 7.12. *Maximum likelihood estimates for the second order trend surface model (7.2): cancer data*

	Constant	(E–W)	(N–S)	(E–W)²	(N–S)²	(E–W)(N–S)
Parameter estimates	97.2	−37.6	75.4	−40.6	−128.4	137.5
Standard deviation	28.3	89.8	57.0	77.6	53.5	87.5
t value	3.43	−0.42	1.32	−0.52	−2.40	1.57

$\tilde{\rho}=0.12$, likelihood ratio test on $\rho=22.97$ (significant as $\chi_1{}^2$), $\tilde{\sigma}=13.9$, $R^2 \times 100 = 41.1$.

Table 7.13. *Leverage values for CMAs*

Observation	Model (7.1)	Model (7.2)
15	0.179	0.272
17	0.305	0.439
20	0.174	—
29	—	0.197
35	0.256	0.263
36	0.208	0.201
37	0.159	0.173
38	0.197	0.292
56	0.140	0.154
57	—	0.142
63	0.144	0.236
71	0.437	0.647
72	0.293	0.428

introduced into the specification in order to downweight the influence of CMAs with small populations. Some experiments were carried out.

Table 7.14 reports the results of fitting the model:

$$y = A\theta + u$$
$$u = \rho Wu + e$$

where $E[e] = 0$ and $E[ee^{\mathsf{T}}] = \sigma^2(W^+)^{-1}$ and where W^+ is diagonal with elements $\{p_i\}$ where p_i is the population of the *i*th CMA. So error variance is inversely related to population size and CMAs with small populations are downweighted in the estimation procedure for both θ and ρ.

The CMAs with large standardised residuals are 42 and 45. The

276

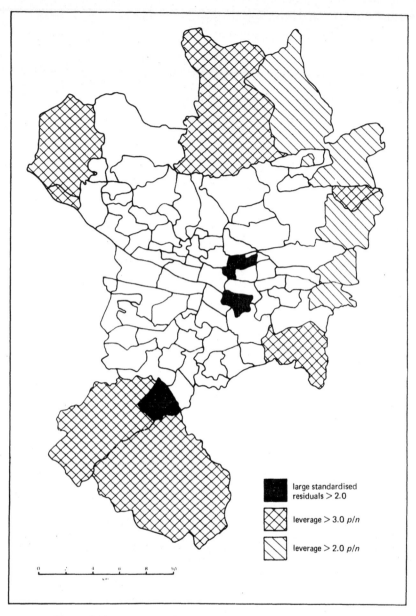

Figure 7.11. Diagnostics from fitting 2nd order trend surface with SAR errors: Glasgow cancer SMR data.

distribution of residuals (ẽ) appear to follow a normal distribution (Figure 7.14(*a*)). There is no evidence of any residual first order spatial correlation (Figure 7.14(*b*)).

CMAs 17, 38, 56, 63, 71 and 72 have high leverages which suggest that the suburban fringe may have a disproportionate influence on model fit. If the rows associated with these observations are deleted from

$$(\mathbf{W}^+)^{\frac{1}{2}}(\mathbf{I}-0.1174\mathbf{W})\mathbf{y} \text{ and } (\mathbf{W}^+)^{\frac{1}{2}}(\mathbf{I}-0.1174\mathbf{W})\mathbf{A}$$

Figure 7.12. Residual plots for final set of (independent) residuals from 2nd order trend surface with SAR errors: (*a*) plot of residuals against the sum of neighbouring residuals, (*b*) plot of residuals against distance from CMA 40.

(*a*)

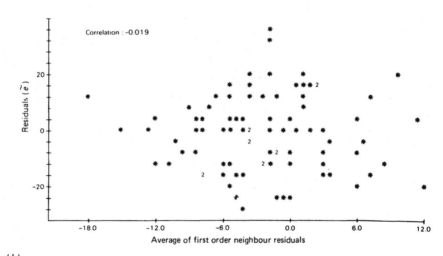

(*b*)

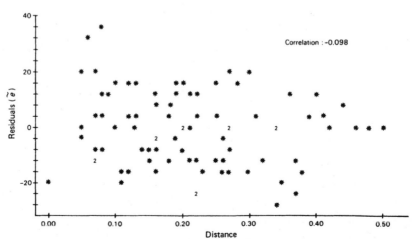

Table 7.14. *Population weighted second order trend surface model (7.2):*
cancer data

	Constant	(E–W)	(N–S)	(E–W)^2	(N–S)^2	(E–W) (N–S)
Parameter estimates	112.40	−72.86	62.77	−13.68	−134.59	155.30
Standard deviation	28.30	87.59	64.85	72.98	57.88	76.86
t value	3.97	−0.83	0.96	−0.18	−2.32	−2.02

$\bar{\rho} = 0.11$, $\bar{\sigma} = 41.61$, $R^2 \times 100 = 40.1$.

and θ re-estimated then only the constant coefficient is still significant in the
second order model. The CMAs with large standardised residuals are 42, 45
and 73.

In conclusion, the results of these analyses confirm significant trends in
cancer mortality rates away from the city centre with particularly high
rates in CMAs 42 and 45, possibly also 43 though this may be due to the low
population level. The low rate in CMA 73 may also be the result of its low
population level. There is evidence, however, that it is the contrast in SMRs
between the suburban CMAs on the boundary and the other CMAs that lies
at the heart of these apparent gradients. This data set will be re-examined in
Chapter 8.

Figure 7.13. Plot of variance of SAR model ($\bar{\rho} = 0.12$) against number of
contiguous CMAs.

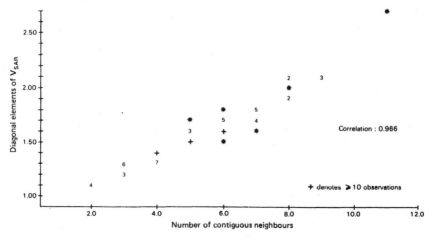

279

Example 7.5: Outlier problems and trend surface analysis: data editing and data transformations

In section 6.1 it was noted that area 1 of the pollution data contained possible outliers. A boxplot of the final set of (independent) residuals from the first order trend surface fit to area 1 (using V_{SAR}) shows that there may be four outliers (Figure 7.15): three possible and one probable. The observations together with their standardised residuals are shown in Table 7.15. The residual at (5,7) was not considered to be sufficiently atypical to

Figure 7.14. (*a*) Rankit plot of (independent) residuals for 2nd order trend surface model with SAR errors. (*b*) Plot of residuals against the average of neighbouring residuals.

(*a*)

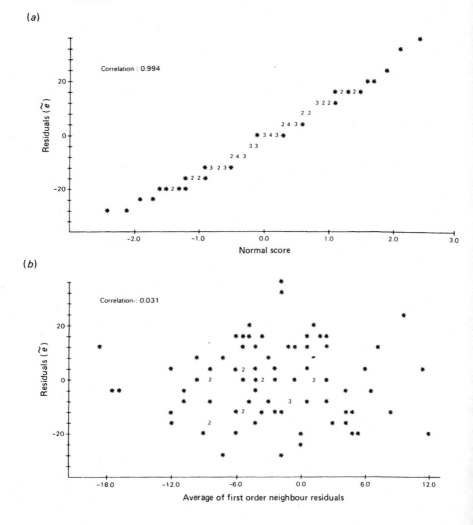

(*b*)

Table 7.15. *Possible outliers: area 1 pollution data*

Observation		Observed	Standardised	Local
Row	Column	value	residual	average
5	7	52	2.03	36.5
7	1	10	−2.56	33.0
8	2	62	3.87	29.0
9	3	0	−2.21	19.3

warrant special treatment. The other three form a diagonal strip in the bottom left corner, two at the boundary. Their values were replaced by the average of their neighbours that share a common side (Table 7.15). The first order model was then re-estimated using V_{SAR} on this outlier adjusted data set. The final set of residuals are better behaved. Figure 7.16(a),(b) shows stem and leaf and rankit plots of residuals when the original data are analysed and when the outlier adjusted data are analysed. Table 7.16 shows that adjusting for outliers in the original data creates a new set of possible outliers. Although there are more possible outliers the standardised residuals do not appear to be seriously high and only one of the values is at a boundary (with a relatively low value). From Figure 7.16(b) it appears that the residuals may follow a heavy tailed distribution so it is worth exploring the effect of a mild Box–Cox transformation of the original data on model fit. The square root ($y^{\frac{1}{2}}$) transformation was applied both to the original data (y) and the outlier adjusted data (y_{adj}). A comparison of these two data sets will show whether a data transformation takes care of both the outlier problem and the distributional problem at the same time. Stem and leaf and rankit plots are shown in Figure 7.16(c)(d). Table 7.17 shows that the square root transformation on y still leaves outliers. On the other hand the square root transformation on y_{adj} leaves better behaved residuals. The plot of the residuals against the sum of their neighbouring residuals suggests there is no residual spatial correlation (Figure 7.17). Table 7.18 gives some

Figure 7.15. Boxplot of (independent) residuals from a first order trend surface model with SAR errors: Area 1 pollution data. * = possible outlier, ○ = probable outlier.

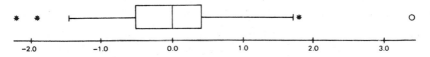

details on the fits of these different models. In all cases the first order trend surface model with V_{SAR} residuals is significant and provides quite a good resistant description of the data. The higher R^2 figures for model fits with outlier adjusted data are a consequence of the method of adjustment.

7.1.2 Non-stationary mean, stationary increments: semi-variogram models and polynomial generalised covariance functions

When surfaces are only stationary in the increments the approach of the previous section is no longer valid and we must turn to the statistical theory developed by Matheron for continuous surfaces and the data analytical

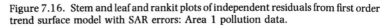

Figure 7.16. Stem and leaf and rankit plots of independent residuals from first order trend surface model with SAR errors: Area 1 pollution data.

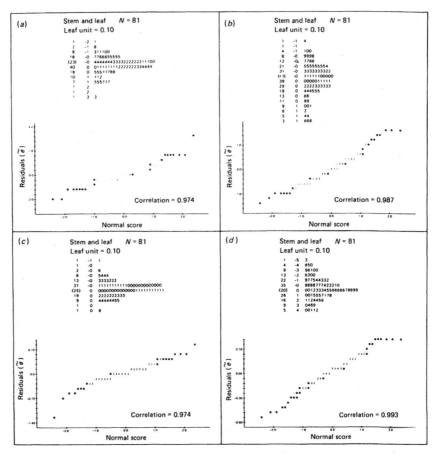

Table 7.16. *Possible outliers: area 1 pollution data after adjusting for initial outliers*

	Observation		
Row	Column	Observed value	Standardised residual
3	2	62	2.36
3	5	55	2.44
4	5	24	−2.09
5	7	52	2.42
6	8	41	2.14
8	1	50	2.20

Table 7.17. *Possible outliers: area 1 pollution data (a) after square root transformation (b) after square root transformation of outlier adjusted data*

	Observation		
Row	Column	Observed value	Standardised residual
(a) 7	1	10	−2.28
8	2	62	2.92
9	1	46	2.07
9	3	0	−3.99
(b) 1	1	32	−2.09
9	8	4	−2.47
9	9	0	−2.42

methods that this theory has given rise to (see section 3.1). From regionalised variable theory two approaches to the description of spatial surfaces have emerged: an approach based on fitting semi-variogram models and another based on fitting polynomial generalised covariance functions. Of the two approaches to date, the former is far more widely used since it is felt to provide a more informative description of spatial variation. The use of regionalised variable theory has been most evident in mining (see for example Journel and Huijbregts, 1978) but has a growing use in many areas of the earth and environmental sciences including hydrology (Kitanidis, 1983), water quality and irrigation studies (Hajrasulika *et al.*, 1980; Hughes and Lettenmaier, 1981), soil sciences (Webster, 1985) and remote sensing (Woodcock and Strahler, 1983; Curran, 1988). The usual aims are to describe spatial variation, estimate surface properties and subsequently interpolate and map. We consider some specific applications where semi-variogram models have been fitted.

Oliver and Webster (1986) remark that 'geomorphic features are essentialy random but spatially dependent and it is appropriate therefore to describe them using stochastic functions' (p. 491). They use semi-variogram models to describe variational properties such as directionality, periodicity and different scales of variation. Some models describe the presence of gentle variation in the mean but it is usually assumed that global trend properties have been previously removed from the data perhaps using a model such as (7.1) or median polish. Figure 7.18 shows several examples of estimated semi-variograms. Figure 7.18(a) shows a transitive semi-variogram model. The data recorded the depth of chalk strata beneath the Chiltern Hills. A quadratic trend was removed from the data and semi-variogram estimates obtained for four directions. An isotropic circular model of the form

$$\gamma(h) = c_0 + c\{1.0 - (2/\pi)\arccos(h/a) + (2h/\pi a)(1.0 - (h^2/a^2))^{\frac{1}{2}} \qquad 0 < h \leqslant a$$
$$= c_0 + c \qquad\qquad\qquad h > a$$
$$= 0 \qquad\qquad\qquad h = 0$$

is fitted with $c_0 = 40.8 \, m^2$; $c = 247.4 \, m^2$; $a = 12.1$ km. Figure 7.18(b) shows semi-variograms of stone content in topsoil measured in four different directions and showing evidence of directionality in surface variation (Burgess and Webster, 1980). Figure 7.18(c) gives an isarithmic plot of the

Figure 7.17. Plot of (independent) residuals against sum of neighbouring residuals from a first order trend surface model with SAR errors and square root transformed outlier adjusted data: Area 1 pollution data.

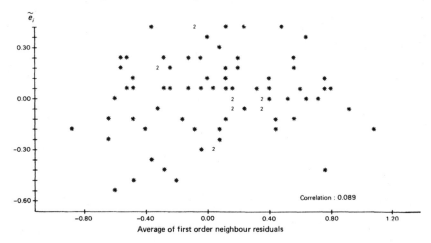

Table 7.18. *Model fits: pollution data (area 1)*

	(1)	(2)	(3)	(4)
$\tilde{\beta}_{00}$	5.09	5.11	2.53	2.56
S.E.$(\tilde{\beta}_{00})$	0.75	0.71	0.27	0.25
$\tilde{\beta}_{01}$	−2.57	−1.83	−1.20	−0.92
S.E.$(\tilde{\beta}_{01})$	0.86	0.79	0.31	0.28
$\tilde{\beta}_{10}$	−1.71	−2.37	−0.68	−1.00
S.E.$(\tilde{\beta}_{10})$	0.86	0.79	0.31	0.28
$\tilde{\rho}$	0.20	0.22	0.20	0.23
$R^2 \times 100$	69.8	78.3	70.8	81.7
GMC$^+$	5.74	6.71	5.60	6.96

$^+$ =Generalised Moran coefficient as standard normal deviate on OLS residuals. (1) Original data, (2) Original data with outlier adjustment, (3) Square root transformed original data, (4) Square root transformed outlier adjusted data.

semi-variogram. The semi-variogram is unbounded and very nearly linear with nuggett variance and the isarithmic plot approximately elliptical. Burgess, Webster and McBratney (1981) fitted the model

$$\gamma(\mathbf{h}, \xi) = c_0 + u(\xi)|\mathbf{h}|$$
$$u(\xi) = [A^2\cos^2(\xi - \phi) + B^2\sin^2(\xi - \phi)]^{\frac{1}{2}}$$

where ϕ is the direction of maximum variation, A the gradient of the semi-variogram in that direction, and B the gradient in the direction $\phi + (\pi/2)$. The proportion A/B defines the anisotropy ratio. Such anisotropy is not unusual in soil survey work particularly in the analysis of water sorted material with structures showing greater similarity in directions parallel to a river than in directions at right angles to the river (Webster, 1985). Figure 7.18(d) shows two semi-variograms of soil pH in the Wyre Forest, together with fitted functions. The subsoil is modelled using a linear function with

$$\gamma(\mathbf{h}) = 0.174 + 0.000228\mathbf{h}$$
$$\gamma(0) = 0$$

The topsoil has two models fitted, a distance function

$$\gamma(\mathbf{h}) = 0.150\mathbf{h}^{0.193}$$

and a less successful exponential model where

$$\gamma(\mathbf{h}) = 0.0263 + 0.0248\{1.0 - \exp(\mathbf{h}/27.7)\}$$
$$\gamma(0) = 0$$

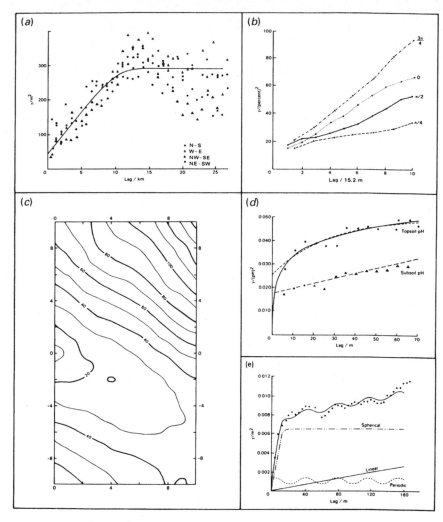

Figure 7.18. Empirical semi-variograms: (*a*) Transitive semi-variogram of the residuals from a regional trend: sub-Upper Chalk surface beneath the Chiltern Hills. (*b*) Sample semi-variograms of stone content in the top soil at Plas Gogerddan in four principal directions. (*c*) Isarithmic representation and sample semi-variograms of stone content at Plas Gogerddan. Border scales are in sampling intervals. (*d*) Unbounded semi-variogram of the pH of the soil in the Wyre forest. (*e*) Semi-variogram of the gilgai land surface on the Bland Plain (New South Wales) showing the four components of the fitted model. (*a*)(*d*)(*e*) After Oliver and Webster (1986). (*b*)(*c*) After Webster (1985).

The first two are examples of unbounded models indicating the presence of non-stationarity in the soil properties (Oliver and Webster, 1986). Finally, Figure 7.18 (*e*) shows a semi-variogram plot of topographic height across gilgai terrain on the Bland plain in central New South Wales and the component models used to describe the empirical plot (Oliver and Webster, 1986). The functional form is a combination of linear, spherical and periodic models plus a nuggett variance, indicating the ability of this approach to handle complex structure including periodicity. The authors argue that in analysing periodic structures, the semi-variogram is just as powerful as the power spectrum (and can be estimated without the need for large regularly spaced data sets). The semi-variogram gives an estimate of the frequency of the periodicity very close to that of the power spectrum. The power spectrum for the data for Figure 7.18(*e*) is shown in Figure 7.19. Further examples can be found in the above mentioned papers and references. Kitanidis (1983) gives examples of the fitting of polynomial generalised covariance functions to precipitation data.

It is usual in empirical work to de-trend the data and then fit a semi-variogram model rather than estimate both elements simultaneously. There are dangers in estimating this way as have been shown in Chapter 4.

Figure 7.19. Power spectrum of the gilgai land surface on the Bland Plain computed from 355 sites using a Bartlett window. After Oliver and Webster (1986).

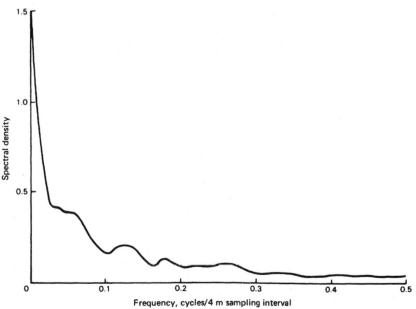

7.1.3   Discrete data

The logistic probability surface model is a natural redefinition of (7.1) for nominal data (Wrigley, 1985, p. 246). Let $\mathbf{A\theta}$ denote the trend component of the surface and let $P\{Y_i = y_i\}$ denote the probability that the observation at site i is y_i. For binary data the model is

$$P\{Y_i = 1\} = \exp[\mathbf{A}_i\mathbf{\theta}]/(1.0 + \exp[\mathbf{A}_i\mathbf{\theta}]) \tag{7.5}$$

with

$$P\{Y_i = 0\} = 1.0/(1.0 + \exp[\mathbf{A}_i\mathbf{\theta}]) \tag{7.6}$$

from which two maps representing (7.5) and (7.6) can be drawn that provide a smoothed description of the surface. $\mathbf{A}_i$ is the ith row of $\mathbf{A}$. This model can be generalised for s unordered categories so that

$$P\{Y_i = r\} = \exp[\mathbf{A}_i\mathbf{\theta}_r]/\Sigma_{t=1,\ldots,m}\,\exp[\mathbf{A}_i\mathbf{\theta}_t] \qquad r = 1, \ldots, s \tag{7.7}$$

Wrigley (1977) reviews the model and provides an empirical example using an $s = 3$ category model to describe levels of aircraft noise disturbance around Manchester's Ringway airport. Figure 7.20 shows the third order probability surfaces for the three categories of annoyance (highly annoyed, moderately annoyed, only a little annoyed).

There are parallels between (7.5) and (7.6) and the autologistic model described in Chapter 3. Indeed, if in (3.19) we set $\alpha_i = \mathbf{A}_i\mathbf{\theta}$ and $\beta_{ij} = 0$ for all i and j the model produces (7.5) and (7.6) as a special case.

It may be of interest to explore a model such as (3.19) in addition to (7.5) and (7.6) in the same spirit as (7.2) was proposed instead of (7.1). In the case of non-lattice data the fit of a model such as (3.19) is likely to be very sensitive to the spatial distribution of sites and the definition of neighbours for each site. Some evidence for this is given in Haining (1983c, Table IIa) in representing a map of adoption of TB controls in Swedish farms.

An autologistic model with a trend surface representation for α_i may provide a more parsimonious description of a surface and indicate the relative importance of different scales of variation. Wrigley (1977) commented on the problem of testing for residual spatial correlation in fitting logistic probability surface models so that estimating autologistic models may also help to address this issue. Little work appears to have been done in this area and the following example is mainly illustrative. Parameter estimates are pseudo maximum likelihood (see section 4.2.2) and were obtained by using a function minimising routine and trying out numerous initial values in the search algorithm.

Figure 7.21 shows a map of farmsteads in an area of Sweden taken from

288

Haining (1983c, figure 4). Farmsteads where TB controls had been adopted by 1935 are black dots. The stimulus to adopt derives from neighbourhood contact and the influence of three regional dairies. The resulting map might therefore be expected to contain rather complicated patterns of variation (adoption/non-adoption). A series of (two category) autologistic and logistic models were fit to this data, that is, parameters estimated and the maximised log likelihood function evaluated. Trend surfaces of up to order two were fitted. Results are summarised in Table 7.19. A site has the value $y_i = 1$ if it is an adopter of TB controls by 1935, 0 if it is not yet an adopter. Two sites are neighbours (in the autologistic model) if their distance apart is less than 0.1 units (the area was scaled to 1.4×1.4 units) so that the average number of neighbours per site was just over 6. Different results are obtained if different distance rules are used. There appears to be a

Figure 7.20. Third order probability surfaces (*a*) highly annoyed, (*b*) moderately annoyed, (*c*) only a little annoyed. After Wrigley (1977).

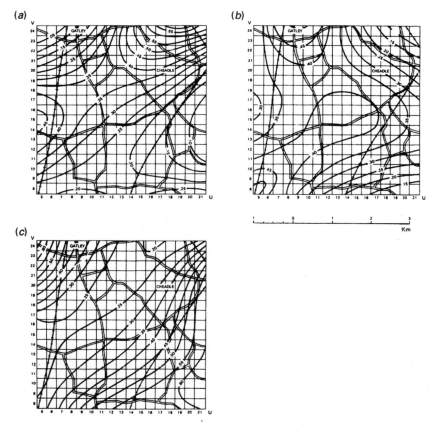

289

strong case for choosing the zero order autologistic model as a summary of the map.

One use for models of this type may be to indicate those areas where the local pressure to adopt is greatest. If for each site the probability that $y_i = 1$ given the set of values in the neighbouring sites is evaluated using the fitted model, then the resulting map suggests where subsequent adoption may be most likely to occur – namely those sites where $y_i = 0$ but where the conditional probability of $y_i = 1$ is relatively high. Figure 7.22 (*a*) and (*b*) taken from Haining (1983c) gives examples using different distance rules.

With some correlated count data it may be possible to transform values so that the data are normal, or nearly normal. Symmetrising Box–Cox transformations should be considered since in some cases these may also make the data more nearly normal. This approach offers an alternative in those situations where the analysis of spatially correlated count data is

Figure 7.21. Distribution of Swedish farms adopting TB controls by 1935.

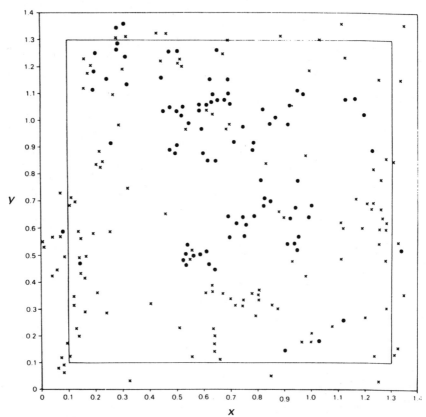

Table 7.19. *Fitting orders of logistic and autologistic model to TB control*
adoption data

	Model order and type				
	O(L)	O(A)	1(L)	1(A)	2(L)
Constant	−0.37	−1.69	−1.97	−2.37	−1.90
E–W			0.24	0.32	0.81
N–S			1.96	0.38	2.09
$(E–W)^2$					−0.58
$(N–S)^2$					0.02
(E–W)(N–S)					−0.56
$\hat{\beta}$		0.46		0.53	
Log likelihood function	−150.78	−95.86	−137.93	−92.88	−135.69

L denotes logistic model ($\beta = 0$), A denotes autologistic model ($\beta \neq 0$), 0, 1, 2 denote order of trend surface component.

difficult or where no suitable spatial model exists. For example, the Poisson and negative binomial distributions frequently arise as models for count data. Unfortunately, spatially correlated forms of these models are restricted to competitive interaction and do not include the case of positive correlation. This problem is mentioned in the case of count data in Haining (1981a) using settlement data from Dacey (1968) where the frequency distribution of houses per quadrat followed a negative binomial distribution but showed evidence of positive spatial correlation.

Given the importance of count data, particularly in the social sciences this is an area that deserves attention (Cressie and Read, 1989).

7.2   Interpolation and estimating missing values

Throughout this section it is assumed that site values are spatially correlated and vary smoothly across the area. It is also assumed that it is not possible or practical to visit all the sites and actually collect the data! For point or area interpolation on a continuous surface the quality of the estimates depends on how well the sampling design has detected all the relevant scales of variation in the underlying surface. Interpolation procedures cannot be expected to provide good estimates if the sampling procedure has missed important elements of surface variation such as periodicity or trend.

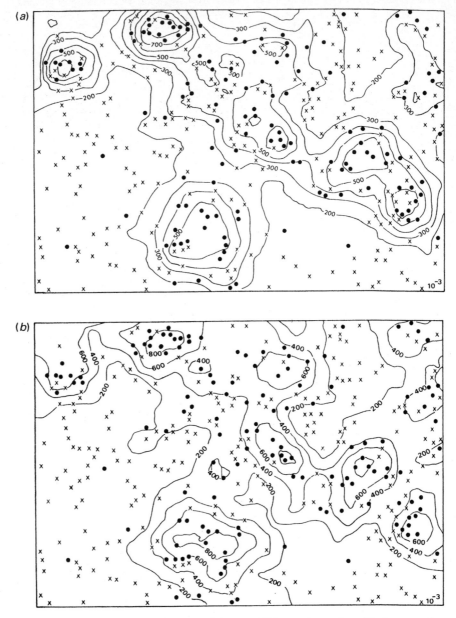

Figure 7.22. Observed pattern of adopters of TB controls together with the expected probability surface (*a*) negative exponential distance weighting, (*b*) nearest neighbour weighting. ● = adopted TB controls, × = did not adopt TB controls. After Haining (1983c).

In the case of missing value estimation it is assumed that observations are missing 'at random', that is, the deletion process does not depend on the form of the surface. Situations where all large or small values have been censored are excluded. If parts of a remotely sensed image are obscured by cloud cover this too may violate the assumption of a random deletion process since cloud cover can be related to the form of the underlying surface. On the other hand sensor failure along a scan line is not usually related to the form of the underlying surface. The situation where observations have been deleted at regular spatial intervals need not raise problems of this nature unless the data contain an important scale of variation that coincides with the interval (or some multiple of the interval).

We now examine different estimation procedures which are distinguished by the amount and type of information that is utilised in the observed data. We follow the classification in Bennett *et al.* (1984).

7.2.1 Ad hoc and cartographic techniques

The simplest interpolation methods are so called 'ad hoc' techniques. These techniques disregard any spatial information in the observed data set. Examples of these are where missing values (in a grid or county system) are replaced by the grand mean or median, a method which is apparently often used and is satisfactory if there are no data trends and no spatial correlation. A special case of this approach widely used in soil surveys is to assign to the site with the missing value the mean value of the characteristic of the soil class within which the site is assumed to lie. A still simpler method is to discard sites with missing observations which is perhaps the safest method but potentially very wasteful of data. Bennett *et al.* note that these methods have usually been developed to meet specific problems and provide satisfactory solutions within those specific terms of reference.

The second category are termed 'cartographic' techniques, widely used in interpolation routines for computerised map drawing from sample data. However, they also provide a basis for point site estimation within a continuous areal framework or a discrete spatial framework such as a set of counties or quadrats. This category of methods exploits the geometric properties of the set of sites where data have been collected. One group are termed spline methods. Ripley (1981, pp. 38–44) discusses these. A simple example is the construction of Dirichlet cells round each sample site with each point in the cell assigned the same value as the sample site. The resulting map may be subsequently smoothed to eliminate the sharp discontinuities at the cell boundaries (see for example Tobler and Lau, 1978).

Distance weighting methods estimate missing values as weighted averages of neighbouring (observed) values. Frequently these weights are chosen as a function of distance, for example:

$$\hat{y}_i = \Sigma_{j=1,\ldots,n} \lambda_{ij} y_j \qquad (7.8)$$

where n is the number of observed sites, $\hat{y}_i$ is the predicted value at location i and λ_{ij} is a weighting scaled so that $\Sigma \lambda_{ij} = 1$. Inverse distance weighting schemes where $\lambda_{ij} = d_{ij}^{-\alpha}$ are often recommended where α is a positive valued parameter and d_{ij} is the distance between the location of site i where the estimate is required, and site j. The value of α can be varied to achieve different surface effects. In general small values of $\alpha (< 1)$ produce rather flat surfaces whilst larger values of α produce more rolling surfaces. Ripley (1981, p. 36) discusses some of the problems with this approach. In particular, estimates are susceptible to clustered data points (Figure 7.23) and values of α should exceed 2 if the interpolated surface is not to show a smooth surface of variation punctuated by peaks or spikes at the observation points. If the surface contains a linear trend the predicted surface will not have this appearance while if the surface is higher order, since $\hat{y}$ is bounded above and below by the maximum and minimum values of the observations, the predicted surface cannot 'follow' gradients. The effects of data point clustering can be ameliorated by using only observed values that are in a close neighbourhood of location i to predict y_i and this will also help to prevent 'spiked' interpolated surfaces. Tobler (1979b) suggests an iterative procedure that also removes spiking in which, after interpolation, observed sites are themselves estimated from the interpolated sites and so on until convergence. However, one effect of this adjustment is

Figure 7.23. Effect of clustering on point estimation.

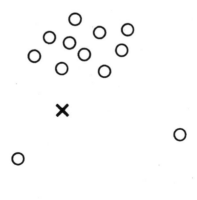

that $\hat{y}_i \neq y_i$ at the sites where observations are recorded. In the case of data collected on an irregular areal partition Kennedy and Tobler (1983) have proposed replacing distance weighting in (7.8) by weights based on the length of the common boundary between county i and each of its neighbours. This can be extended to second, third and higher order neighbours (Tobler and Kennedy, 1985). Again, the final map could be smoothed to remove discontinuities at the boundary. Upton (1985) indicates problems with the boundary contact method in relation to distance weighting (see Figure 7.24) which he also argues is computationally easier.

Although different distance or boundary weightings as well as other embellishments such as directional and gradient adjustments (Shepard, 1984) can be introduced to reflect surface properties of known importance these techniques depend largely on geometrical relationships between the sites. They are unlikely to perform well on surfaces with marked areal non-stationarity and some evidence for this has been given by Upton (1985) who compared the Kennedy–Tobler boundary weighting approach with the distance weighting approach in (7.8) using U.S. state population density data. Results are given in Table 7.20 albeit for the worst cases but they indicate how poorly these types of methods can perform on this kind of data.

Some of the problems of using (7.8) may be overcome by distance weighted least squares. In this method, reviewed by Ripley (1981, p. 37), $\hat{y}_i = A_i \hat{\theta}_i$ where A_i is a row vector of locational co-ordinates for site i corresponding to a trend surface model and

$$\hat{\theta}_i = (A^T \Lambda_i{}^2 A)^{-1} (A^T \Lambda_i{}^2 y) \tag{7.9}$$

Figure 7.24. Polygon (B) hidden from neighbouring polygon (A) when boundary contact weights used.

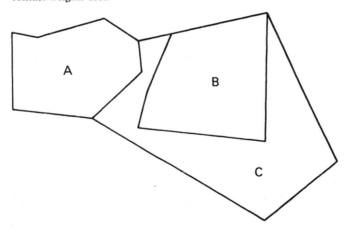

Table 7.20. *Estimated population density based on all other states (worst cases only)*

State	Actual density	Kennedy–Tobler estimate	Estimate using unscaled latitude & longitude with [b]		Estimate using scaled latitude & longitude with [b]	
			d^{-4}	d^{-6}	d^{-4}	d^{-6}
Arizona	15.6	34.3	22.2	16.6	25.4	18.2
California	127.6	14.5	5.6	4.5	5.4	4.5
Colorado	21.3	14.0	11.8	6.5	15.2	9.2
Idaho	8.6	14.6	22.3	19.1	22.0	19.2
Illinois	199.4	89.4	119.6	132.2	126.5	137.0
Maine	32.1	81.7	305.9	244.9	257.9	174.4
Montana	4.8	10.1	11.0	6.3	12.7	7.5
Nevada	4.4	17.6	111.3	124.2	111.8	124.2
New Jersey	953.1	307.8	336.3	299.3	342.4	303.4
New Mexico	8.4	32.7	25.6	21.5	28.4	22.4
North Dakota	8.9	18.6	15.6	10.1	20.5	12.3
Ohio	260.0	122.5	120.1	96.2	132.2	105.6
Oregon	21.7	27.8	47.5	49.8	41.6	45.0
South Dakota	8.8	24.7	19.0	16.7	22.1	17.7
Vermont	47.9	220.7	172.1	105.9	128.6	89.9
Washington	51.2	23.1	21.4	21.3	20.0	20.3
West Virginia	72.5	138.3	175.8	151.9	184.4	154.6
Wyoming	3.4	15.7	17.1	16.9	18.0	16.6

[a] Source: Kennedy and Tobler (1983), [b] d is a measure of distance.

where A is the matrix of location co-ordinates for the observations (**y**) and Λ_i is a diagonal matrix with entries $[\lambda_{i,1}^{\frac{1}{2}}, \ldots, \lambda_{i,n}^{\frac{1}{2}}]$. Note that θ_i is estimated for each site, i, to be interpolated. This method appears to be intermediate computationally between the geometrical or cartographic methods discussed so far and the next set of procedures that exploit other properties of the observed data values. In (7.9) the order of trend surface may be selected to reflect data properties.

7.2.2 *Distribution based techniques*

These methods proceed by fitting a model to the observed data which is then used to estimate the unobserved point sites or areas. We examine two approaches.

7.2.2 (a) Sequential approaches (sampling a continuous surface)

The idea underlying the sequential approach to interpolation and missing value estimation is indicated in Figure 7.25. Following the results in Chapter 4 the two components of surface variation are estimated simultaneously. The arrow returning from model evaluation to model specification is to indicate that it may be necessary to respecify the model in the light of information obtained at the evaluation stage, before proceeding to interpolate. Although all the methods in this final category involve much more computation they have the advantage of providing an estimate of interpolation errors which in some applications (and if the model is a good fit) may be of considerable importance.

Assume that n observations have been drawn from a model given by (7.2) but where V is known. The problem is to predict Y_0, a single value at a given

Figure 7.25. Interpolation and missing data estimation: sequential analysis.

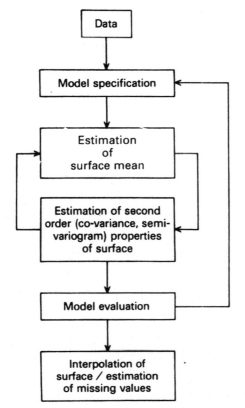

site. The true value must be

$$Y_0 = A_0\theta + e_0$$

where A_0 is the vector of location co-ordinates for the interpolation point and e_0 is the true but unknown value of the prediction disturbance. We assume

$$E[e_0] = 0; \; E[e_0{}^2] = \sigma_0{}^2$$

$$E[ee_0] = \begin{bmatrix} E[e_1e_0] \\ \vdots \\ E[e_ne_0] \end{bmatrix} = \Gamma = \text{Cov}(e,e_0) = \text{Cov}(Y,Y_0)$$

We define a linear predictor $\hat{Y}_0 = c^T Y$ where c is chosen to

$$\text{minimise } \sigma_y{}^2 = E[(\hat{Y}_0 - Y_0)^2]$$
$$\text{subject to } E[\hat{Y}_0 - Y_0] = 0$$

The constraint implies that $c^T A = A_0$. It can then be shown that the prediction error

$$\hat{Y}_0 - Y_0 = c^T e - e_0 \tag{7.10}$$

and the prediction or estimation variance

$$\begin{aligned} \sigma_y{}^2 = E[(\hat{Y}_0 - Y_0)^2] &= E[(c^T e - e_0)(c^T e - e_0)^T] \\ &= c^T Vc - 2c^T \Gamma + \sigma_0{}^2 \end{aligned} \tag{7.11}$$

The interpolation problem is to minimise (7.11) subject to (7.10), that is minimise

$$\psi = c^T Vc - 2c^T \Gamma + \sigma_0{}^2 + 2(A_0 - c^T A)\pi \tag{7.12}$$

where π is a $k \times 1$ vector of Lagrangian multipliers (k is the rank of A)[5]. After differentiating with respect to c and π and setting the two equations to zero we obtain

$$\begin{bmatrix} \hat{c} \\ \hline -\hat{\pi} \end{bmatrix} = \begin{bmatrix} V & A \\ A^T & 0 \end{bmatrix}^{-1} \begin{bmatrix} \Gamma \\ \hline A_0 \end{bmatrix}$$

from which it follows that

$$\left. \begin{aligned} \hat{c} &= V^{-1}[I - A(A^T V^{-1} A)^{-1} A^T V^{-1}]\Gamma + V^{-1} A(A^T V^{-1} A)^{-1} A_0{}^T \\ \hat{\pi} &= (A^T V^{-1} A)^{-1}(A_0 - A^T V^{-1}\Gamma) \end{aligned} \right\} \tag{7.13}$$

so that the best linear unbiased estimator for Y_0 is

$$\begin{aligned} \hat{Y}_0 &= c^T Y \\ &= A_0 (A^T V^{-1} A)^{-1} A^T V^{-1} Y + \Gamma^T V^{-1} Y - \Gamma^T V^{-1} A (A^T V^{-1} A)^{-1} A^T V^{-1} Y \end{aligned}$$

Now if **Y** is MVN then if **θ** is unknown, the ML estimator is

$$\tilde{\boldsymbol{\theta}} = (\mathbf{A}^\mathsf{T}\mathbf{V}^{-1}\mathbf{A})^{-1}\mathbf{A}^\mathsf{T}\mathbf{V}^{-1}\mathbf{Y}$$

so

$$\hat{Y}_0 = \hat{\mathbf{c}}^\mathsf{T}\mathbf{Y} = \mathbf{A}_0\tilde{\boldsymbol{\theta}} + \boldsymbol{\Gamma}^\mathsf{T}\mathbf{V}^{-1}[\mathbf{Y} - \mathbf{A}\tilde{\boldsymbol{\theta}}] \tag{7.14a}$$

$$= \mathbf{A}_0\tilde{\boldsymbol{\theta}} + \boldsymbol{\Gamma}^\mathsf{T}\mathbf{V}^{-1}\tilde{\mathbf{e}} \tag{7.14b}$$

where $\tilde{\mathbf{e}}$ is the vector of residuals from the trend surface.

The error variance (or minimised estimation variance) is

$$\sigma_e{}^2 = \sigma_0{}^2 - \boldsymbol{\Gamma}^\mathsf{T}\mathbf{V}^{-1}\boldsymbol{\Gamma} + [\mathbf{A}_0 - \mathbf{A}^\mathsf{T}\mathbf{V}^{-1}\boldsymbol{\Gamma}]^\mathsf{T}\tilde{\boldsymbol{\pi}} \tag{7.15}$$

which is the minimised value of (7.11) subject to (7.10) and can be obtained from (7.12) using (7.13).

The formal similarity between (7.14) and (7.8) shows this method to be an estimation procedure based on weighted local averaging. One of the important differences is that whereas in (7.8), and the other cartographic techniques, the weights are specified a priori and usually without reference to the statistical properties of the data, in (7.14) they are specified in terms of first and second order data properties in order to yield an unbiased minimum variance estimator. The other and possibly more important difference is that estimation variances can now be computed.

The estimator (7.14) assumes **V** is known, which is generally not the case. So we require an estimate of **V**, $\tilde{\mathbf{V}}$. A consequence of this is that (7.14) is no longer unbiased although it is consistent. Where computationally feasible, it seems preferable to simultaneously estimate **θ** and **V** in (7.2) by maximum likelihood (see Chapter 4) using the n observed values, and then substitute these into (7.14). If the variables are normal, (7.14) is then the maximum likelihood estimator for Y_0 and the conditional specification of the process (Martin, 1984, pp. 1279–80; Dowd, 1984). The parameter $\sigma_0{}^2$ must be estimated and the ML estimator is

$$\tilde{\sigma}_0{}^2 = n^{-1}(\mathbf{Y} - \mathbf{A}\tilde{\boldsymbol{\theta}})^\mathsf{T}\tilde{\mathbf{V}}^{-1}(\mathbf{Y} - \mathbf{A}\tilde{\boldsymbol{\theta}})$$

Note that the unobserved site does not enter into parameter estimation at any stage.

The estimator (7.14) is essentially the universal (point) kriging estimator of Y_0. However, in kriging, trend (or drift) is usually removed *prior* to analysis and the residuals from the trend used to estimate the semi-variogram (rather than the covariances **V**). Methods of estimating and modelling the semi-variogram have been discussed in Chapters 3 and 4.

The kriging estimator, with estimated trend ($\hat{\mu}$) is given by:

$$\hat{Y}_0 = \hat{\mu}_0 + \hat{\Psi}^\top \hat{\gamma}^{-1} [Y - \hat{\mu}] \tag{7.16}$$

where $\hat{\mu}_0$ is the trend evaluated at the interpolation point and $\hat{\gamma}$ is the $n \times n$ matrix of semi-variogram estimates (usually in the form of an analytic function) and $\hat{\Psi}$ is the $n \times 1$ vector of semi-variogram estimates between the point to be interpolated and the n data points, that is

$$\hat{\Psi}^\top = [\hat{\gamma}(\mathbf{x}_0, \mathbf{x}_1), \ldots, \hat{\gamma}(\mathbf{x}_0, \mathbf{x}_n)]$$

where $\mathbf{x}_0$ is the location co-ordinate of the point to be interpolated and $\mathbf{x}_i$ the location of the ith data point. (7.16) should be compared with (7.14).

The minimum estimation (or kriging) variance is

$$\sigma_e^2 = \hat{\Psi}_*^\top \hat{K}^{-1} \hat{\Psi}_* - \sigma_0^2 \tag{7.17}$$

where

$$\hat{K} = \begin{bmatrix} \hat{\gamma} & 1 \\ 1^\top & 0 \end{bmatrix} \qquad \hat{\Psi}^{*\top} = [\hat{\Psi}^\top 1]$$

and σ_0^2 is interpreted as the nugget variance. Equation (7.17) should be compared with (7.15). The derivation of (7.16) and (7.17) can be developed as for V and Γ above, recalling that, when $c(h)$ is defined, $c(h) = c(o) - \gamma(h)$. The additional unit term in $\hat{\Psi}^*$ is for the (single) Lagrangian multiplier.

$\hat{Y}_0$ is obtained by re-introducing the trend evaluated at the point of interpolation. This might be done simply by evaluating $A_0 \hat{\theta}$ if the trend was removed by least squares, or by interpolating planes between a median polish fit (see Cressie, 1986). The kriging variance at a point depends on the semi-variogram (or covariances) and the distribution of points, not on the local structure of observations. Variances will tend to be larger for areas where sampling is relatively sparse and at the boundaries where there are fewer neighbours (Figure 7.26).

Like (7.8), estimators (7.14) and (7.16) interpolate, that is, if we predict at a site at which there is an observation, $\hat{y}_i = y_i$, since $\hat{c} = (0, \ldots, 0, 1, 0, \ldots, 0)$ where the 1 is in the ith position. (Note that the simple trend surface estimator $A\hat{\theta}$ does not interpolate.) A problem with this however is that unless σ_0^2 is small, the interpolated surface may be smooth with sharp peaks at the observation points. This means that the interpolated surface will depend on the distribution of sample points. Block kriging offers a solution to this problem and is also a better method of interpolating area data (such as core samples of soil). In the case of block kriging the prediction is for an area A of size $|A|$ centred at the point $\mathbf{x}_0$ and (disregarding trend):

$$\hat{Y}(\mathbf{x}_0) = \hat{s}^\top \hat{\gamma}^{-1} Y \tag{7.18}$$

where $\hat{s}^T = [\int_A \gamma(x_1,x)p(x)dx, \ldots, \int_A \gamma(x_n,x)p(x)dx]$ and $\int_A p(x)dx = 1$ with

$p(x) = 1/|A|$ if $x\varepsilon A$ and 0 otherwise.

The minimised estimation variance for area A is

$$\sigma_e^2 = \hat{s}^{*T}\hat{K}^{-1}\hat{s}^* - \iint_A \hat{\gamma}(x,y)p(x)p(y)dxdy$$

where $\hat{s}^*$ contains the additional unit term and the second term is the average variance between any pair of points in A, called the within block variance. Unlike σ_0^2 this is never zero. Figure 7.27 shows two maps that have been interpolated by point and block kriging.

If the observations are drawn from a multivariate normal distribution then the kriging estimator $\hat{Y}_0$ is the conditional expectation of Y_0 given the $\{Y_i\}$ and will be conditionally unbiased (if V or γ are known) as well as

Figure 7.26. Error map of sodium content at Plas Gogerddan for block kriging over areas of $920\,m^2$. Isarithms are in units of $(meq/10\,kg)^2$, from the innermost, 1.0, in steps of 0.2 to the outermost, 1.6. After Burgess and Webster (1980).

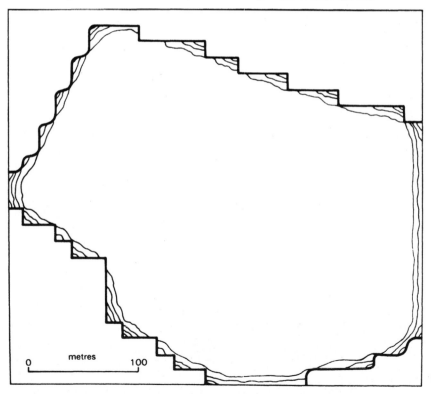

metres

0   100

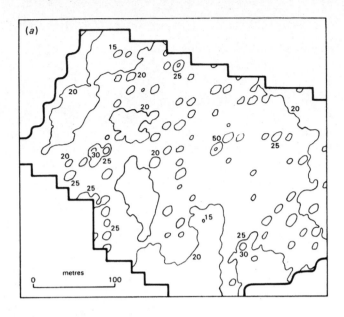

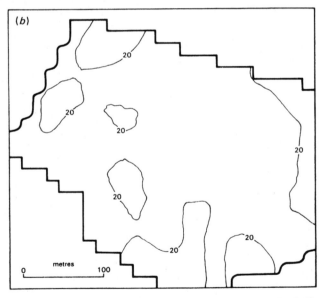

Figure 7.27. (*a*) Isarithmic map of sodium content in Cae Ruel, Plas Gogerddan, produced by point kriging. Isarithms are at intervals of 5 meq/10 kg. (*b*) Map of sodium content at Plas Gogerddan made by block kriging over areas of 920 m². Only the isarithm for 20 meq/10 kg appears. After Burgess and Webster (1980).

Table 7.21

ordered data values	1.0	1.1	3.6	4.0	5.2	6.1	7.7
weights	0.1	0.3	0.05	0.15	0.1	0.25	0.05
cumulative weights	0.1	0.4	0.45	0.60	0.7	0.95	1.00

globally unbiased. If the underlying distribution is not normal the estimator will not be unbiased and will be sensitive to the effects of outliers. Methods that have sought to get round this by transforming the data to normality (for example log normal kriging) 'rest on dubious assumptions which in practice either do not hold or are very difficult to verify' (Dowd, 1984, p. 101). However, resistant methods have been developed to counter the effects of outliers to which the kriging estimator is sensitive.

In Chapter 6 resistant methods of estimating trend, the semi-variogram, covariances and model fitting were discussed and since the kriging estimator depends on all these quantities they contribute to the development of resistant kriging. However, the presence of Y in (7.14), (7.16) and (7.17) means that the problem of outliers resurfaces at the stage of determining $\hat{Y}_0$.

One approach to resistant kriging is to minimise $E|\hat{Y}_0 - Y_0|$ rather than $E[(\hat{Y}_0 - Y_0)^2]$. Dowd (1984) discusses this and other approaches but notes that 'enthusiasm . . . has been somewhat dampened by . . . computational and theoretical difficulties' (p. 101).

Hawkins and Cressie (1984) and Dowd (1984) suggest several similar approaches to resistant kriging based on a weighted median approach described in Henley (1981). One method is as follows:
(i)   Obtain a resistant estimate of the semi-variogram.
(ii) Remove data point Y_i and use standard kriging to estimate the kriging weights for $\hat{Y}_i$. Compute the kriging variance ($\sigma_{e(i)}^2$).
(iii) Use these weights to get a resistant estimate of Y_i. This might be done as follows. Sort the $\{Y_j\}_{j \neq i}$ into ascending order and interpret their associated kriging weights as relative frequencies of occurrence. The estimator of Y_i, known as a weighted median estimator and denoted $\tilde{Y}_i$ is then the Y_i value (or some interpolated value), with the cumulative frequency of 0.5. Henley's example is shown in Table 7.21.

Linear interpolation gives $\tilde{Y}_i = 3.7$. (It is argued by Dowd (1984, p. 102) that significant negative weights would be unlikely to arise in situations where resistant kriging is needed.)

(iv)   Edit Y_i by replacing it by the Winsorised version

$$Y_i^{(s)} = \begin{aligned} &= \tilde{Y}_i + c\sigma_{e(i)}^2 & \text{if } Y_i - \tilde{Y}_i > c\sigma_{e(i)}^2 \\ &= Y_i & \text{if } |Y_i - \tilde{Y}_i| \leqslant c\sigma_{e(i)}^2 \\ &= \tilde{Y}_i - c\sigma_{e(i)}^2 & \text{if } Y_i - \tilde{Y}_i < -c\sigma_{e(i)}^2 \end{aligned}$$

a value of $c = 2.0$ is suggested by Hawkins and Cressie (1984).

(v)   Repeat this for all i.

(vi)   To krige any point or block, estimate the weights using the robust estimator of the semi-variogram on the original data, *but* predict the point or block value by replacing $\mathbf{Y}$ with $\mathbf{Y}^{(s)}$, the vector of edited values.

Hawkins and Cressie (1984) state that the usual kriging variances, based on the robust estimate of the semi-variogram are adequate approximations here. There is further discussion and modifications in the previously cited papers. The procedure can be shortened by eliminating the editing phase.

The underlying model for these resistant procedures is a contaminated normal distribution. If data are non-normal a transformation is usually recommended, before adopting the resistant procedures. Armstrong (1984) also points out that resistance to outliers is only one of the problems to be faced and stresses the sensitivity of current kriging methods to the choice of semi-variogram model and the distribution of sample sites across the area. She gives an example to indicate how bad things can be as a result of small changes in model specification and the distribution of points. These issues reflect other aspects of robustness not merely departure from the normal distribution.

7.2.2   (b) Simultaneous approaches

Figure 7.28 represents the simultaneous approach to missing value estimation. Model specification is based on the properties of the observed data and parameter values and missing values are then estimated simultaneously. Suppose a region has n areas and observations on a variable are missing from k of these. The observation vector $\mathbf{y}$ may be partitioned, after suitable permutation, so that $\mathbf{y}^T = (\mathbf{y}_0{}^T | \mathbf{y}_m{}^T)$ where $\mathbf{y}_0$ denotes the $(n - k)$ dimensional vector of observed values and $\mathbf{y}_m$ is the k dimensional vector of missing values. Let the covariance matrix, $\sigma^2\mathbf{V}$, be similarly partitioned (after permutation) so that

$$\mathbf{V} = \begin{bmatrix} \mathbf{V}_{oo} & \mathbf{V}_{om} \\ \mathbf{V}_{mo} & \mathbf{V}_{mm} \end{bmatrix} \qquad \mathbf{V}^{-1} = \begin{bmatrix} \mathbf{V}^{oo} & \mathbf{V}^{om} \\ \mathbf{V}^{mo} & \mathbf{V}^{mm} \end{bmatrix}$$

so that $\sigma^2\mathbf{V}_{ij}$ is the covariance matrix for the sub-vectors $\mathbf{Y}_i$ and $\mathbf{Y}_j$ of the partitioning. $\mathbf{V}$ depends on the parameter vector $\mathbf{z}$.

In the case of normal data, the log likelihood function for the case where k observations are missing but the missing values are to be estimated is (Martin, 1984)

$$L(\boldsymbol{\theta},\sigma,\mathbf{z},\mathbf{y}_m\,|\,\mathbf{y}_o) = -(n-k)\ln(2\pi)/2 - (n-k)\ln(\sigma^2)/2 - \ln(\,|\,\mathbf{V}^{mm}\,|\,/\,|\,\mathbf{V}^{-1}\,|\,)/2 \\ -(\mathbf{y}-\mathbf{A}\boldsymbol{\theta})^{\mathsf{T}}\mathbf{V}^{-1}(\mathbf{y}-\mathbf{A}\boldsymbol{\theta})/2\sigma^2$$

The ML estimators are as follows

$$\tilde{\mathbf{y}}_m = \mathbf{A}_m\tilde{\boldsymbol{\theta}} + \tilde{\mathbf{V}}_{mo}(\tilde{\mathbf{V}}_{oo})^{-1}(\mathbf{y}_o - \mathbf{A}_o\tilde{\boldsymbol{\theta}}) \tag{7.19}$$

$$\tilde{\boldsymbol{\theta}} = (\mathbf{A}^{\mathsf{T}}\tilde{\mathbf{V}}^{-1}\mathbf{A})^{-1}(\mathbf{A}^{\mathsf{T}}\tilde{\mathbf{V}}^{-1}\mathbf{y}) \tag{7.20}$$

$$\tilde{\sigma}^2 = (\mathbf{y}-\mathbf{A}\tilde{\boldsymbol{\theta}})^{\mathsf{T}}\tilde{\mathbf{V}}^{-1}(\mathbf{y}-\mathbf{A}\tilde{\boldsymbol{\theta}})/(n-k) \tag{7.21}$$

and $\mathbf{z}$ is estimated by minimising

$$|\,\mathbf{V}_{oo}\,|^{(1/(n-k))}(\mathbf{y}-\mathbf{A}\tilde{\boldsymbol{\theta}})^{\mathsf{T}}\mathbf{V}^{-1}(\mathbf{y}-\mathbf{A}\tilde{\boldsymbol{\theta}}) \tag{7.22}$$

A numerical procedure might be to start with an initial estimate for $\mathbf{y}_m$ (perhaps using local or global means of the observed data), then estimate $\boldsymbol{\theta}$

Figure 7.28. Interpolation and missing data estimation: simultaneous analysis.

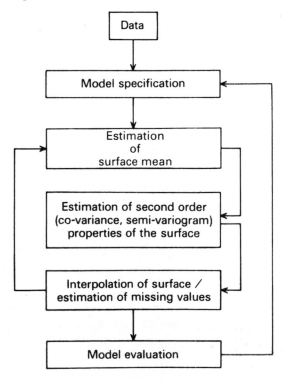

(setting $V = I$) then z and σ^2. At the second iteration (7.19) to (7.22) are evaluated using estimates from the first iteration and the cycle continues until convergence. Note that unlike the sequential approach the 'data' vector contains both observed values and estimates of the missing values. Mardia and Marshall (1984) refer to this as unified universal kriging. The procedure is close to that suggested by Orchard and Woodbury (1972). Martin (1984) discusses this estimation procedure in detail and suggests a number of approximations. He observes that it is similar to the EM algorithm.

Haining *et al.* (1989) consider the effects of missing data on the estimation of the other parameters of the model, particularly the mean which may be of substantive interest. They examine the effects of differing numbers and configurations of missing values on a surface where $V = (I - \tau W)^{-1}$ and $\tau \geqslant 0$. The estimation variance for the mean increases as the number and compaction of missing values increase and as the level of spatial correlation (measured by τ) increases. They also examine the effect of missing values on the estimation of τ. Where, as in remote sensing, numerous images of the same scene are available with possibly different levels and patterns of deletion, the different levels of information loss (in estimating parameters) may provide helpful criteria in deciding which image to analyse. The results of Haining *et al.* (1989) have been generalised by Martin (1989) who shows that information loss on the estimation of τ also depends on where in the region (relative to the boundary) the missing observations are located.

Earlier it was suggested that this simultaneous approach might seem more appropriate where data arise from a spatial grid or matrix where missing values represent gaps in this matrix. Missing values in remotely sensed data usually occur either in linear sequences when a detector fails along a scan line or in clustered patches arising from scattered cloud cover. Current remote sensing practice that uses only a single waveband takes an average of neighbouring observed values (Bernstein *et al.*, 1984). Table 7.22 contrasts this averaging technique ($\bar{N}$) with estimates obtained from (7.19) on an 11×11 data set with artificially deleted values. Results in Haining *et al.* (1989) indicate that a square window encircling the area to be interpolated, of size at least 9×9 and large enough so that the missing values do not exceed 10% of the total data matrix, provides a useful rule of thumb. Two different models were used for V: 1. $V = (I - \tau W)^{-1}$ and 2. $\sigma^2 V = \{C(i,j)\}$ where $C(i,j)$ is the empirical spatial covariance evaluated for sites i and j and depending only on the distance between the two sites. So 2 assumes local stationarity whilst 1 is non-stationary. $A_m \tilde{\theta}$ is the estimate of the surface mean, RMSE is the root mean square error between the estimated values and the (known) obliterated pixels. Estimates for θ, σ^2 and τ are given by (7.20) to (7.22).

Three cases of missing data are examined: an 11×1 linear sequence; a 3×3 clustered set and a 2×2 clustered set. The original data are highly correlated but show evidence of trend which was modelled as a quadratic surface. Estimated spatial correlations obtained using the residuals from the trend surface suggested that a symmetric structure for V was plausible and that a CAR model was reasonable. Of the three methods the simple averaging method performed least well although whether the improvements achieved by the other methods justify the considerable extra computation may be a consideration. Of the other two methods, that based on the CAR model converges more rapidly and also gives generally better estimates.

This method has also been applied to the estimation of missing meteorological records (Haining *et al.*, 1984) and census data (Griffith *et al.*, 1989). Glasbey and O'Sullivan (1988) consider the problem of estimating soil resistance values missing due to the presence of stones. The example given in Martin (1984) indicates how very highly model dependent missing value estimation is.

7.2.3 *Extensions*

7.2.3 *(a) Obtaining areal properties*

The above interpolation methods can be used to compute totals for the variable over some specified area A. If A is small a simple procedure is to interpolate by block kriging to a central point and treat this as representative of the area. In the case of large areas McBratney and Webster (1983) suggest computing a number of point estimates across the area A to form a regular grid. Associate to each estimate its Dirichlet tile. Let $\hat{Y}(\mathbf{x}_i)$ be the estimate at $\mathbf{x}_i$ with tile area V_i, then

$$\hat{Y}_A = \Sigma_i V_i \hat{Y}(\mathbf{x}_i)$$

A similar problem is to identify the proportion of an area over which the variable is less than a certain critical value (say Z). If the sample density is high enough and uniform enough Dirichlet tiles can be constructed and their areas evaluated (Green and Sibson, 1978). If the sampling density is low or the sample coverage has gaps so that the tiles are large then, as before, the area may be kriged prior to constructing the tiles. A different

307

Table 7.22. *Missing value estimates: different forms for* $\hat{V}$

Case (a) CAR model. Case (b) estimated empirical covariances.

												RMSE
11 × 1 linear												
True value	4.5	5.0	4.3	3.3	2.4	3.8	4.4	4.2	3.9	2.3	3.6	
$\bar{N}$	4.5	4.9	3.1	2.85	4.3	3.7	4.6	3.1	2.6	2.75	2.7	0.920
Case (a)												
$A_m \tilde{\theta}$	4.10	3.91	3.74	3.57	3.43	3.29	3.17	3.07	2.98	2.90	2.84	
$\tilde{y}_m$	4.48	4.49	3.53	3.33	4.00	3.92	4.14	3.29	2.70	2.22	2.63	0.780
Case (b)												
$A_m \tilde{\theta}$	3.99	3.92	3.84	3.76	3.67	3.58	3.48	3.38	3.28	3.17	3.06	
$\tilde{y}_m$	5.44	5.36	3.70	3.20	4.84	3.75	4.49	3.69	2.62	2.35	2.74	0.953
3 × 3 clustered												
True value	5.5	3.7	4.0	2.4	3.8	4.4	3.1	3.7	5.2			
$\bar{N}$	4.7	4.2	3.5	3.3	3.675	4.2	2.2	3.0	3.25			0.888
Case (a)												
$A_m \tilde{\theta}$	3.59	3.52	3.47	3.54	3.42	3.32	3.36	3.19	3.04			
$\tilde{y}_m$	4.24	3.97	3.70	3.54	3.65	3.76	2.84	3.25	3.40			0.879
Case (b)												
$A_m \tilde{\theta}$	3.67	3.64	3.61	3.56	3.48	3.40	3.32	3.19	3.06			
$\tilde{y}_m$	4.55	4.11	4.03	4.08	3.40	4.02	2.52	2.78	3.04			1.055

approach is given in Switzer (1977) based on a weighted estimator of the form

$$\Sigma_i w_i I(Y_i \leqslant Z)$$

where $w_i > 0$ for all i, $\Sigma_i w_i = 1$ and I is the indicator function. The weights are chosen as a function of the correlation structure in the data to downweight the contribution of clustered sites and Switzer gives an example using pollution data. As Ord notes in the discussion to Switzer's paper a comparison of these methods would be interesting. The Dirichlet method has no optimality properties when sample spacings are irregular but the weightings are always positive (unless two sample points coincide) and can be computed once and for all.

Ripley (1981, pp. 64–72) states that prediction standard errors are of little help in assessing the variability of areal properties such as these and suggests the use of simulation to determine the fluctuations of the unknown parts of the surface. He describes how this should be done.

7.2.3 (b) Reconciling data sets on different areal frameworks

The need sometimes to place data sets from different agencies on the same areal framework was noted in Chapter 2. It is a problem that may assume particular importance as part of the development of computerised data banks held within geographical information systems. The problem is to convert data recorded on one areal partition (source zones) to another (target zones). If the source areas are small then they can be represented as points and point interpolators used to estimate values at a new set of grid points which can then be summed to yield values for the new target zones. Lam (1983) describes this as poor practice: properties depend on the type of point interpolation method used; there may be considerable ambiguity in assigning values to grid points; the method is not volume preserving which means that the sum of values over the new target zones may not correspond to the sum over the original source zones. An alternative approach is simply to parcel up the values of those source zones that overlap any given target zone. In the case of source zones that overlap two or more target zones, adjust or weight values in proportion to the size of overlap. A variant of this, the pycnophylactic method, incorporates neighbouring values in the estimate of each target zone in an iterative procedure in which after obtaining target zone values, values are then averaged using adjoining neighbours, adjusting values also to preserve volumes. Tobler (1979c) describes the method and Lam (1983) gives a simple worked example. A method using information on other variables is described by Flowerdew and Green (1989).

309

7.2.3 (c) Categorical data

The methods of interpolation described here assume interval scale data. Interpolation of categorical maps, in contrast, seems to have received relatively scant treatment although some suggestions have been given in Switzer (1975), Guptill (1978) and Tobler (1979a). The use of conditional expectations to interpolate interval scale data suggests that a suitable analogue for categorical data is provided by the 'auto' models described in Chapter 3 and also discussed as models for categorical data in section 7.1. Besag (1986) examines the use of these models for cleaning contaminated image data.

7.2.3 (d) Other information for interpolation

None of the methods for interpolation discussed here can be used uncritically for they all make assumptions about the nature of the phenomenon under study and the nature of the unobserved parts of the surface which may or may not be true. The choice of distance exponent (α) in (7.8) can determine the degree of spikiness of a surface and, despite earlier comments, a spiked surface may be appropriate in certain cases such as atmospheric smoke concentrations for which dispersion theory suggests there should be a pattern of very localised peaks superimposed on a flat regional field. In the case of kriging estimators, results hinge on the choice of semi-variogram model and here too theoretical knowledge may be a factor in choosing one model rather than another (Webster, 1985). Methods should be compared in terms of how they use information contained in the observed data, their accuracy, amount of computational effort and how flexible they are in allowing theoretical and process information to be incorporated into the interpolator. When interpolating over large areas (weather maps, remotely sensed images) the curvature of the earth's surface may need to be allowed for. Projection onto a flat surface should only be done after interpolating and contouring. Earth surface distance and directional relationships between points are not accurately preserved on a flat surface (Willmott *et al.*, 1985).

Where complete information on other variables is available the methods of this section can be used with the mean modelled by a regression model. The problems of selecting a regression model are not the same as those of choosing a trend surface model as will be evident in the next chapter. Nonetheless the approaches discussed here can be extended in this way and for many situations, particularly in the case of census data, this may offer the possibility of greatly improved estimates and much reduced prediction errors.

NOTES

1. The minimisation routine used to obtain ρ in the SAR model placed bounds on the value of ρ and used first derivatives as well as the function value (NAG routine EO4BBF). A routine based on only function values and used to obtain the parameter estimates reported in Haining (1987c) gave a value of ρ that in fact attained the stationary maximum. There are small differences in other parameter estimates depending on which routine was used.

2. In the regression model defined by (7.1) the least squares estimator for θ, is $\hat{\theta} = (A^TA)^{-1}A^Ty$. Leverage measures the influence of the ith data point ($i = 1, \ldots, n$) on the predicted value at site i and is the ith diagonal element of H, denoted h_{ii} where

$$H = A(A^TA)^{-1}A^T \qquad (F1)$$

This can be seen by noting that

$$\hat{y} = A\hat{\theta} = A(A^TA)^{-1}A^Ty = Hy$$

A large value for $h_{ii} (> 2.0p/n$ where p is the number of regression coefficients) signifies that $\hat{y}_i$ depends strongly on y_i. In the case of model (7.2), for the regression element of the model

$$H = A(A^TV^{-1}A)^{-1}A^TV^{-1} \qquad (F2)$$

In both F1 and F2 the trace of H is p.

3. The (externally) studentised residual

$$e_i^* = \hat{e}_i / s(i)(1 - h_{ii})^{\frac{1}{2}}$$

where h_{ii} is the leverage of observation i, $\hat{e}_i = y_i - \hat{y}_i$ is the model residual, and

$$s(i) = \{(1/(n-p-1))[(n-p)s^2 - \hat{e}_i^2/(1-h_{ii})]\}^{\frac{1}{2}}$$

where $s^2 = \Sigma_i \hat{e}_i^2/(n-p)$ and p is the number of regression coefficients estimated.

The standardised residual, which does not remove the effect of the ith observation in the estimation of the error variance, is:

$$e_i^+ = \hat{e}_i / s(1 - h_{ii})^{\frac{1}{2}}$$

An absolute value larger than 2.0 is often considered large. The DFITS of observation i is

$$\text{DFITS} = [h_{ii}/(1 - h_{ii})]^{\frac{1}{2}} e_i^*$$

and an absolute value larger than $2(p/n)^{\frac{1}{2}}$ is often considered large.

For a discussion of these diagnostics see Belsley, Kuh and Welsch (1980).

4. If O_i denotes the observed number of deaths in area i due to a specified disease then providing the disease is non-infectious O_i can be considered binomial with parameters n_i and p_i where n_i is the number of individuals in area i and p_i is the probability of any individual dying from the disease. If $n_i \to \infty$ and $p_i \to 0$ such that $n_i p_i \to \lambda_i$ then O_i is Poisson with parameter λ_i. So $E[O_i] = \text{Variance}[O_i]$.

5. This specification does not ensure positive weights. Negative weights are sometimes considered undesirable because 1. they can produce negative

estimated values for Y_o (which may be physically impossible), 2. they can produce estimates higher than any observed sample value (which need not necessarily be a bad thing), and 3. they give rise to considerable variation over short distances.

To ensure positivity a second constraint would be needed, namely $c \geqslant 0$. The solution to this minimisation problem uses the Kuhn–Tucker theorem (see Barnes and Johnson, 1984).

8

Analysing multivariate data sets

This chapter examines the problems of analysing data where several sets of spatially referenced variables have been measured. Data refer to groups rather than individuals (ecological analysis). Section 8.1 considers parametric and non-parametric correlation measures of point to point association and indices for describing spatial association between two variables. Section 8.2 considers the problems of building regression models for spatially referenced data.

Since areal units or point sites are not closed entities, effects cannot be assumed to operate only on an intra-area basis, so that to the usual problems associated with analysing variate relationships we have the additional problems that arise from the spatial nature of the system and in particular the possible existence of spatial relationships between the observational units. Data may refer to areas that are modifiable. It is important to realise that not only are results scale dependent (so that there is no single correlation coefficient or regression model) but even at a given scale estimates could vary depending on the particular areal partition.

8.1   Measures of spatial correlation and spatial association

8.1.1   Correlation measures

The Pearson product moment correlation coefficient, $\hat{r}$, provides a measure of association between two variables measured at the interval scale. If the two variables are denoted Y_1 and Y_2 then

$$\hat{r} = \frac{\Sigma_{i=1,\ldots,n}(y_{1,i}-\bar{y}_1)(y_{2,i}-\bar{y}_2)}{(\Sigma_{i=1,\ldots,n}(y_{1,i}-\bar{y}_1)^2\Sigma_{i=1,\ldots,n}(y_{2,i}-\bar{y}_2)^2)^{\frac{1}{2}}} \qquad (8.1)$$

where the bar denotes the sample mean. In the case of ranked data, the

Spearman rank correlation coefficient, $\hat{r}_s$, provides a measure of association and

$$\hat{r}_s = 1 - [6\Sigma_{i=1,\ldots,n} \, d_i^2/(n^2 - n)] \tag{8.2}$$

where d_i is the difference in the ranks of the ith observation for Y_1 and Y_2; (8.2) can be derived from (8.1) (Siegal, 1956, pp. 202–4). Both (8.1) and (8.2) measure 'point to point' association or covariation between Y_1 and Y_2 and are unaffected by the spatial distribution of observations. But standard procedures for testing for significance assume independent observations and are affected by the presence of trend or correlation in the set of observations. In the case of time series data, the effect of trend and serial correlation in the distribution of $\hat{r}$ are discussed in Yule (1926) and Bartlett (1935). The sampling variance of $\hat{r}$ may be seriously underestimated by the usual estimator when bɔth Y_1 and Y_2 are dependent (positive correlation). In the case of two serially correlated variables where both follow a first order autoregressive model:

$$Y_{1,t} = \rho_1 Y_{1,t-1} + e_t$$
$$Y_{2,t} = \rho_2 Y_{2,t-1} + e_t$$

then

$$\mathrm{Var}(\hat{r}) = n^{-1}[(1 + \rho_1\rho_2)/(1 - \rho_1\rho_2)] \tag{8.3}$$

which shows why *both* variables must be correlated for the problem to arise. If only one of the variables is dependent or if both variables show only weak dependence then conventional test procedures are usually safe to use.

The effect of spatial correlation on the asymptotic variance of $\hat{r}$ is examined in Richardson and Hemon (1981; 1982). They assume a sample from a bivariate normal distribution (constant mean) where observations are spatially correlated and stationary. Table 8.1 shows the asymptotic variance of $\hat{r}$ when both Y_1 and Y_2 follow a stationary first order CAR model (Table 8.1(a)) and a stationary first order SAR model (Table 8.1(b)).[1] The effect is serious for parameter values close to the stationary maximum.

A study by Bivand (1980) using Monte Carlo methods and simulating spatial data using a first order symmetric SAR model, varying the spatial parameter in order to generate different levels of spatial dependence, examined the seriousness of the problem in the case of small non-stationary data sets. His results confirm these general findings but in addition he suggested that the problem was more serious for non-lattice data than for lattice data (see also Clifford *et al.*, 1989). Figure 8.1 compares simulated sampling distributions of $\hat{r}$ when Y_1 and Y_2 are independent and both follow

Table 8.1.

(a) Asymptotic value of $n\,Var(\hat{r})$ when Y_1 and Y_2 are first order CAR models

$R_2(0,1)$					$R_1(0,1)$				τ_2
	0.00	0.100	0.200	0.300	0.400	0.500	0.600	0.700	
0.000	1.00								0.000
0.100	1.00	1.042							0.095
0.200	1.00	1.088	1.195						0.168
0.300	1.00	1.138	1.321	1.558					0.2123
0.400	1.00	1.192	1.469	1.866	2.444				0.2355
0.500	1.00	1.248	1.637	2.251	3.270	4.989			0.2456
0.600	1.00	1.307	1.824	2.720	4.419	7.917	15.903		0.2491
0.700	1.00	1.368	2.025	3.256	5.883	12.432	33.523	140.603	0.2499
τ_1	0.00	0.095	0.168	0.212	0.235	0.245	0.249	0.249	

(b) *Asymptotic value of n Var(r̂) when Y₁ and Y₂ are first order SAR models*

R₂(0.1)	R₁(0.1)										ρ₂
	0.00	0.100	0.200	0.300	0.400	0.500	0.600	0.700	0.800	0.900	
0.000	1.00										0.000
0.100	1.00	1.041									0.049
0.200	1.00	1.085	1.178								0.094
0.300	1.00	1.130	1.282	1.458							0.133
0.400	1.00	1.178	1.394	1.655	1.960						0.165
0.500	1.00	1.227	1.515	1.878	2.320	2.864					0.190
0.600	1.00	1.279	1.647	2.130	2.743	3.530	4.540				0.209
0.700	1.00	1.332	1.788	2.411	3.237	4.347	5.844	7.894			0.225
0.800	1.00	1.387	1.940	2.729	3.823	5.369	7.581	10.826	15.893		0.236
0.900	1.00	1.444	2.094	3.096	4.541	6.710	10.045	15.403	24.838	44.846	0.244
ρ₁	0.000	0.049	0.094	0.133	0.165	0.190	0.209	0.225	0.236	0.244	

Values for $\tau_1 > \tau_2$ and $\rho_1 > \rho_2$ are obtained by symmetry. $R_i(0.1)$ are the first order spatial correlations corresponding to the parameter values of the model.
From Richardson and Hemon, 1982.

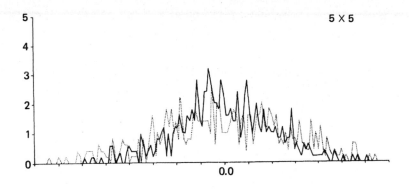

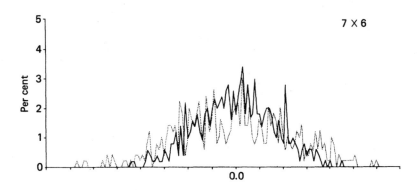

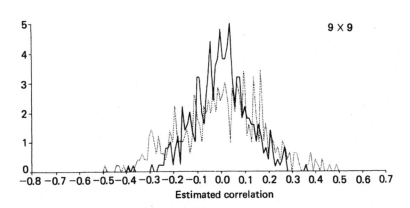

Figure 8.1. Empirical sampling distribution of the correlation coefficient for autocorrelated and non-autocorrelated data: — $\rho = 0.0$, - - - $\rho = 0.2$. After Haining (1980).

a single parameter non-stationary SAR model. The greater spread of $\hat{r}$ in the case of autocorrelated data is evident.

Statistical significance can be tested using Fisher's Z transformation which behaves as a standard normal deviate (see for example Freund, 1962). Alternatively obtain:

$$(n-2)^{\frac{1}{2}}|\hat{r}|(1-\hat{r}^2)^{-\frac{1}{2}} \tag{8.4}$$

Under the null hypothesis (H_0) that the true correlation is zero, (8.4) is t distributed with $(n-2)$ degrees of freedom. But both these testing procedures assume independent approximately normal observations with finite variance. For spatially correlated data one solution might be to pre-whiten the data by obtaining independent residuals (see Chapter 3). It is then the two sets of residuals that are correlated. Clifford *et al.* (1985) suggest obtaining an adjusted value for $n(N')$ which they call the 'effective sample size'. This measures the 'equivalent' number of independent observations. Under fairly mild conditions

$$N' \approx 1 + \left[\frac{\Sigma_k N_k C_1^*(k) C_2^*(k)}{n^2 s_1^2 s_2^2}\right]^{-1} \tag{8.5}$$

where N_k is the number of pairs of observations separated by distance lag k, $C_i^*(k)$ is the estimated spatial covariance at lag k for $Y_i(i=1,2)$ with edge correction (see Chapter 4) and s_i^2 is the sample variance for $Y_i(i=1,2)$. All lags are used. Clifford and Richardson (1985) approximate (8.5) with

$$N' \approx 1 + n^2 (\text{trace}(\hat{\mathbf{R}}_1 \hat{\mathbf{R}}_2))^{-1} \tag{8.5a}$$

where $\hat{\mathbf{R}}_1$ and $\hat{\mathbf{R}}_2$ are the estimated correlation matrices for the two processes. The second terms on the right of (8.5) and (8.5a) are estimators for the inverse of the variance of $\hat{r}$ when the data are spatially correlated.

The hypothesis of independence between Y_1 and Y_2 is rejected if

$$(N'-2)^{\frac{1}{2}}|\hat{r}|(1-\hat{r}^2)^{-\frac{1}{2}} \tag{8.6}$$

exceeds the critical value of the t statistic with $N'-2$ degrees of freedom.

Clifford and Richardson (1985) report simulation results for first order SAR processes on 12×12, 16×16 and 20×20 lattices. $\hat{\mathbf{R}}_1^*$ and $\hat{\mathbf{R}}_2^*$ are obtained using estimated correlations up to lag 4. The type I error for a 5% test ranged from 3.5% to 7% (using 500 simulations). Type I errors in their tests when using (8.4) ranged from 8.2% in the least correlated case to 52% when $R(0,1)=0.8$. The adjustment appears to be very important. (See also Clifford *et al.*, 1989, where results for lattice and non-lattice situations are given using all distance lags.)

Table 8.2 shows results based on smaller lattices (11×11 and 7×7).

Table 8.2. *Type I errors for a 5% test on the Pearson product moment correlation coefficient based on 300 simulations of first order SAR processes. The upper figure is the unadjusted test and the lower figure is the test based on the Clifford and Richardson (1985) adjustment.*
(a) 11×11

	ρ_1					
ρ_2	0.0	0.10	0.15	0.20	0.225	0.24
0.0	0.0566			0.0500		0.0400
	0.0600			0.0533		0.0466
0.10		0.0700			0.1000	
		0.0600			0.0400	
0.15			0.10000		0.1500	
			0.0500		0.0366	
0.20				0.1900		0.3100
				0.0433		0.0400
0.225					0.3366	
					0.0433	
0.24						0.5000
						0.0566

(b) 7×7

	ρ_1					
ρ_2	0.0	0.10	0.15	0.20	0.225	0.24
0.0	0.0500			0.0648		0.0631
	0.0433			0.0477		0.0421
0.10		0.0769			0.1245	
		0.0535			0.0535	
0.15			0.1103		0.1772	
			0.0535		0.0702	
0.20				0.1973		0.2307
				0.0602		0.0735
0.225					0.2575	
					0.0702	
0.24						0.3110
						0.0769

Table 8.3. *Properties of N' for different levels of spatial dependence in* Y_1
and Y_2. *The Pearson product moment correlation coefficient*

(1) 11×11 $(n = 121)$

ρ_1	R(1,0)	ρ_2	R(1,0)	Mean (N')	St. Dev. (N')
0.00	0.0	0.00	0.0	121.38	4.82
0.20	0.523	0.00	0.0	128.29	21.16
0.24	0.762	0.00	0.0	140.42	40.91
0.10	0.211	0.10	0.211	106.78	9.31
0.225	0.650	0.10	0.211	86.51	19.67
0.15	0.343	0.15	0.343	86.47	12.21
0.225	0.650	0.15	0.343	66.94	16.12
0.20	0.523	0.20	0.523	56.96	12.89
0.24	0.762	0.20	0.523	43.36	13.31
0.225	0.650	0.225	0.650	39.40	11.18
0.24	0.762	0.24	0.762	28.21	8.88

(2) 7×7 $(n = 49)$

ρ_1	R(1,0)	ρ_2	R(1,0)	Median (N')	IQR
0.0	0.0	0.0	0.0	49.74	4.33
0.20	0.510	0.0	0.0	50.74	14.99
0.24	0.721	0.0	0.0	52.07	20.41
0.10	0.210	0.10	0.210	45.04	7.81
0.225	0.625	0.10	0.210	37.35	12.99
0.15	0.339	0.15	0.339	38.86	12.02
0.225	0.625	0.15	0.339	31.71	12.02
0.20	0.510	0.20	0.510	28.32	10.77
0.24	0.721	0.20	0.510	23.36	9.64
0.225	0.625	0.225	0.625	22.01	9.25
0.24	0.721	0.24	0.721	17.66	7.69

IQR denotes inter-quartile range.

Again first order SAR processes are simulated with unobserved border values set to zero. $\hat{R}_1$ and $\hat{R}_2$ are obtained using estimated correlations (C^*) up to lag 6. It would be useful to have an approximate adjustment for n without having to evaluate (8.5). Although such an uncritical adjustment should be used with caution the results in Table 8.3 may be useful in providing some simple bounds for the test. In the case of the 7×7 lattice estimated correlations are rather unstable and some large adjustments to n are indicated when (8.5) is evaluated. In practice many of these would probably be deemed inappropriate (particularly those when $N' \gg n$). For this reason we report the median properties of N'. Note that the adjustment will give values of $N' > n$ if there is negative spatial correlation.

Equation (8.4) is used to test the significance of the Spearman rank correlation coefficient in the case of independent observations and $n > 10$ (Siegal, 1956, p. 212). The Clifford and Richardson adjustment (8.5) was also examined therefore as a means of adjusting (8.4) for $\hat{r}_s$ in the case of correlated ranked data. Table 8.4 shows results based on the same 11×11 and 7×7 lattices. First order SAR processes are simulated and the observations rank ordered. $\hat{R}_1$ and $\hat{R}_2$ were obtained using estimated correlations up to lag 6 from the ranked data. Table 8.5 gives values for N' that parallel the results in Table 8.3.

The adjustment to n proposed by Clifford and Richardson applies to stationary correlated data with constant variance. Data with trend may need to be de-trended (see Chapter 7) prior to using their method. The important issue here is not to use conventional procedures to test for the significance of the correlation coefficient, and to recognise that a large $\hat{r}$ (or $\hat{r}_s$) value may be due to spatial correlation effects or where the variables are only associated through a common dependence on a third variable that has a spatial trend. The risks of inferring association between variables that is nothing other than the product of the spatial characteristics of the system are real and call for caution on the part of the user (Haining 1991a). One implication of this work is for comparative analyses between different areas or time periods. If the spatial dependency structures are different the correlation estimates are being drawn from different sampling distributions. Differences therefore may be due to differences in the spatial properties of the surface (as well as any scale effects associated with different areal partitions) rather than to any differences in the associations between the variables. Applications together with a comparison between the method of pre-whitening and the Clifford and Richardson approach are discussed by Haining (1991a) and Clifford *et al.* (1989).

In the case of descriptive techniques based on a matrix of correlations (such as principal components) again correlations may confound inter-variate patterns of association with spatial patterns present in the variables in the particular study area.

Bivariate correlations can also be computed between pairs of variables separated by a fixed distance or lag – the bivariate analogue of the correlogram. In geostatistics the cross semi-variogram is preferred and used in kriging (McBratney and Webster, 1983b).

Table 8.4. *Type I errors for a 5% test on the Spearman rank correlation coefficient based on 300 simulations of a first order SAR process converted to rank order values.*
The upper figure is the unadjusted test and the lower figure is the test based on the Clifford and Richardson (1985) adjustment

(a) 11×11

	ρ_1					
ρ_2	0.0	0.10	0.15	0.20	0.225	0.24
0.0	0.0566			0.0400		0.0333
	0.0566			0.0566		0.0333
0.10		0.0700			0.0966	
		0.0533			0.0333	
0.15			0.0866		0.1466	
			0.0500		0.0333	
0.20				0.1833		0.2966
				0.0366		0.0466
0.225					0.3300	
					0.0433	
0.24						0.4633
						0.0666

(b) 7×7

	ρ_1					
ρ_2	0.0	0.10	0.15	0.20	0.225	0.24
0.0	0.0466			0.0580		0.0596
	0.0466			0.0546		0.0526
0.10		0.0668			0.1237	
		0.0401			0.0702	
0.15			0.1036		0.1471	
			0.0501		0.0501	
0.20				0.1739		0.2107
				0.0568		0.0702
0.225					0.2441	
					0.0602	
0.24						0.2876
						0.0635

322

Table 8.5. *Properties of N' for different levels of spatial dependence in* Y_1 *and* Y_2. The Spearman rank correlation coefficient

(a) $11 \times 11(n=121)$

ρ_1	ρ_2	Mean (N')	St. dev. (N')
0.00	0.00	121.51	4.95
0.20	0.00	127.65	19.43
0.24	0.00	139.52	37.79
0.10	0.10	107.93	9.12
0.225	0.10	88.72	19.91
0.15	0.15	86.67	12.28
0.225	0.15	69.46	16.63
0.20	0.20	59.45	13.45
0.24	0.20	45.51	14.31
0.225	0.225	41.42	11.89
0.24	0.24	29.74	9.49

(b) $7 \times 7(n=49)$

ρ_1	ρ_2	Median (N')	IQR
0.0	0.0	49.57	5.03
0.0	0.2	50.43	13.77
0.0	0.24	52.13	20.16
0.10	0.10	45.79	7.71
0.10	0.225	38.50	13.61
0.15	0.15	39.47	10.24
0.15	0.225	32.55	12.62
0.20	0.20	29.37	11.77
0.20	0.24	24.22	10.23
0.225	0.225	22.58	10.51
0.24	0.24	18.41	8.71

IQR denotes interquartile range. St. dev. = standard deviation.

8.1.2 Measures of association

The following discussion is based on Hubert and Golledge (1982). Consider two variables which we label F and G. They are both measures at the ordinal scale and there are no tied ranks. There are n observation sites and the location of rank i on variable F is $l_F(i) = (x_F(i), y_F(i))$, whilst the location of rank i on variable G is $l_G(i) = (x_G(i), y_G(i))$. Measures of spatial association deal with the degree to which similar values on the two variables are spatially close to one another. Tjøstheim (1978) proposed a measure of association which (in an unnormalised form) can be expressed as

$$\Lambda = \Sigma_{i=1,\ldots,n} \, d(l_F(i), l_G(i)) \tag{8.7}$$

where $d(\,,\,)$ is a measure of spatial separation. Thus, (8.7) measures the degree to which identical rank values on F and G occupy positions that are close together in space. Tjøstheim (1978) used straight line distance measures so that after removing terms that do not affect association:

$$d(l_F(i), l_G(i)) = x_F(i)x_G(i) + y_F(i)y_G(i)$$

If the x and y co-ordinates are standardised to mean zero and unit variance, then Tjøstheim's index becomes:

$$A = \frac{\Sigma_{i=1,\ldots,n}(x_F(i)x_G(i) + y_F(i)y_G(i))}{\Sigma_{i=1,\ldots,n}[(x_F(i))^2 + (y_F(i))^2]} \tag{8.8}$$

$$= \tfrac{1}{2}[r_{x_F,x_G} + r_{y_F,y_G}]$$

where r_{x_F,x_G} and r_{y_F,y_G} are the correlations between the x and y co-ordinates respectively for identically ranked observations on F and G. Under the assumption that F and G are independent and the n observations on F and G are (spatially) independent (Tjøstheim, 1978):

$$E[A] = 0$$

$$\mathrm{Var}[A] = \frac{(\Sigma x_i^2)^2 + 2(\Sigma x_i y_i)^2 + (\Sigma y_i^2)^2}{(n-1)(\Sigma x_i^2 + \Sigma y_i^2)^2} = [2n^2/(n-1)](1 + r_{xy}^2)$$

where (x_i, y_i) is the location of the ith observation, and r_{xy} is the Pearson correlation between the set of (x, y) co-ordinates over the n locations. For large n, A can be tested as a normal variate in the case of systematic sampling designs, for small n, sampling experiments are required to determine the distribution.

The moments and reference distribution for A are obtained by assuming that one set of ranks (on the variable F, say) and their spatial locations are fixed, and the ranks of the other variable (in this case, G) are assigned at random to the n spatial locations. Each permutation (ρ) generates a value for the index $(A(\rho))$. This raises two problems. First, the inference framework disregards the spatial structure of the data. (Tjøstheim assumes F and G are spatially independent.) When the data refer to point observations on a continuous surface the assumption that the observations are independent need not raise difficulties since it is often possible to arrange for the sampling interval to be large enough so that the assumption is met. When data refer to areas that partition a region it may not be possible to meet this requirement. Second, each permutation generates a value for an aspatial measure of association such as Spearman's rank $(r_s(\rho))$. This is because each permutation gives rise to a new pairing of values on F and G at each of the n locations. As Hubert *et al.* (1985b) note 'the inference model for Tjøstheim's measure is spatially contaminated with an association statistic that is aspatial in design' (p. 38). Since spatial association as defined above is conceptually distinct from point-to-point association they argue that what is needed is 'a strategy for evaluating the relative size of spatial association *per se, conditional* on the fixed level of point-to-point association that is present in the data' (p. 39). They suggest treating this problem as a special case of the cross product statistic for spatial autocorrelation discussed in Chapter 6.

The cross product statistic compares the correspondence between two matrices which we define now as $\{d_{ij}\}$, where d_{ij} is some spatial separation measure between sites i and j; and $\{c_{ij}\}$ where c_{ij} is some bivariate function measuring the similarity between F measured at site i and G measured at site j. The statistic is defined as

$$\Gamma = \Sigma_{i,j} c_{ij} d_{ij} \tag{8.9}$$

Tjøstheim's index arises when d_{ij} is measured in terms of straight line distances and

$c_{ij} = 1$ if the rank of F at location i = rank of G at location j
$\quad\quad = 0$ otherwise

However, the reference distribution, $\Gamma(\rho)$, is now obtained by permuting *pairs* of observations on F and G treating the two values at any location $i(f_i$ and $g_i)$ as indivisible units. In this way the sampling distribution preserves the aspatial measure of point-to-point association. As an example (Hubert *et al.*, 1985a) let d_{ij} be symmetric and for $i \neq j$

$$c_{ij} = (|f_i - g_j| + |f_j - g_i|)/2 \tag{8.10}$$

Defined in this way $\Gamma(\rho)$ is a generalised spatial autocorrelation statistic. The methods of section 6.2 can be used to obtain the reference distribution for the statistic. The null hypothesis is that the *n pairs* show no evidence of spatial pattern; the alternative hypothesis is that the *n pairs* of observations display pattern (or structure or spatial predictability).

Hubert *et al.* (1985b) report the following study. Figure 8.2 shows rank order values of per capita residential burglary in 64 police reporting districts in Santa Barbara, California in 1980 and 1981. The Spearman rank correlation between the 64 observations is 0.66 which is significant at less than the 0.01 level using conventional test procedures. The Figures 8.2(*a*) and (*b*) also show evidence of spatial association using (8.10) and the Euclidean distance between district centroids in (8.9). A significance level of less than 0.01 is obtained based on either a Type III approximation or Monte Carlo methods. The results indicate that areas have similar rankings in terms of residential burglary in 1980 and 1981 and that areas of similar rank are found together. So, given a knowledge of the rank of a district in 1980 it should be possible to make a good prediction of its ranking in 1981 and of its neighbours' ranks in 1980 and 1981. Figures 8.3 and 8.4 show the effects of maximising and minimising respectively the raw spatial association index Γ. In both these figures $\hat{r}_s$ remains the same and the two figures show clearly how *strong* positive or negative association might look.

Different inference frameworks in sections 8.1.1 and 8.1.2 raise a possible source of confusion here. Hubert *et al.* (1985b) examined pattern properties in the burglary data and under the randomisation hypothesis the sampling distribution of the Spearman rank correlation coefficient is unaffected by the spatial properties of the two data sets since the spatial distribution of values is only one of *n*! possible spatial patterns. In the earlier discussion of the Spearman rank correlation coefficient the spatial distribution of values was important and expressed an important attribute of the data. In section 8.1.1, correlation measures were tested *conditional* on the structure of spatial dependence in the two variables. There is a difference of objective here which underlies the purpose for which statistical analysis is being carried out. Use of the measure of spatial association in Hubert *et al.* (1985b) is conditional on a given level of point-to-point association but their use of the measure of point-to-point association is not conditional on a given level of spatial structure. Since they reject a measure of spatial association that is contaminated by different levels of point-to-point association, it is at least arguable in a study of spatial data that any measure of point-to-point association should allow for the spatial structure in the data.

Returning to the Santa Barbara burglary data, Table 8.6 shows the spatial correlations for the two sets of data where the first order neighbours of an area are those areas that share a common border with it, and higher order neighbours are defined in the usual way. The number of pairs of

(a)

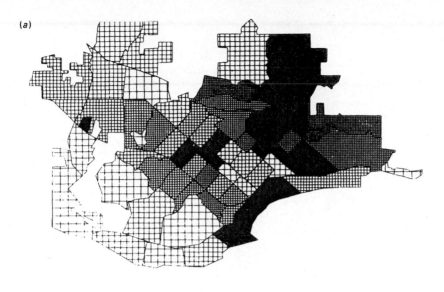

(b)

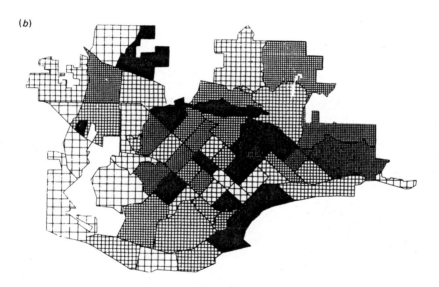

Figure 8.2. Rank order of per capita residential burglary (a) 1981, (b) 1980. After Hubert *et al.* (Darker shading denotes higher rates.)

(a)

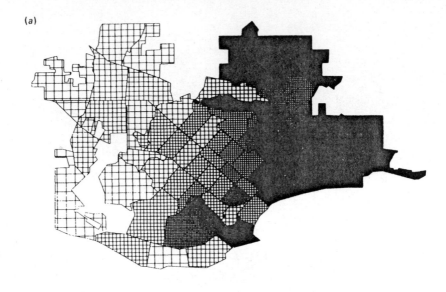

(b)

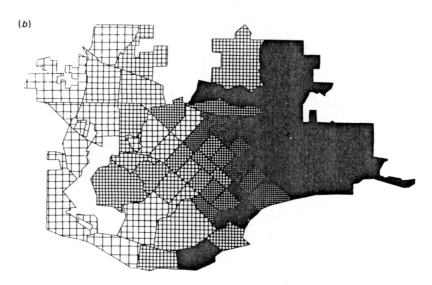

Figure 8.3. Maximised spatial association, Residential burglary (*a*) 1981, (*b*) 1980. After Hubert, Golledge, Costanzo and Gale (1985). (Darker shading denotes higher rates.)

(a)

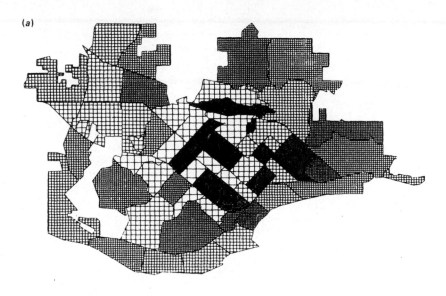

(b)

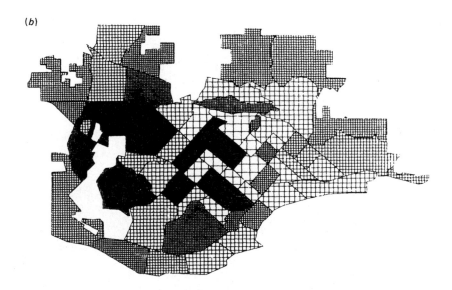

Figure 8.4. Minimised spatial association, Residential burglary (a) 1981, (b) 1980. After Hubert, Golledge, Costanzo and Gale (1985). (Darker shading denotes higher rates.)

Table 8.6. *Spearman's rank correlation measure for the 1980 and 1981
burglary data adjusted for the spatial structure of the data*

	Lag correlations						
	R(0)	R(1)	R(2)	R(3)	R(4)	R(5)	R(6)
1980	1.0	0.275	0.111	0.054	0.015	−0.067	−0.273
1981	1.0	0.055	−0.000	0.059	0.010	−0.044	−0.111
No. of data pairs	64	146	286	384	390	329	249

Spearman rank correlation $= 0.6638$, $n = 64$, Adjusted n $(N') = 47.37$,
t transform: 6.989 (no adjustment), 5.979 (Clifford–Richardson adjustment).

(ranked) data values used to compute the estimates are also shown. The
adjusted t transform for the Spearman rank correlation coefficient is given
together with the adjusted value of $n(N')$. In this case there is still a strong
(area to area) association between the two sets of data.

Where data on many variables have been collected a matrix of cross
product statistics (8.9) can be constructed measuring spatial associations
between each variable and lagged values of the remaining variables. (The
diagonal elements of such a matrix would correspond to the conventional
univariate measures described in Chapter 6.) Wartenberg (1985) has
suggested carrying out a principal components analysis (PCA) of this
matrix. (The analogy is drawn with the usual form of PCA applied to a
matrix of Pearson correlation coefficients although problems arise – such as
negative eigenvalues – which are discussed in the paper.) He discusses
examples in which the different components are identified with pattern
properties of the multivariate data set.

8.2   Regression modelling

Through regression modelling the analyst tries to account for variation in a
set of observations on a response variable Y in terms of a set of explanatory
variables or covariates $X_1, \ldots, X_k$. A functional expression is specified and its
fit to the data assessed. Apart from trying to identify a good model a further
objective may be to obtain good estimates of the regression coefficients. In
some situations there may be an additional objective which is to provide a
prediction equation. We do not intend to deal with this issue explicitly here,
however spatial interpolation and missing value estimation have been

discussed in Chapter 7 and the formal procedures described there for the trend surface case (where $\mu = A\theta$) would seem to be applicable to the regression case (where $\mu = X\beta$), providing there are no missing values in X. The main difference lies in the problem of choosing X. Weisberg (1985, Chapter 10) provides an introduction to missing data estimation in the standard regression model.

The standard regression model is defined by the equation

$$y = X\beta + e \qquad (8.11)$$

where y is the $(n \times 1)$ column vector of (interval or ratio scale) observations on Y over the n observational units, X is an $(n \times (k+1))$ matrix with a first column of 1s and the remaining entries x_{ij} denoting the value of variable $j(j = 1, \ldots, k)$ at site $i(i = 1, \ldots, n)^2$. The vector $\beta^T = (\beta_0, \ldots, \beta_k)$ specifies the regression coefficients to be estimated from the data and e is the $(n \times 1)$ vector of errors or disturbances and it is assumed that $E(e) = 0$ and $E(ee^T) = \sigma^2 I$ where I is the identity matrix.

There is a formal similarity between (8.11) and the usual trend surface model (7.1), although there are also important points of difference. In (7.1) A specifies only the location co-ordinates, and powers of those co-ordinates, and model fitting proceeds in a strict order sequence from constant to linear to quadratic to cubic, etc. The main problem is to determine the best order of model for describing the data. Only full order models should be fitted so that stepwise regression (starting with all the terms for some high order model) is not an appropriate fitting procedure. With high order models (cubic and above) there are inherent multicollinearity problems because powers of lower order terms are used in the model. This places constraints on the type of computational algorithms that can be used.

In the case of regression modelling multicollinearity problems may be present but are not inherent and stepwise regression may be an appropriate fitting procedure. In non-experimental applications the main problems to be confronted are those of model specification and variable selection. In Chapter 2 we reviewed Leamer's (1978) classification of specification searches in regression. Subject matter theory may be particularly important either in specifying a model (or small set of models) to be assessed against the data or in helping the analyst construct a sensible model for the data in an inductive search.

Table 8.7 summarises the main problems encountered in using (8.11). It can be seen that problems divide into two groups – those due to a failure to satisfy the assumptions of least squares which is the procedure used to fit and make inferences with (8.11) and those due to the nature of the data. The reader is referred to Weisberg (1985) and Wetherill *et al.* (1986) for a

Table 8.7. *Problems that may arise in fitting the regression model using ordinary least squares*

	Consequences	Check	Remedial action
(a) Problems due to assumptions of least squares			
(i) Residuals Not normal	Inferential test procedures based on F test may be invalid.	Rankit plot; Shapiro W test (and others).	Transform y values (Box–Cox transformation). Use of different error models (Generalised linear modelling).
Heteroscedastic	Biased estimation of error variance and hence inferential test procedures may be invalid.	Plot residuals against y, x's & other variables. Anscombe's test (and others).	Transform y variable ($y^{\frac{1}{2}}$,$\log(y)$,y^{-1}; Box–Cox transformation). Weighted least squares.
Not independent.	Inferential test procedures may be invalid. Underestimate true sampling variance of regression estimates. Inflated R^2.	Residual plots. Some tests (e.g. time: Durbin–Watson; space: Moran).	Iterated generalised least squares.
(ii) Non-linearity of functional relationship	Poor fit; meaningless results; non-independent residuals.	Scatterplots of y against x's. Added variable plots.	Transform x's and/or y variables.

(b) Problems due to the nature of the data

(i) Multicollinearity amongst explanatory variables.	$(X^TX)^{-1}$ unstable hence an unstable fit. Variances of regression estimates inflated.	Correlation measures. Tests based on eigenvalues of (X^TX).	Transform explanatory variables. Delete variables. Ridge regression.
(ii) Many explanatory variables.	Difficulties in performing: (i) efficient analysis, (ii) sifting out variables.		Added variable plots for variable selection. Transform x's and/or y to simplify model. Stepwise regression.
(iii) Outliers and leverage effects.	May severely distort model fit. Model fit is dependent on a few values.	Exploratory data analysis methods. Sensitivity analyses.	Robust, resistant regression. Data deletion.
(iv) Inaccurate data.	Meaningless results.	Exploratory data methods may highlight errors.	Delete or replace inaccurate values.
(v) Incomplete data.	Missing at random: could be wasteful of other information if this has to be discarded. Not missing at random: suspect inferences.		Estimate missing values (missing at random). Reduce data matrix to the cases with full information.
(vi) Categorical response variable.	'Normal' linear regression model inappropriate.		Generalised linear model (e.g. logistic regression).

discussion of these issues. Note that some tests may be inappropriate when more than one assumption does not hold. Here we consider the special problems that may be encountered when the observational units refer to spatially referenced sites or areas.

8.2.1 Problems due to the assumptions of least squares not being satisfied

Two of the problems noted in Table 8.7 are particularly likely to occur with spatially referenced data. If areal units differ in size or population then where the response variable measures a density feature or rate (e.g. where the denominator is the population of each area) then error variances may not be constant and it may be necessary to allow for the situation where

$$E[\mathbf{ee}^\top] = \sigma^2 \mathbf{D}$$

where $\mathbf{D}$ is diagonal but values are not identical (see Chapter 2). This problem is noted by Anselin and Can (1986) in the fitting of urban density functions and is discussed in Pocock *et al.* (1981) in the context of modelling mortality rates across a set of geographical areas of different population size.

However, the assumption that $\mathbf{D}$ is diagonal may also be unrealistic in the case of spatial data. There could be a number of reasons for this but most frequently cited is model misspecification due to excluded variables that are spatially autocorrelated, misspecification of the functional form or spatial heterogeneity. Model misspecification due to not including the effects of different types of spatial mechanisms operating between the spatial units such as interaction, transfer and diffusion mechanisms may also need to be considered.

It may be very difficult, statistically, to distinguish these sources. In Chapter 4 the generalised Moran coefficient was described as a test for spatial autocorrelation in the residuals. This is a test that indicates whether it is safe to assume $E[\mathbf{ee}^\top]$ is diagonal. However this does not absolve us of the need to explore the possible reasons for the residual pattern, nor is there a 'standard' solution to the problem. Other diagnostics should be explored in order to suggest a way forward.

Maps of residuals may be informative both in judging whether this assumption has been violated and indicating a possible reason. Figure 8.5 shows a map of positive and negative residuals from an urban population model of southwest Wisconsin (Haining, 1981b). The reported Moran test showed there to be significant nearest neighbour spatial autocorrelation. The map of residuals suggested that a possible cause of this pattern was the presence of small but unspecified amounts of non service employment across the set of towns and villages which the original model had not included. Figures 8.6(*a*),(*b*) show another example depicting residuals from an urban income model for an area of Pennsylvania (Haining, 1987a). The residuals from the model in the case of first order (small) urban places show significant nearest neighbour spatial autocorrelation. However, there is a clear concentration of negative residuals (per capita income levels lower than expected) in the northeast. A similar cluster is evident in the case of residuals for the second order (larger) urban places. A formal test is not feasible because of the number of observations but the map of residuals is suggestive of spatial heterogeneity. Paelinck and Klaassen (1979) refer to the 'pumping' effect that large metropolitan areas have on the income levels of nearby smaller urban areas in the sense of diverting consumer

Figure 8.5. Distribution of positive and negative residuals from the urban population model: 1970. $\oplus$ = positive residuals for towns, $+$ = positive residuals for villages, $\ominus$ = negative residuals for towns, — = negative residuals for villages. After Haining (1981b).

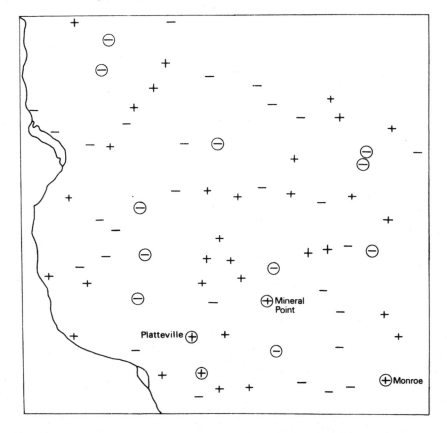

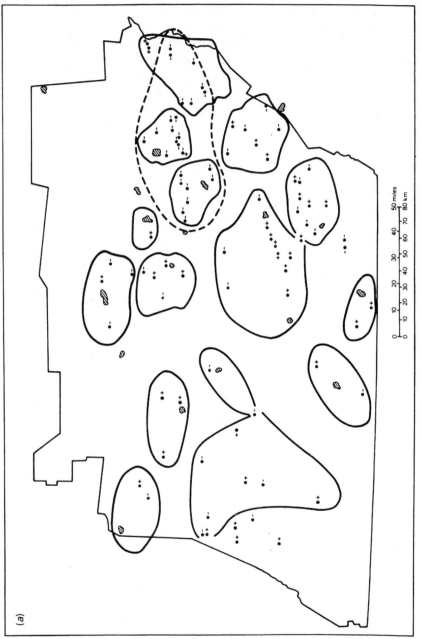

Figure 8.6. Residual maps of urban income model: (a) First order urban places. Solid line delimits edge of market region of 2nd order places; dotted line identifies area with a concentration of negative residuals.

(a)

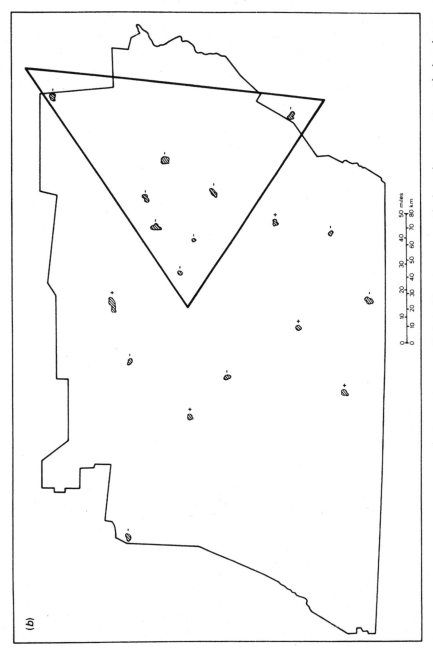

(b) Second order urban places. Triangle encloses 2nd order places all of which have negative residuals. + denotes positive residual. − denotes negative residual. After Haining (1987a).

expenditure away from such smaller places. The cluster of negative residuals is sandwiched between several large metropolitan areas so there is theoretical and empirical support for a model of spatial parameter variation. In both these cases the map of residuals can be used to guide the analyst towards a better model in a substantive sense apart from helping to avoid violating statistical assumptions. Scatterplots may also be helpful. Figure 8.7 shows the effects of two types of spatial heterogeneity on model fit. Region A will tend to have positive residuals and B negative residuals. Plots of residuals against their neighbours may indicate spatial correlation. The importance of including new variables can be explored using added variable plots which are discussed in Weisberg (1985).[3]

Figure 8.7. The effects of spatial parameter heterogeneity on the residuals of a homogeneous spatial model: (*a*) Differences in the constant coefficient, (*b*) Differences in the slope coefficient. × = sub-region A, ○ = sub-region B, —— = relationship for A and B, —— = model fit.

(*a*)

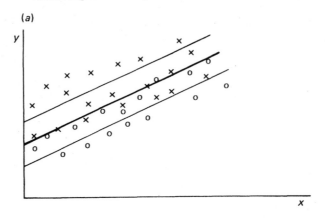

(*b*)

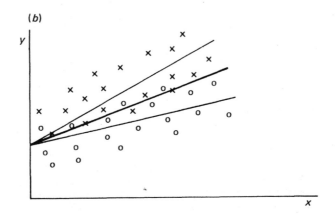

8.2.2 Problems of model specification and analysis

Since areal units are not closed entities we may wish to respecify (8.11) to reflect underlying interactions. One approach is to respecify the error model for (8.11) so that

$$y = X\beta + u$$
$$E[u] = 0; \quad E[uu^T] = V \tag{8.12}$$

where V (non-diagonal) is a model for spatial variation. If (8.12) is fit the non-independence of the error structure is allowed for in the fitting procedure but is not otherwise explained. In experimental situations (such as agricultural uniformity trials) where the set of predictor variables and their levels are determined by the experiment, a correlated errors model is a natural choice when residuals are found to be correlated. Attention focuses then on the choice of error model. In non-experimental situations the justification for this model is less clear cut since the set of relevant predictor variables is not defined. In such cases it is usual to consider fitting models like (8.12) if the residuals are correlated and no further variables can be identified or adding further variables does not remove the residual correlation. Residual correlation in such situations might, for example, reflect the combined effects of a large number of influences which are difficult to specify but which together display spatial persistance or continuity. The study by Loftin and Ward (1983) into the effects of population density on fertility rates includes a correlated errors model. Their study shows how such a re-definition of the error model (from 8.11) can alter model inferences substantially.

An alternative route is to include lagged explanatory variables in (8.11). For example:

$$y = GX\beta + e$$
$$E[e] = 0; \quad E[ee^T] = \sigma^2 I \tag{8.13}$$

where $G = \{g_{ij}\}$ and $g_{ii} = 1$ and $|g_{ij}| < 1$. G is fixed. Alternatively we might wish to specify different parameters associated with the lag terms. For example, as a special case

$$y_i = \beta_0 + \beta_1 x_i + \gamma\Sigma_{j=1,\ldots,n} w_{i,j} x_j + e_i \tag{8.14}$$

or

$$y = 1\beta_0 + X\beta_1 + \gamma WX + e$$

where γ is a parameter and w_{ij} is an element of the fixed connectivity matrix W that specifies spatial relationships between the sites. This model arises in the study of the housing market and in particular the modelling of spatial variation in house prices in which price is not only a function of structural characteristics of the house, its location (with respect to the city centre for example) and area characteristics (including environmental and demographic characteristics) but also the characteristics of *neighbouring* areas. Anas and Eum (1984) remark: 'the spillovers among neighbouring and otherwise substitutable submarkets can be taken into account to specify models in which market information from *other* submarkets becomes capitalised into housing prices' (p. 105). The estimate of each coefficient implicitly measures the price of that attribute.

A further option is to include lagged response variables into (8.11):

$$y = X\beta + Ky + e$$
$$E[e] = 0; \; E[ee^T] = \sigma^2 I$$

(8.15)

where $K = \{k_{ij}\}$ and $k_{ii} = 0$ and typically $k_{ij} = \rho w_{ij}$ where $\{w_{ij}\} = W$ as before and ρ is a parameter to be estimated. A form of (8.15) has been used in the study of urban retail spatial price variation where price setting at any location depends in part on local patterns of competition so that the price set at site *i* is partly a response to prices set at sites that are considered to affect the market share of the retailer at site *i*. Haining (1983b, 1984) uses (8.15) as a model for spatial variation in petrol pricing. A form of (8.15) has also been proposed as a model for spatial income variation. If Y denotes total income accruing to the residents of an area or community and if X denotes exogenous expenditures (export and investment income and government outlays) and C denotes endogenous expenditures (local consumption) by community residents then

$$Y = X + C$$

(8.16)

Since local consumption is a function of local income

$$C = cY$$

(8.17)

where c is the income creating local propensity to consume. From (8.16) and (8.17)

$$Y = (1-c)^{-1} X$$

This is a simple Keynesian income model. If we now consider this as a model for a set of towns in a region then, for any town, X can be decomposed into exogenous income originating from outside the region (X_1) and exogenous income originating from other towns in the region due for example to consumer expenditure in the town by inhabitants of other towns (X_2):

$$Y = (1-c)^{-1}(X_1 + X_2)$$

where

$$X_2 = \Omega Y$$

and $\Omega = \{\omega_{ij}\}$ is fixed with $\omega_{ii} = 0$ and ω_{ij} describes the patterns of movements between towns for purposes of consumer expenditure, being zero between towns where no such transfers occur (Haining, 1987a). Griffith (1981) and Anselin and Can (1986) use a model similar to (8.15) to represent intra-urban variation in population density while Doreian (1980) uses the model for describing geographical variation in political support and Doreian and Hummon (1976) in modelling patterns of political control.

We now consider some of the problems that arise if the class of possible regression structures for a set of data must be extended to include these other models.

8.2.2 (a) Model discrimination

An important issue is whether it is always possible to choose, on statistical grounds, between these models. Consider for example (8.12) with $V = \sigma^2[(I - \rho W)^T (I - \rho W)]^{-1}$ where W is a fixed connectivity matrix. Then

$$\left. \begin{array}{l} y = X\beta + u \\ u = \rho W u + e \end{array} \right\} \Rightarrow \begin{array}{l} y = X\beta + (I - \rho W)^{-1} e \\ = X\beta + \rho W X\beta + \rho W y + e \end{array} \tag{8.18}$$

Therefore it is not possible to discriminate, on purely statistical grounds, between these forms although the substantive interpretation of the models would be quite different. The same situation may arise in choosing between a model of spatial heterogeneity (where regression parameters vary between sub regions or segments of the study area) and one of the above spatial models. Presumably in the case of spatial heterogeneity a map of residuals from (8.11) will tend to show a more spatially blocked structure but the evidence is likely to be ambiguous.

8.2.2 (b) Specifying W

When dealing with irregular areal partitions or point sites spatial relationships between the observational units are specified by a connectivity matrix (W). This raises a number of problems because usually the analyst has to resort to fairly *ad hoc* criteria both for specifying which entries in W are non-zero (the joins) and the values for the non-zero entries (the weights). It is difficult to make this aspect of model specification (particularly the specification of the weights) any less arbitrary, and yet regression results will be sensitive to the choice of **W**. The way **W** is specified depends on whether the analyst is testing a specific hypothesis or trying to develop a model for the data. There are currently four approaches in general use. The first is to use geometrical properties of the set of sites or areas and/or interaction data (where available) to specify joins and weights for **W**. This approach has been described in detail in Chapter 3. The second is to use theoretical arguments to specify a form for **W**. Figure 8.8 shows two join systems for an urban population model (Haining, 1981b) and an urban

Figure 8.8. Systems of linkages for different inter-urban relationships (*a*) Orders and linkages between central places in southwest Wisconsin. After Haining (1981b). (*b*) Linkages between central places in Pennsylvania. After Haining (1987a).

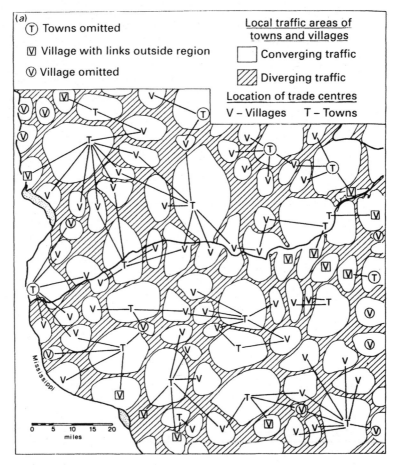

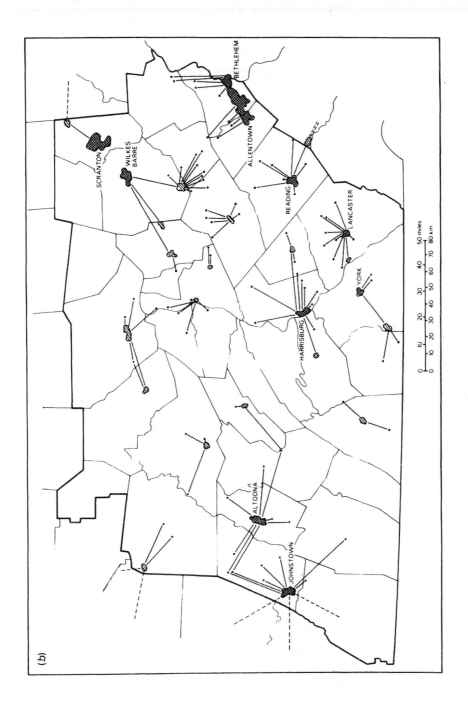

(b)

SCRANTON

WILKES BARRE

BETHLEHEM

ALLENTOWN

READING

LANCASTER

YORK

HARRISBURG

ALTOONA

JOHNSTOWN

0   10   20   30   40   50 miles
0   10   20   30   40   50   60   70   80 km

income model (Haining, 1987a) in which non-zero joins are based on a theory of inter-urban relationships and (in the case of Figure 8.8(*a*)) traffic flow data. In addition the joins were treated as directional, since income levels in large towns are dependent on income levels in towns in the market hinterland, but the relationship is not symmetrical. If the observational units are areas then if the areas are zones within a city an inverse distance weighting between all pairs of areas may be appropriate, reflecting the high degree of intra-urban mobility, whereas if the areas are, say, American states or even English counties then a nearest neighbour contiguity criterion may be adequate. These approaches are often adopted when the analyst has in mind some mechanism that might be responsible for the observed pattern of variation.

A third approach compares alternative **W** matrices within a formal non-nested hypothesis testing framework. Anselin (1986) evaluates a number of non-nested testing procedures. A drawback appears to be that it may be very difficult to choose between alternative **W** matrices using these tests particularly if the matrices are rather similar. Anselin notes that the 'statistical elegance of the tests . . . is not reflected in clearer insights as to which are the proper theoretical specifications' (p. 282).

A final approach, and the one adopted in subsequent examples in the absence of strong theoretical grounds for the choice, is to consider fitting the regression model under several different forms of **W**. The purpose is to gauge the sensitivity of regression results to the form of **W** (since the particular form of **W** is not often of major interest). If the aim of statistical analysis is the construction of better fitting models with better behaved residuals that form of **W** should be chosen which helps these targets to be met. This may mean that **W** is specified after inspecting the residuals of the fit of (8.11). In this case little substantive significance should be attached to the **W** matrix that provides the best fit.

8.2.2 (c) Parameter estimation and inference

Parameter estimation and inference for (8.12) has been extensively discussed in Chapter 4. Model (8.13) has been discussed by Draper and Guttman (1980). The estimation of β is by weighted least squares in which if **G** is symmetric

$$\tilde{\beta} = (X^T G^2 X)^{-1} (X^T Gy)$$

so that the regression problem can be solved using standard unweighted least squares on **GX** and **Y**.

In the case of (8.14) where there is a separate parameter associated with the lagged explanatory variables, standard least squares regression can be used treating $\mathbf{WX}$ as another explanatory variable. If, however, $\mathbf{X}$ is spatially correlated the problem of multicollinearity (between $\mathbf{X}$ and $\mathbf{WX}$) will arise.

Maximum likelihood estimation and inference for (8.15) where $\mathbf{K} = \rho\mathbf{W}$ has been examined by Ord (1975) and reviewed in Upton and Fingleton (1985). The estimators are given by

$$\tilde{\boldsymbol{\beta}} = (\mathbf{X}^\mathsf{T}\mathbf{X})^{-1}\mathbf{X}^\mathsf{T}\tilde{\mathbf{Z}}$$
$$\tilde{\sigma}^2 = (1/n)\tilde{\mathbf{Z}}^\mathsf{T}\mathbf{M}\tilde{\mathbf{Z}}$$

where $\tilde{\mathbf{Z}} = (\mathbf{I} - \tilde{\rho}\mathbf{W})\mathbf{y}$; $\mathbf{M} = \mathbf{I} - \mathbf{X}(\mathbf{X}^\mathsf{T}\mathbf{X})^{-1}\mathbf{X}^\mathsf{T}$ and $\tilde{\rho}$ is that value of ρ which minimises

$$-2n^{-1}\ln|(\mathbf{I} - \rho\mathbf{W})| + \ln\tilde{\sigma}^2$$

Note that the estimate of ρ does not depend on $\tilde{\boldsymbol{\beta}}$ so that unlike (8.12) an iterative procedure is not required.

The upper triangle of the asymptotic covariance matrix for the estimates is:

$$V(\sigma^2,\rho,\boldsymbol{\beta}) = \sigma^4 \begin{bmatrix} n/2 & \sigma^2\mathrm{tr}(\mathbf{A}^{-1}\mathbf{W}) & \mathbf{0}^\mathsf{T} \\ & V(\rho,\rho) & \sigma^2\mathbf{X}^\mathsf{T}\mathbf{A}^{-1}\mathbf{WX\boldsymbol{\beta}} \\ & & \sigma^2\mathbf{X}^\mathsf{T}\mathbf{X} \end{bmatrix}^{-1}$$

where:

$$V(\rho,\rho) = \sigma^2[\sigma^2\mathrm{tr}(\mathbf{W}^\mathsf{T}(\mathbf{A}^{-1})^\mathsf{T}\mathbf{A}^{-1}\mathbf{W}) + \boldsymbol{\beta}^\mathsf{T}\mathbf{X}^\mathsf{T}\mathbf{W}^\mathsf{T}(\mathbf{A}^{-1})^\mathsf{T}\mathbf{A}^{-1}\mathbf{WX\boldsymbol{\beta}}]$$
$$+ \sigma^4\Sigma_{i=1,\ldots,n}(\omega_i/(1-\omega_i))^2$$

$\mathbf{A} = (\mathbf{I} - \rho\mathbf{W})$, $\{\omega_1, \ldots, \omega_n\}$ are the n eigenvalues of $\mathbf{W}$ and tr denotes trace.

It is tempting with these lagged models (8.14 and 8.15) to fit long lags and then decide on the significance of the coefficients. Apart from the computational difficulties in the case of (8.15) the estimates at different lags will be highly intercorrelated and imprecise so that there will be great difficulty in making any useful inferences. Estimation is probably best approached by building from the bottom, adding variables one at a time and carrying out careful diagnostic checks.

Other estimation procedures for these models have been suggested. Anselin (1988b) discusses instrumental variables estimation. Hepple (1979) carried out a Bayesian analysis using (8.12) where Y was the second hand value of cars in 49 states of the U.S.A. and X_1 was new car price differentials attributable to transport charges and sales tax. The approach is similar to examining the whole likelihood function and Figure 8.9 shows various posterior distributions for the regression parameter (β_1) and the

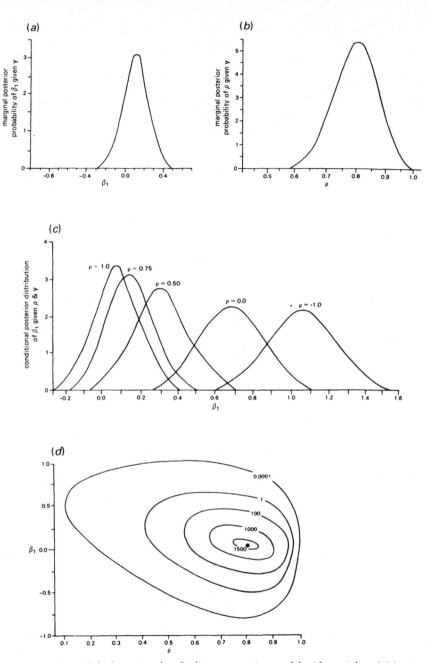

Figure 8.9. Bayesian plots for linear regression model with spatial model for the errors. (*a*) marginal posterior probability of β_1 given **y**, (*b*) marginal posterior probability of ρ given **y**, (*c*) conditional posterior distribution of β_1 given ρ and **y**, (*d*) bivariate posterior distribution for β_1 and ρ. After Hepple (1979).

346

spatial parameter ρ in a SAR model for V. Figure 8.9(c) shows the sensitivity of the regression parameter to values of ρ and 8.9(d) shows the joint sensitivity of ρ and β_1. This information is useful in identifying how clear cut the maximum is and indicating the effects of modelling the spatial aspects of the data on the estimate of the regression component of the model.

For (8.18), maximum likelihood estimates can be obtained by adding macros to commercial packages like MINITAB (Griffith, 1988). Where ρ is known or fixed (8.18) can be estimated by regressing $(\mathbf{I}-\rho\mathbf{W})\mathbf{y}$ on $(\mathbf{I}-\rho\mathbf{W})\mathbf{X}$ to obtain the minimum residual sum of squares fit. For (8.15) with $\mathbf{K} = \rho\mathbf{W}$, if ρ is known or fixed, the procedure is to regress $(\mathbf{I}-\rho\mathbf{W})\mathbf{y}$ on $\mathbf{X}$. Links between the standard regression model and the 'spatialised' forms above are discussed by Haining (1990).

8.2.2   (d)   Model evaluation

A problem of considerable interest is whether, after fitting any of the models (8.12), (8.14) or (8.15), there is still spatial autocorrelation in the model residuals. Testing with the generalised Moran coefficient is no longer valid. There are, currently, two options. A simple check is to plot the regression residuals against sums or weighted averages of their neighbouring residuals across different lags. Such informal plots should provide a useful guide and can be quite easily implemented.

A different approach has been developed by Anselin (1988a) as part of a general program of discriminating between alternative models with different correlation structures and different degrees of heteroscedasticity. Different models are presented as special cases of a more general model:

$$\mathbf{y} = \rho\mathbf{W}_1\mathbf{y} + \mathbf{X}\boldsymbol{\beta} + \boldsymbol{\varepsilon}$$
$$\boldsymbol{\varepsilon} = \lambda\mathbf{W}_2\boldsymbol{\varepsilon} + \boldsymbol{\varphi}$$

where $E[\boldsymbol{\varphi}] = 0; E[\boldsymbol{\varphi}\boldsymbol{\varphi}^\top] = \boldsymbol{\Omega}$ ($\boldsymbol{\Omega}$ diagonal).

By imposing constraints on the system different special cases emerge:

$$\text{(8.11):}\ \rho = 0;\ \lambda = 0;\ \boldsymbol{\Omega} = \sigma^2\mathbf{I}$$
$$\text{(8.12) } (\mathbf{V} = \mathbf{V}_{SAR})\text{: } \rho = 0;\ \boldsymbol{\Omega} = \sigma^2\mathbf{I}$$
$$\text{(8.15): } \lambda = 0;\ \boldsymbol{\Omega} = \sigma^2\mathbf{I}$$

Anselin (1988a) gives the Lagrange multiplier tests for each of these models, of which the generalised Moran coefficient described in Chapter 4 is one example. The interested reader should consult the original paper for details.

8.2.3   Interpretation problems

The open nature of the observational units raises important questions about what interpretation can be placed on the results of a regression analysis which identifies significant covariates. Where inter-area mobility is high the values of the explanatory variables for area i may provide poor estimates of the extent to which the response variable in i is an expression of exposure to the explanatory variable levels in i. This of course is one reason why new regression models with lagged terms were specified. However, even that adjustment may only be a surrogate for the complex patterns of movement (in time) which underlie the evolution of population characteristics. In the case of modelling intra-urban mortality rates, for example, the residents of any area are exposed to a complex set of risk factors, and levels of those factors, over time. Yet measurements for the area (including neighbouring areas) may provide a poor estimate of these true exposure levels for the residents.

The interpretation of regression results presents further difficulties when two or more explanatory variables possess similar spatial patterns. As a consequence it may be impossible to disentangle the contribution of each explanatory variable to variation in the response variable. This might be due to variate dependence but could also be due to the organisation of the spatial system. For example, levels of stress and pollution may both decline with distance from the city centre. Assessing the contribution of each on measures of health may be impossible within a non-experimental regression framework. Since social class is also often related to distance from the city centre this may add an additional variable with a spatial structure that is confounded with other potentially significant explanatory variables.

8.2.4   Problems due to data characteristics

In fitting (8.11) the estimates of the regression parameters may be sensitive to outliers and extreme values. Procedures for checking the sensitivity of model fit to particular data values are described for example in Belsley, Kuh and Welsch (1980) and are available in some packages (e.g. MINITAB). In models such as (8.12), (8.14) and (8.15) estimates of both the regression and spatial parameters will be sensitive to outliers and it seems likely that because of the inclusion of spatial parameters, model fitting may also be sensitive to spatial *clusters* of outliers (see Appendix). Residual variance may be greater for boundary observations than for those in the interior (see Chapter 2). There is as yet no recommended procedure for assessing the sensitivity of model fit to extreme values in the case of models such as (8.12) and (8.15). If sensitivity of estimates of the spatial parameters to extreme values is ignored then diagnostics on the regression parameters can be obtained by deleting from the vectors $(I - \rho W)y$ and $(I - \rho W)X$ (for 8.18) and $(I - \rho W)y$ and X (for 8.15) and re-estimating the regression parameters. Martin (1984) gives estimators for both sets of parameters of a regression model with general correlated errors where observations are missing (or treated as missing).[4] Evaluating these equations deleting each observation in turn (and perhaps some clusters) may prove impractical and may introduce implicit assumptions into the model fit concerning the value of any deleted observation since observations round any deleted case are now on a boundary. An alternative approach might be to 'smooth' or 'perturb' values that are suspected of being influential, rather than deleting them.

If the distribution of high leverage values is associated with one part of the map, (such as high levels of per capita income in a small number of suburban areas of a city) the fit of the model may be strongly influenced by the characteristics of these areas.[5] If high leverages in a regression model (such as 8.15) are associated with border cells, results could be particularly sensitive to the type of assumptions used to fit the model at the boundary. Draper and Guttman (1980) suggest that border effects raise fewer problems in a 'deterministic' model such as (8.13) than in 'stochastic' models such as (8.12) and (8.15) for example, since in the case of a natural boundary (edge of a land mass for example) the matrix G is simply defined to reflect the absence of influences from this direction.

Mention has already been made of the possible sensitivity of results to the zoning system when dealing with aggregate data and the desirability of performing analyses on other plausible zoning systems of equivalent scale (see Chapter 2). Where several equally plausible zoning systems are being compared a further criterion could be to select that aggregation for which residual correlation is not a serious problem (if such exists).

8.2.5 Numerical problems

Model (8.12) calls for an iterative routine in which the spatial and regression parameters are iteratively estimated until convergence. (The estimation of (8.12) is discussed in Chapter 4.) There is also a double inversion problem in estimating β, that is, evaluation of $(X^T V^{-1} X)^{-1}$. The problems of ensuring a proper inversion of $(X^T X)$ in fitting (8.11) (see also (8.18)) are discussed in Wetherill *et al.* (1986). A suitable choice for V^{-1} can eliminate the need to invert a large matrix (see Chapter 3). The iterative cycle is further extended if robust regression is carried out, using the final set of model residuals to specify a weights matrix to downweight the influence of observations with large residuals. In practical implementations the author has noted two problems for certain forms of (8.12) in particular where a moving average model has been used to represent V. First, if the moving average parameter is not restricted to lie in the invertible range (see Chapter 3) there may be many local maxima and searching the surface for the global maximum can be difficult. A second, and closely related problem, is that in the iterative cycle the total sum of squares may be apportioned over the regression and error model elements of (8.12) in a rather unstable way. At successive iterations the regression residual sum of squares may rise and fall sharply. It seems sensible when including the error model not to allow the regression residual sum of squares to deviate sharply from the case where no error model is included in the specification. The analyst should monitor convergence carefully as well as keeping track of numerical accuracy. (Further details are given on pages 127–9 and footnote.)

In the case of fitting (8.14) an unrestricted stepwise regression may be unsuitable since the analyst may not wish to include WX without X. Correlation between WX and X arising from spatial correlation in X may produce unreliable regression parameter estimates associated with the inversion of a near singular matrix. Correlation between two or more explanatory variables arising from identical spatial trends may raise similar numerical problems.

8.3 Regression applications

Example 8.1. Model diagnostics and model revision: (a) New explanatory variables

Weisberg (1985) gives data on motor fuel consumption (gallons/head) for 48 states of the U.S.A. for the period 1971/2. Inter-state variations are accounted for by a regression on:

X_1: tax (cents per gallon)
X_2: % of state population with drivers licence
X_3: average per capita income in $000
X_4: thousands of miles of road

Variable X_4 is not significant and if Y is regressed on X_1, X_2 and X_3 using ordinary least squares this gives the results reported in Display 8.1(a). Some diagnostics are given in part (b) of the display. The GMC statistic which retains the null hypothesis of no spatial autocorrelation is computed using a binary contiguity matrix (W) where entries $w_{ij} = 1$ if states i and j share a common boundary and are 0 otherwise. The lack of spatial correlation up to lag 6 is evident also. So spatial residual correlation does not appear to be a

Display 8.1. *Analysis of US fuel consumption data*

(a) Model fit

$$\hat{Y} = 305.89 - 29.51X_1 + 13.78X_2 - 68.03X_3$$
$$\quad\ (156.5)\quad (10.56)\quad (1.83)\quad\ (16.97)$$

Figures in brackets are standard errors. $R^2 \times 100 = 67.6$, R^2(adjusted) $\times 100$ = 65.4.

(b) Diagnostics
  (i) Spatial correlation
  GMC = 0.0455   E[GMC] = −0.0457
            St. Dev. [GMC] = 0.0995
  GMC as standard normal deviate: 0.958

Lag correlations

	0	1	2	3	4	5	6
Correlation	1.0	0.025	−0.086	−0.008	−0.004	−0.053	0.089
No. of pairs		107	175	214	198	152	118

  (ii) Residual values
  Rank order of largest positive and negative standardised residuals

+ residuals	− residuals
3.62 Wyoming	1.67 Rhode Island
1.87 N. Dakota	1.22 Mississippi
1.35 S. Dakota	
1.27 Virginia	
1.24 Nevada	

(c) Revised model

$$\hat{Y} = 274.3 - 24.82X_1 + 12.13X_2 - 42.87X_3 - 0.11X_5$$
$$\qquad\quad (10.39)\quad (1.92)\quad (20.15)\quad (0.05)$$

$R^2 \times 100 = 70.7$ R^2(adjusted) $\times 100 = 68.0$
Figures in brackets are standard errors.

(d) $\hat{Y} = 504.4 - 46.32X_1 + 10.92X_2 - 2.96X_6$
$$=\qquad\ (10.86)\quad (1.81)\quad (0.69)$$

$R^2 \times 100 = 68.8$ R^2(adjusted) $\times 100 = 66.7$
Figures in brackets are standard errors.

problem. We now turn to the residuals where apart from Wyoming there are no large values. If the regression is re-run deleting Wyoming this does not affect the significance of the variables, although parameter estimates are affected, as Figure 8.10 shows. Figure 8.11 shows rankit plots of the residuals using all 48 states (*a*) and excluding Wyoming (*b*). These plots suggest that the residuals are well behaved. However, if we plot residual values against population size there is evidence that residual variance may be related to population size (Figure 8.11(*c*)). Since the analysis is based on averages we might expect states with larger populations to have smaller variances. We might also argue that values derived from states with small populations are probably less reliable and should be downweighted in the fit. On both counts we might justify examining a weighted least squares estimate of (β_1, β_2, β_3), downweighting states with small populations. A concern we might have in adopting this strategy is that the apparently greater spread of residual values at low population levels may be a consequence of the four states marked on Figure 8.11(*c*) and the relatively small number of very large states. The evidence is ambiguous for a purely statistical manipulation of the data. On the other hand there does seem to be something of a pattern to the states with large and small standardised residuals. States with large positive residuals generally have small dispersed populations whilst Rhode Island has a small but very concentrated population. It may be better therefore (and geographically more interesting) to add another explanatory variable that measures intra-state population

Figure 8.10. Estimates of (β_1, β_2) obtained by deleting each case in turn. After Weisberg (1985).

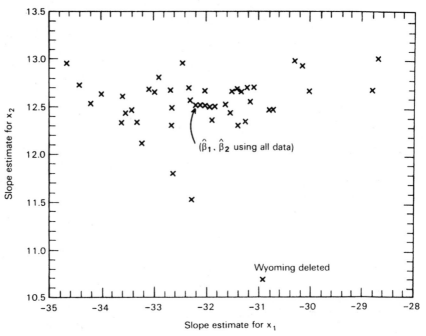

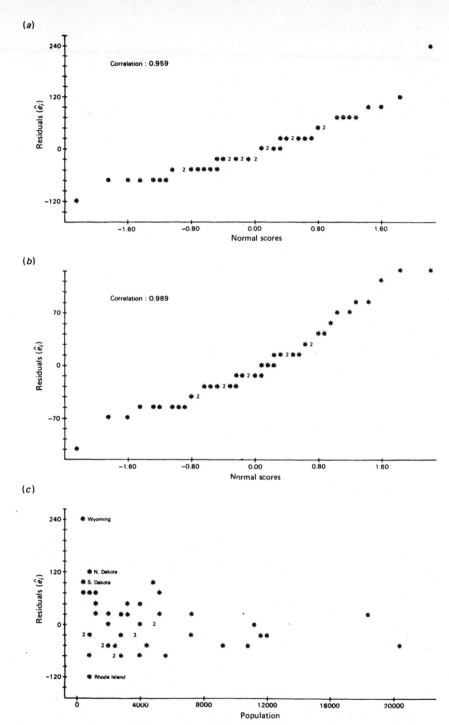

Figure 8.11. Rankit plots for U.S. fuel consumption model residuals (a) residuals from model fitted with 48 states, (b) residuals from model fitted without Wyoming, (c) plot of residuals against population size: U.S. fuel consumption data.

Table 8.8. *Population density data (1970) for the 48 states in Weisberg*
(1985) table 2.1

State	1	2	3	4	5	6	7
Pop. Den.	32.1	81.7	48.0	727.0	905.4	623.7	381.4
,,	8	9	10	11	12	13	14
	953.5	262.4	260.1	143.9	199.3	156.3	81.1
,,	15	16	17	18	19	20	21
	48.0	50.5	67.8	8.9	8.7	19.4	27.5
,,	22	23	24	25	26	27	28
	276.5	396.7	116.9	72.5	104.2	85.7	79.0
,,	29	30	31	32	33	34	35
	125.6	81.2	95.0	67.9	46.9	37.0	81.1
,,	36	37	38	39	40	41	42
	37.2	42.7	4.8	8.6	3.4	21.3	8.4
,,	43	44	45	46	47	48	
	15.7	12.9	4.4	51.3	21.7	127.7	

Pop. Den. = population density.

dispersal. Certainly we might expect the level of state fuel consumption to be dependent on the spatial dispersal of the population. Table 8.8 contains 1970 population density data for the 48 states. (The numbering follows Weisberg, 1985, Table 2.1.) Population density is obviously not ideal since it does not characterise the degree of spatial concentration of the population at the intra-state level. Part (c) of Display 8.1 records the result of including population density (X_5). The model has been slightly improved and all the variables are significant. If instead of X_5 we use the percentage of people living in urban places (X_6) which might be a better measure of spatial concentration, this variable is also significant. However, the correlation between X_3 and X_6 is 0.624 and X_3 now ceases to be significant. Since the sign of the coefficient for X_3 appears rather anomalous (Display 8.1) this might be a preferable model. The results for this fit are given in part (d) of Display 8.1.

Example 8.2. Model diagnostics and model revision:
(b) developing a spatial regression model

Cliff and Ord (1981, p. 209, p. 237) report the results of an analysis of spatial variation in the percentage, in value terms, of the gross agricultural output of each county in Ireland consumed by itself (Y) as a function of a measure of county accessibility on the arterial road network (X). We shall use their data set to exemplify some of the earlier points on the utility of exploratory techniques and the need to consider alternative specifications of spatial relationships. Figure 8.12 gives a scatterplot of the data.

Table 8.9 reports the results of an ordinary least squares analysis of the data together with an analysis of the residuals from the fit. It is evident that there is significant spatial autocorrelation. The GMC statistic is evaluated for three different types of W matrix. The binary matrix sets $w_{ij} = 1$ if counties i and j share a common border, 0 otherwise and the standardised binary matrix sets row sums to unity. The construction of the weighted matrix (row sums equal to unity) is discussed in detail in Cliff and Ord (1981, p. 209). There also appears to be some evidence of negative spatial correlation at lag 3.

The approach taken here, to dealing with the problem of residual correlation, is to consider adding spatially lagged forms of the original variables. Figure 8.13 shows added variable plots to try and determine whether WX or WY or both should be added to the model. It is the shape of the plots we are interested in (see note 3). Since this test is being used informally we have fitted the various models by ordinary least squares, which is not really correct for regressing WX on X. Correlation values have been added as a simple summary, but should only be interpreted broadly. The evidence of these plots irrespective of how W is constructed is that adding WY is preferable to adding WX; that if WX is added WY should be added; that if WY is added WX is not needed.

Table 8.10 reports the results of fitting both the lagged response and the lagged explanatory variable models. A problem with the lagged explanatory

Figure 8.12. Plot of percentage, in value terms, of the gross agricultural output of each county consumed by itself (Y) against road accessibility index (X).

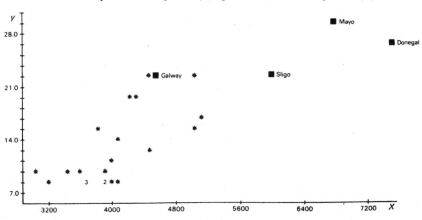

Table 8.9. *Ordinary least squares analysis of Irish data*

$\hat{Y} = -8.44 + 0.0053X$
  (-2.65) (7.42) $\qquad\qquad$ (*t* values in brackets)
$R^2 \times 100 = 69.7$
R^2 (adjusted) $\times 100 = 68.4$
$n = 26$; residual variance: 13.58.

(*a*) Lag correlations computed for the residuals

lag	0	1	2	3
Correlation	1.00	0.387	−0.089	−0.218
No of pairs		58	93	94

(*b*) Autocorrelation tests on the residuals

	GMC	E[GMC]	St.Dev. [GMC]	St. Normal Dev.
Binary matrix	1.726	−0.238	0.535	3.67
Standardised binary matrix	0.315	−0.057	0.126	2.95
Weighted matrix	0.429	−0.057	0.146	3.32
(Cliff and Ord, 1981, p. 230)				

St.Dev. = standard deviation.

variable model is that for two definitions of the W matrix, X and WX are strongly correlated. The lagged response variable model provides the better fit and Figure 8.14 which shows plots of the final set of independent residuals ($\tilde{e} = (\tilde{e}_1, \ldots, \tilde{e}_n)^T$) against W$\tilde{e}$ and rankit plots suggests that there is no serious residual first order spatial correlation and that the residuals are well behaved. Of the two forms for W it seems that the standardised binary weights matrix produces the better fit.

The next step might be to add WX to the lagged response variable model. We have already developed insights into the data to suggest that this might be both unwise and not very rewarding. There remains the problem then as to whether we should accept the lagged response variable model. To this author, the model is substantively unconvincing. Model (8.12) seems better (that the spatial correlation in the data is the outcome of numerous local factors that operate over relatively small inter-county scales) and the fact that both WY and WX could be added to the standard regression model certainly does not rule out fitting (8.12). Table 8.11 shows the results of fitting (8.12) with $V = V_{SAR}$. Of the three forms of W, the weighted W matrix seems to provide the best fit and also appears best in terms of residual correlation plots and rankit plots (Figure 8.15). The fit seems slightly poorer than the lagged dependent variable model. Table 8.11 also shows the results of using a moving average and a conditional autoregressive model

Figure 8.13. Added variable plots: adding WX or WY or both to the original regression model (a) standardised binary weights, (b) Cliff and Ord (1981, p. 209) weights.

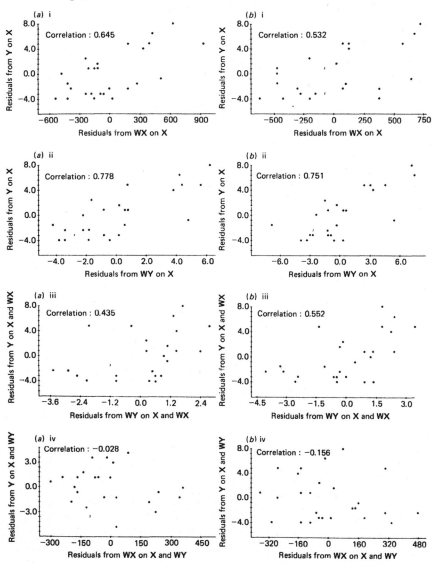

Table 8.10. *Fitting different spatial regression models to Irish data*

(*a*) Lagged explanatory variable: $Y = \alpha + \beta X + \gamma WX + e$

	Binary **W**	Standardised binary **W**	Weighted **W** (Cliff and Ord, 1981, p. 230)
$\tilde{\alpha}$	−14.13	−23.97	−20.59
	(−3.55)	(−5.24)	(−4.22)
$\tilde{\beta}$	0.0056	0.0026	0.0030
	(8.25)	(3.13)	(3.25)
$\tilde{\gamma}$	0.0002	0.0063	0.0050
	(2.15)	(4.05)	(3.02)
$r_{\mathbf{x,wx}}$	0.06	0.57	0.58
Residual variance	11.80	8.28	10.16

t values in brackets under the coefficient estimates, $r_{\mathbf{x,wx}}$ is the Pearson correlation between X and WX.

(*b*) Lagged response variable: $Y = \alpha + \beta X + \rho WY + e$

	Standardised binary **W**	Weighted **W** (Cliff and Ord, 1981, p. 230)
$\tilde{\alpha}$	−6.24	−6.71
	(−3.10)	(−3.21)
$\tilde{\beta}$	0.0024	0.0028
	(4.38)	(5.11)
$\tilde{\rho}$	0.731	0.646
	(6.38)	(5.13)
Residual variance	5.25	5.67

t values in brackets under coefficients.
(Note: These results agree with those given in Anselin, 1988a and Bivand, 1984.)

for the error structure using the binary connectivity matrix for W. The moving average model performed badly. The parameter estimate when constrained to the invertible range attained the upper maximum (lag one correlations in the residuals are close to the maximum allowed – see Table 8.9); when not constrained the surface appeared to contain many local turning points and in the iterative cycle the apportionment of the total sums of squares to the regression and error elements of the model was rather volatile. The estimate of the constant coefficient was very unstable. The conditional autoregressive model fit well but did not improve on the SAR model.

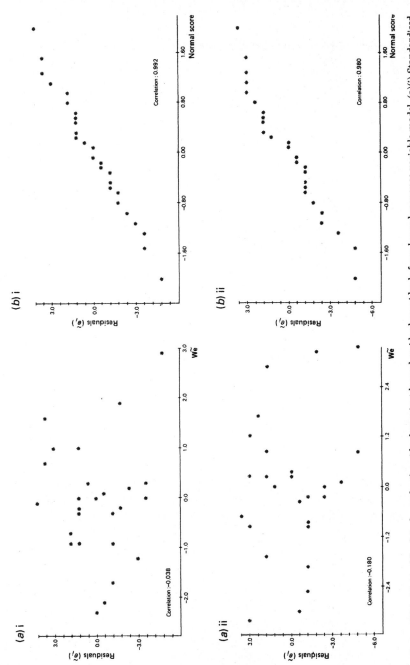

Figure 8.14. Plots of final (independent) residuals against lagged residuals: residuals from lagged response variable model. (a)(i) Standardised binary weights, (ii) Standardised weights, (b)(i) Standardised binary weights, (ii) Standardised weights.

Table 8.11. *Fitting model (8.12) to the Irish data*

(*a*) SAR errors

$$E[\mathbf{u}\mathbf{u}^T] = \mathbf{V} = \sigma^2[(\mathbf{I} - \rho\mathbf{W})^T(\mathbf{I} - \rho\mathbf{W})]^{-1}$$

	Binary **W**	Standardised binary **W**	Weighted **W** (Cliff and Ord, 1981, p. 230)
$\tilde{\alpha}$	1.155	4.670	1.359
	(0.33)	(1.04)	(0.36)
$\tilde{\beta}$	0.0032	0.0024	0.0030
	(5.84)	(3.79)	(4.74)
$\tilde{\rho}$	0.177*	0.843[+]	0.780[+]
	(14.29)	(9.44)	(7.06)
Residual variance	5.36	5.89	5.73

*max value is 0.194, [+] max value is 1.00, *t* value in brackets beneath estimate.

(*b*) Other spatial error models

		Moving average		Conditional auto-regressive
		Invertible range	Unrestricted	
$\tilde{\alpha}$		−1.290	1.766	−3.725
$\tilde{\beta}$		0.0037	0.0034	0.0041
	$\tilde{v}=$	0.193	0.944	$\tilde{\tau}=$ 0.184
Residual variance		8.92	8.07	7.47

Table 8.12 shows leverage values (as percentages) for the 26 counties under (8.11) and (8.12) with $\mathbf{V} = \mathbf{V}_{SAR}$ and **W** given by the weighted form. Under both models, Donegal (E) and Mayo (P) have high leverage values and because of the tendency of (8.12) to downweight the influence of observations in the interior, leverage values for these counties and the other boundary sites, particularly those in the west with few neighbours, tend to be larger for this model (Figure 8.16(*a*),(*b*)). This is unfortunate. As Cliff and Ord (1981, p. 207) note the values of X of these western counties are inflated and worse than in reality because of the way the index was derived. So the fit of the model is now more affected by these 'unreliable' values. There is a case here for using 'bounded influence' regression but we adopt a rather simpler approach. Display 8.2 reports the results of re-estimating

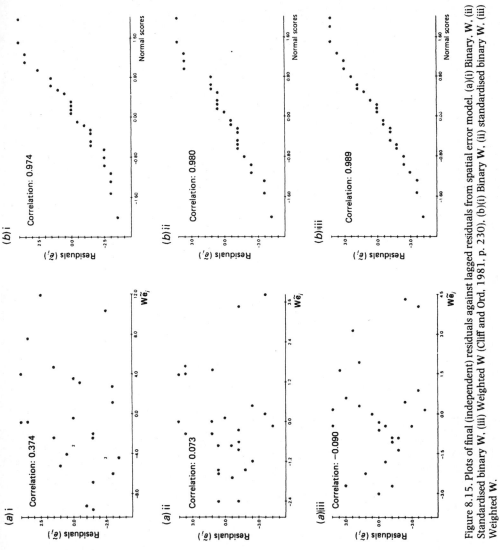

Figure 8.15. Plots of final (independent) residuals against lagged residuals from spatial error model. (a)(i) Binary **W**, (ii) Standardised binary **W**, (iii) Weighted **W** (Cliff and Ord, 1981, p. 230), (b)(i) Binary **W**, (ii) standardised binary **W**, (iii) Weighted **W**.

Table 8.12. *Leverage values (as % of total leverage) for model (8.11) and model (8.12) with* $V = V_{SAR}$, W *the weighted form*

County	A	B	C	D	E	F	G	H
(8.11)	2.87	2.63	1.92	2.05	19.97	5.07	1.96	2.99
(8.12)	2.14	2.70	1.95	2.71	25.05	4.67	2.53	4.68
,,	I	J	K	L	M	N	O	P
	4.50	2.76	3.51	2.63	2.16	1.95	2.27	12.92
	3.75	2.11	3.06	1.92	2.29	2.04	2.32	12.98
,,	Q	R	S	T	U	V	W	X
	2.18	1.94	2.06	1.94	6.77	2.30	2.80	2.40
	1.95	2.08	2.04	2.79	3.08	3.02	2.50	2.59
,,	Y	Z						
	2.28	3.05						
	1.92	1.99						

High leverage $> 11.5\%$.

Display 8.2. *Evaluating the effects on parameter estimates of deleting two counties (Donegal and Mayo) with high leverages*

Ordinary least squares

$\hat{Y} = -11.253 + 0.0059X$ $R^2 \times 100 = 54.4$
   (-2.30) (5.12) $R^2(\text{adjusted}) \times 100 = 52.3$
                $n = 24$

t values in brackets
GMC $= 0.317$ E[GMC] $= -0.0635$
             St. Dev.[GMC] $= 0.1290$ St. Normal dev.: 2.94
(Using standardised binary weights matrix)

Spatial error model (using standardised 0/1 matrix)
$Y = \alpha + \beta X + u$
$u = \rho W u + e$
$\hat{\alpha} = 6.206$ $\hat{\beta} = 0.0019$ $\hat{\rho} = 0.833$
 (3.28) (4.24) (8.17)
Residual variance $= 6.17$
t values in brackets

Diagnostics for Sligo

	OLS	Spatial model	Cut off level
leverage	0.383	0.398	0.249
DFITS	-0.677	-0.928	$+0.577$

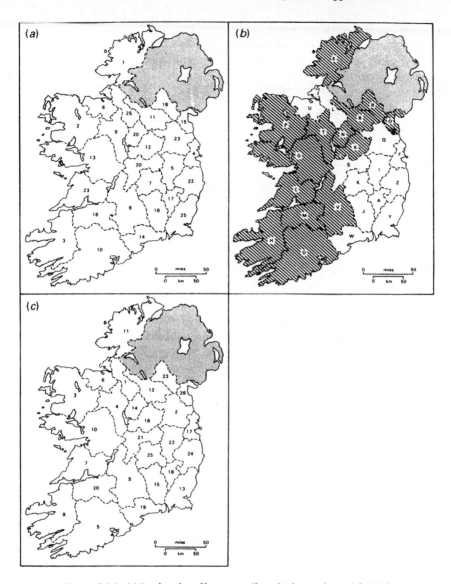

Figure 8.16. (*a*) Rank order of leverages (from highest to lowest) for Irish counties and spatial error model, (*b*) Counties with higher leverages (shaded) under the spatial error model than the OLS model, (*c*) Rank order of residuals (from largest to smallest in absolute value) from spatial model with SAR errors. Generalised weights matrix.

both (8.11) and (8.12) deleting Donegal and Mayo which have the largest leverage scores. (To avoid re-computing the weights, the standardised binary matrix was used for (8.12).) The conclusions are not greatly altered, although Sligo now has a leverage effect and the DFITS measure is also large, which is a combined measure of leverage and residual effects. (Galway, another western county, comes close to the DFITS critical limit.)

There appear to be no serious outlier problems for the final set of regression residuals from (8.12) with the weighted form of **W** although the larger residuals tend to be associated with the western boundary counties (Figure 8.16(*c*)). Also the plot of the residuals from the regression element of the model (**ũ**) against their neighbourhood average (**Wũ**), which underlies the estimation of ρ in (8.12), does suggest that there may be an outlier problem with Sligo in estimating the spatial parameter. Figure 8.17 shows this plot and the suspected outlier is obtained by analysing an ordinary least squares fit of **ũ** on **Wũ**. However, if a resistant, *R*-estimator fit (Li, 1985, p. 331), of **ũ** on **Wũ** is obtained the slope parameter does not change by very much. (The *R*-estimator fit is available in MINITAB.) It was decided that although it would be useful to examine the fit when **β** is estimated robustly, ρ would be estimated by the usual maximum likelihood method. Table 8.13 reports the results of fitting (8.12) using the Tukey bi-weight with $B = cS$ where $c = 6$ and S is the MAD. The final set of (independent) residuals (**ẽ**) at each iteration are used to specify the weights. The use of these weights does tend to downweight the western boundary counties in the estimate of **β** but the downweighting is in respect of their residual properties not their leverage scores.

Figure 8.17. Plot of final set of regression residuals against lagged residuals: Irish data.

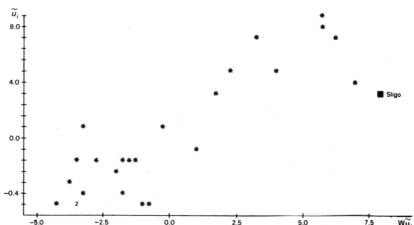

Regression applications

Table 8.13. *Robust estimation of (8.12) for* β *using Tukey's bi-weight with final set of Tukey weights (in alphabetical order across rows)*

$Y = 1.9941 + 0.0028X + (I - 0.794W)^{-1}e$. Residual variance $(\bar{\sigma}^2) = 5.68$.

0.989	0.962	0.918	0.866	0.971
0.980	0.959	0.908	0.997	0.972
0.999	0.821	0.991	0.970	0.999
0.829	0.847	0.999	0.990	0.873
0.893	0.928	0.990	0.973	0.970
0.999				

MAD = 1.715, B = 10.293.

Example 8.3: Regression modelling with census variables: Glasgow health data

We now examine the problem of finding a set of covariates for the cancer mortality data examined in Chapters 6 and 7. The CMAs were constructed by aggregating enumeration district level data so as to preserve social and economic homogeneity. Data on 15 relevant social and economic variables for the 87 CMAs are available from the census.

Logarithmic transforms (to the base e) of the SMR data are used. Following Pocock *et al.* (1981) let $Y_i = \log_e[(O_i \times 100)/E_i]$ where O_i and E_i are the observed and expected numbers of cancer deaths for area i (see Chapter 6 for the definition of E_i). As described in Chapter 7, the observed number of deaths for an area (i) is assumed to be a drawing from the Poisson distribution with parameter $\lambda_i = n_i p_i$. The logarithmic transform helps stabilise the variance and in addition we can write:

$$Y_i = \mu_i + \phi_i$$

where ϕ_i is approximately $N(0, O_i^{-1})$, $\mu_i = p_i$ so that Y_i is $N(\mu_i, O_i^{-1})$. Now $\mu_i = E[Y_i]$ and is assumed to be a function of a set of covariates $(X_1, \ldots, X_k)$. Suppose

$$\mu_i = \Sigma_{j=1,\ldots,k} \, \beta_j X_{i,j} + e_i$$

The first term is the explained variation in Y_i while e_i is the unexplained variation and assumed to be normal with mean zero and variance σ^2. Providing e_i and ϕ_i are independent then

$$Y_i = \Sigma_{j=1,\ldots,k} \, \beta_j X_{i,j} + \eta_i$$

365

where η_i is an independent normal error term with mean zero and variance $\sigma^2[1 + 1/\sigma^2 O_i]$. The model assumes non-constant error variance and so the variance of Y differs between areas. The range of values of O_i (from 20 to 288) is responsible for this but the degree of variability is not large. For this reason a standard stepwise regression was carried out first making no allowance for this property. The explanatory variables are (in percentages):

1.   living in local authority housing;
2.   living in overcrowded conditions;
3.   lacking amenities;
4.   low socio-economic grouping;
5.   migrants;
6.   ethnic community;
7.   elderly;
8.   pensioners living alone;
9.   single adult families;
10.   young children;
11.   in social class 4 and 5;
12.   with no car;
13.   living in owner occupied housing;
14.   large families;
15.   in communal establishments.

The result is summarised in Display 8.3 and Table 8.14 gives the data for the two significant variables X_8 and X_{11}. The R^2 for this model could not be raised by adding additional variables.

Display 8.3

$$\text{Log}_e(\text{S}\tilde{\text{M}}\text{R}) = \hat{Y} = 4.23 + 0.014X_8 + 0.017X_{11}$$
$$\phantom{\text{Log}_e(\text{S}\tilde{\text{M}}\text{R}) = \hat{Y} = 4.23 + } (3.95) \qquad (13.07)$$

t values in brackets $\qquad R^2 \times 100 = 67.9 \qquad R^2(\text{adjusted}) \times 100 = 67.1$
$\hat{\sigma} = 0.100$

Table 8.15 lists diagnostics for the fit summarised in Display 8.3. Figure 8.18(a) identifies those CMAs with large residuals and Figure 8.18(b) maps the full set of residuals and indicates clusters of residuals of similar sign. Application of the Moran test using a binary weights matrix, specifying joins between CMAs that share a common boundary, rejected the null hypothesis of no spatial autocorrelation at the 10% level.

A regression model with correlated errors was fit to the data but did not improve the fit. Nor was the fit improved by lagging any of the variables for which there also seemed little substantive justification. However, when the model was augmented with second order trend surface terms there was a

Table 8.14. *Glasgow CMA data*

CMA	X_8	X_{11}	CMA	X_s	X_{11}
1	7.26	17.33	45	6.84	27.62
2	4.19	15.49	46	7.04	16.85
3	6.14	8.00	47	10.60	12.54
4	8.03	11.18	48	2.08	26.58
5	10.94	11.81	49	8.35	19.38
6	11.80	3.83	50	6.67	21.51
7	5.99	7.83	51	9.47	16.30
8	6.65	4.66	52	4.23	10.11
9	8.48	16.74	53	3.08	26.99
10	9.86	10.26	54	2.90	19.84
11	6.93	18.58	55	2.68	26.25
12	1.69	28.04	56	1.67	32.14
13	3.60	22.57	57	3.54	11.16
14	4.93	16.85	58	9.44	25.48
15	7.91	18.13	59	7.34	16.36
16	1.60	19.66	60	4.97	7.99
17	5.36	12.26	61	8.45	11.16
18	2.68	3.77	62	4.55	15.80
19	3.56	3.73	63	3.91	15.46
20	5.29	7.23	64	10.63	8.08
21	8.22	20.98	65	3.55	31.42
22	5.61	23.53	66	3.37	27.85
23	3.74	25.47	67	5.06	2.31
24	8.23	18.55	68	5.56	4.19
25	5.45	29.21	69	4.60	6.49
26	5.24	20.45	70	5.26	3.35
27	8.06	25.18	71	3.67	5.99
28	4.17	19.28	72	3.90	7.20
29	3.73	19.23	73	3.32	2.28
30	4.59	9.57	74	5.16	6.03
31	2.99	10.89	75	7.84	5.00
32	3.17	5.24	76	8.79	11.30
33	1.81	4.24	77	9.27	11.06
34	2.16	11.81	78	8.93	11.31
35	3.94	11.00	79	9.76	12.64
36	2.94	9.96	80	6.79	25.19
37	4.39	15.00	81	3.77	27.01
38	2.04	19.31	82	2.35	24.22
39	3.32	8.21	83	9.44	7.05
40	17.57	16.85	84	6.79	25.90
41	13.33	17.72	85	3.86	24.43
42	6.19	25.87	86	2.41	25.24
43	10.62	17.86	87	5.59	8.37
44	9.61	23.17			

Table 8.15. *Diagnostics for the fit in Display 8.3*

CMA	Standardised residuals	DFITS	Leverage
12	−2.59	−0.659	
16	−3.09	−0.633	
34		0.377	
40		−0.916	0.189
42	2.17	0.400	
43	2.57	0.560	
78	2.21		

DFITS cut off: $2(p/n)^{\frac{1}{2}} = 0.371$, $p = 3$, $n = 87$, leverage cut off: $3p/n = 0.103$.

significant improvement. The inclusion of these additional terms is consistent with properties of the cancer data identified in Chapter 7. The fit is summarised in Display 8.4(1) where both X_8 and X_{11} remain significant and in addition the higher order trend surface terms are also significant.

Figure 8.19 shows (*a*) the rankit plot for the residuals from the model together with (*b*) the residual against observed number of deaths plot. The residuals seem generally well behaved but there is evidence of a higher residual variance when numbers of deaths are low but this could just be due to the larger number of cases. Further, CMAs 17, 71 and 72 have high leverage values as a consequence of fitting the trend surface terms.

Display 8.5(2) summarises the fit using the model derived earlier with non-constant error variance. The fit is by weighted least squares with the weighting given by $[\sigma^2 + 1/O_i]^{-\frac{1}{2}}$ (see Pocock *et al.*, 1981). Further regressions deleting CMAs 17, 71 and 72 resulted in X_8 ceasing to be significant. Finally, Display 8.4(3) reports the estimation of the model using the resistant *R*-estimator (Li, 1985, p. 331) which fits a regression model using a rank based criterion. This Display provides evidence of the resistance of the fit to the (few) large residuals.

As noted earlier, the observed number of deaths is a Poisson variate. As a final check on the 'robustness' of the findings reported so far, a stepwise Poisson log-linear regression analysis was run on the observed number of deaths including E_i as an offset variable. A summary of the fit (using GLIM) is given in Display 8.4(4).

The evidence of these regression results points to the important role of social class in accounting for spatial variation in Cancer mortality rates. There also appears to be an age/loneliness factor, although the significance

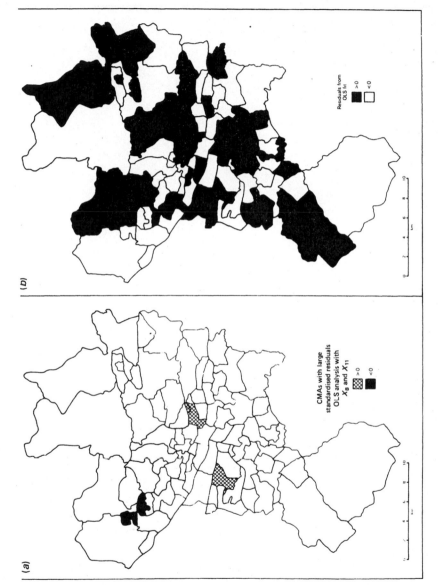

Figure 8.18. (a) Glasgow CMAs with large standardised residuals under OLS analysis of cancer data. (b) Plot of positive and negative residuals from OLS analysis.

Analysing multivariate data sets

Display 8.4. *Fitting models to the cancer SMR data*

1. Fitting with additional quadratic trend: least squares fit

$$\hat{Y} = 4.27 - 0.165X_E + 0.202X_N - 0.610X_E^2$$
$$(-0.38) \quad (0.65) \quad (-1.65)$$
$$- 0.710X_N^2 + 1.296X_EX_N + 0.010X_8 + 0.017X_{11}$$
$$(-2.35) \quad (3.24) \quad (2.55) \quad (12.27)$$

$R^2 \times 100 = 73.7 \qquad R^2(\text{adjusted}) \times 100 = 71.4 \qquad \hat{\sigma} = 0.093$

t values in brackets

X_E denotes E–W co-ordinate of CMA

X_N denotes N–S co-ordinate of CMA

Figure 8.19. Diagnostics from regression model with quadratic trend surface terms (*a*) rankit plot, (*b*) residuals against observed number of deaths plot.

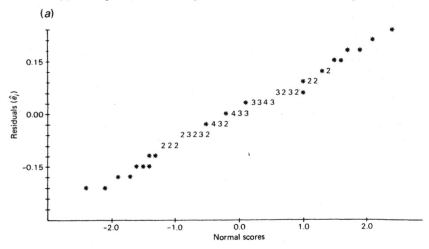

(*a*)

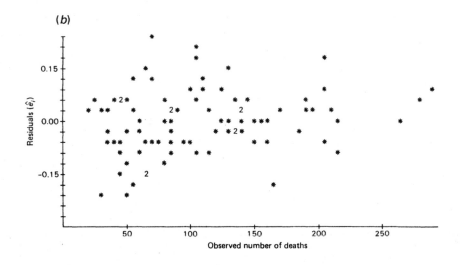

(*b*)

370

2. Fitting with non-constant error variance (weighted least squares)
$$\hat{Y} = 4.26 - 0.142X_E + 0.242X_N - 0.588X_E^2$$
$$\quad\quad (-0.32) \quad (0.17) \quad\quad (-1.57)$$
$$\quad - 0.704X_N^2 + 1.210X_E X_N + 0.009X_8 + 0.016X_{11}$$
$$\quad (-2.19) \quad\quad (3.03) \quad\quad\quad (2.49) \quad\quad\quad (12.27)$$
$\hat{\sigma} = 0.093$. Figures in brackets are t values.

3. Fitting with additional quadratic trend: resistant R-regression
$$\hat{Y} = 4.34 - 0.375X_E + 0.135X_N - 0.461X_E^2$$
$$\quad\quad (0.412) \quad (0.294) \quad (0.347)$$
$$\quad - 0.682X_N^2 + 1.373X_E X_N + 0.009X_8 + 0.017X_{11}$$
$$\quad (0.284) \quad\quad (0.375) \quad\quad\quad (0.003) \quad\quad (0.001)$$
R^2(adjusted) $\times 100 = 46.0 \quad\quad \hat{\sigma} = 0.087$
Figures in brackets are standard error estimates.

4. Stepwise Poisson log linear regression

Step 1 Null model (Variance of Observed number of cancer deaths)
Scaled deviance: 2882.9

Step 2 Add $\log_e E$ (expected number of deaths given age and sex composition of the area)
Scaled deviance: 229.86

Step 3 Add X_{11} (% in social class 4 and 5)
Scaled deviance: 86.764 $\quad\quad \hat{\sigma} = 1.033$

Step 4 Add $\log_e X_7$ (% elderly)
Scaled deviance: 70.726 $\quad\quad \hat{\sigma} = 0.8521$

Step 5 Add $\log_e X_{15}$ (% in communal establishments)
Scaled deviance: 66.298 $\quad\quad \hat{\sigma} = 0.8085$

Step 6 Add 2nd order Trend Surface

Scaled deviance: 59.134 $\quad\quad \hat{\sigma} = 0.768$

Parameter estimate	Standard error	Variable
−0.440	0.160	
1.006	0.022	$\log_e E$
0.018	0.001	X_{11}
0.075	0.030	$\log_e X_7$
0.005	0.003	$\log_e X_{15}$ (n.s.)
−0.176	0.469	⎫
0.254	0.425	⎪
−0.363	0.389	⎬ 2nd order trend surface
−6.07	0.380	⎪
0.948	0.440	⎭

n.s. denotes not significant.
Degrees of freedom 77. $\chi^2_{77, 0.05} \approx 100$

of this is less clear cut. CMAs with large numbers of single elderly people and those in social class 4 and 5 will tend to have high rates. The trend surface element declines towards the outer fringes of the urban area (Figure 8.20) and since this too is significant it suggests an additional 'inner urban area' factor as an added risk factor for people who live in the centre of Glasgow. (The elongation of the surface could be due to the high leverage effects of CMAs 71 and 72 as well as the orientation of the distribution of areas.) To what extent such a factor might be associated with the environmental characteristics of such an area (such as higher levels of stress or pollution) and to what extent this factor is associated with other characteristics of the population (such as diet, exercise, etc.) of the inner areas of a major city cannot be evaluated here with the available data. Figures 8.21(*a*) and (*b*) identify those CMAs with large standardised residuals from the fits summarised in Display 8.4(1) and (4).

Further discussion of this example is given by Haining (1991b) and a more detailed example of the development of spatial regression models for SMR data, applied to a sampled national data set including the simultaneous treatment of spatial correlation and heteroscedastic disturbances can be found in Cook and Pocock (1983) and Cook, Pocock and Shaper (1982). Lovett *et al.* (1986) provide an extended discussion of the use of log linear regression modelling with geographical mortality data.

Example 8.4. *Identifying spatial interaction and heterogeneity: Sheffield petrol price data*

In this final example we examine the problem of accounting for variation in the Sheffield petrol price data. Several monthly data sets were subject to an exploratory analysis in Chapter 6. Here we consider only the March data set. Display 8.5 summarises results from regressing March prices on February prices for 79 of the sites with recorded prices for both months. The inclusion of January prices was not significant. Since the slope coefficient is less than one this provides further evidence of the price fall. Figure 8.22(*a*) shows a scatterplot of the March price data against the February price data, and identifies observations with large standardised residuals and leverages. The rankit plot is shown in Figure 8.22(*b*). We should assess the sensitivity of results to these extreme observations. Some retail sites did not engage in fierce price cutting and a sensitivity analysis might help to identify competitive effects rather better by downweighting the contribution of these retailers to the overall fit. It is also of interest to examine whether or not local patterns of competition have had an effect on the spatial pattern of prices. We shall consider this first.

372

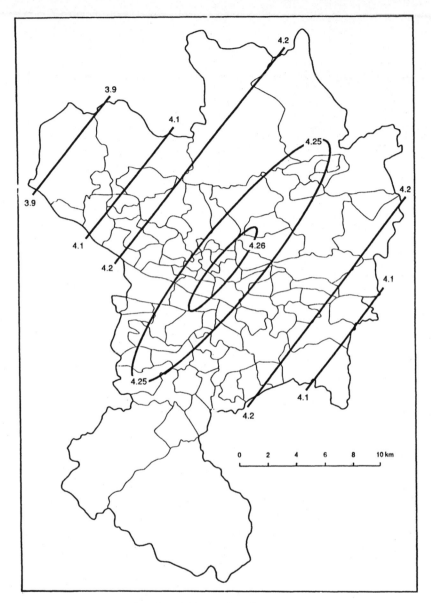

Figure 8.20. Plot of quadratic trend surface component.

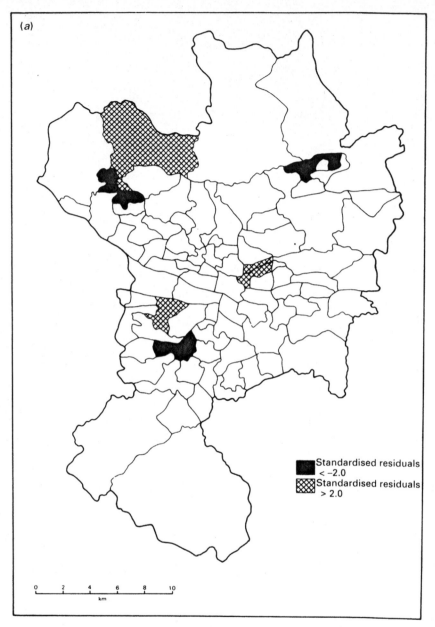

Figure 8.21. Glasgow CMAs with large standardised residuals (*a*) Normal regression model (Display 8.4(1)), (*b*) Poisson regression model (Display 8.4(4)).

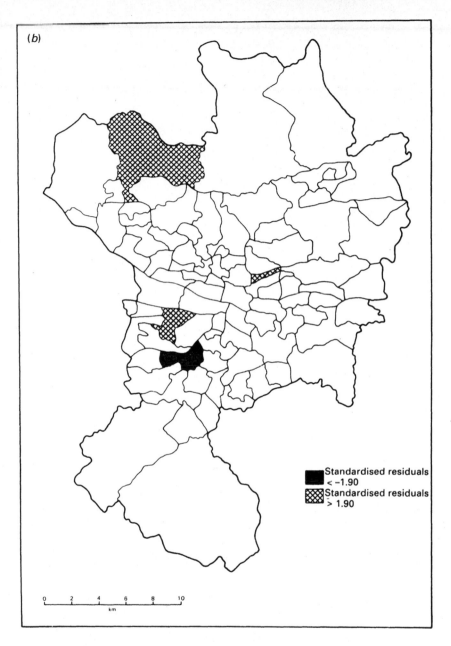

(b)

Standardised residuals
< −1.90

Standardised residuals
> 1.90

0   2   4   6   8   10
            km

Display 8.5. *Analysis of petrol price data: temporal model I*

$$\hat{Y}_{\mathrm{MARCH}} = \quad 3.64 + 0.925 Y_{\mathrm{FEB}}$$
$$(11.49) \quad (0.075)$$

Standard errors in brackets $\quad R^2 \times 100 = 66.3 \quad R^2(\text{adjusted}) \times 100 = 65.8$

$\tilde{\sigma}^2 = 2.87 \qquad n = 79$

Figure 8.22. (*a*) Scatterplot of March prices data against February prices: Sheffield petrol data. $\bigcirc$ = observations with large negative standardised residuals, $\bullet$ = observations with large positive standardised residuals, $\square$ = observations with large leverages, $\blacklozenge$ = observations with large DFITS.

(*a*)

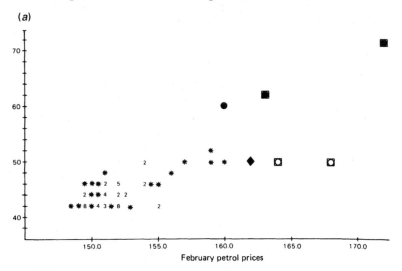

(*b*)

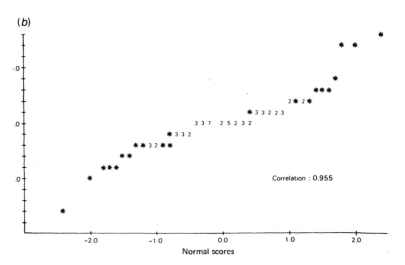

Of the 79 sites, 61 of these sites are close enough to each other to allow examination of spatial interaction effects. Figure 8.23 shows the reduced set ($n=61$) and a hypothesised set of linkages used to specify the connectivity matrix **W**. Sites close together or next to one another on a main route are linked. Several sites close together at crossroads are all interlinked with one another. Row sums of **W** are standardised to one, on the grounds that the more local competitors there are, the less influential any *one* is. Display 8.6 summarises results from regressing March prices on February prices for these 61 sites and the GMC test uses the above **W** matrix. The GMC test

Figure 8.23. Reduced set of petrol stations ($n=61$) and linkages used to test for interaction effects. □ = sites with large standardised residuals (>0), ◨ = sites with large standardised residuals (<0), × = sites with large leverages. Circled stations are not on a main road.

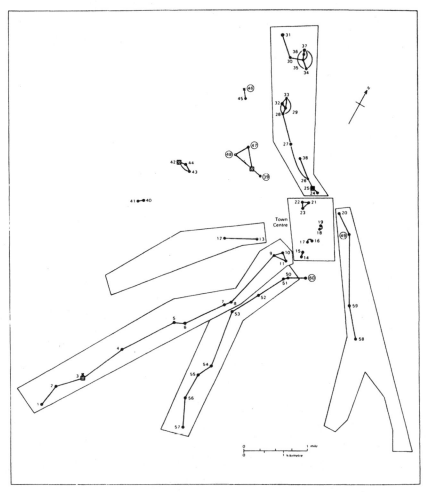

Display 8.6. *Analysis of petrol price data: temporal model II*

$\hat{Y}_{MARCH} = 19.00 + 0.822 Y_{FEB}$
       (14.98) (0.098)

Standard errors in brackets $R^2 \times 100 = 54.0$ R^2(adjusted) $\times 100 = 53.3$
$\hat{\sigma}^2 = 2.22$ $n = 61$
GMC $= 0.262$ E[GMC] $= -0.018$ St.Dev.[GMC] $= 0.139$
GMC (as standard normal deviate) $= 2.02$

Figure 8.24. Residual correlation plot and added variable plots for petrol data. (*a*) Plot of OLS residuals ($\hat{e}$) against (W$\hat{e}$). (*b*) Added variable plot for the inclusion of (Wy$_{MARCH}$) into the regression model.

(*a*)

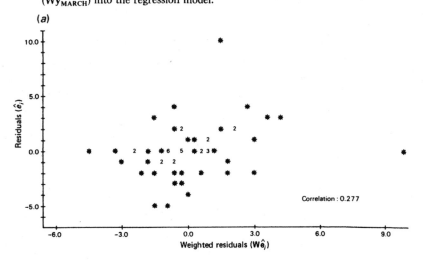

(*b*)

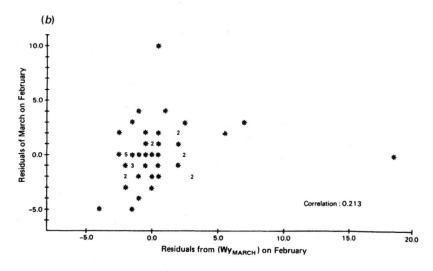

provides evidence of spatial pattern in the residuals.[6] Figure 8.24 shows correlation plots of the regression residuals and an added variable plot for the inclusion of (Wy_{MARCH}). The former plot is used to indicate whether a regression model with a spatial error model should be fit, the latter to indicate whether a regression model with a spatially lagged response variable should be used. The evidence supports both and slightly favours the former. Display 8.7 summarises results for both models and Figure 8.25 shows some diagnostic plots for Display 8.7(a). Statistically there seems little to choose between them but substantively whereas the model of Display 8.7(b) implies only additional spatial interaction, the model of

Display 8.7. *Analysis of petrol price data: spatial models*

(a) Spatial error model ($V = V_{SAR}$)
$$Y_{MARCH} = 20.721 + 0.810Y_{FEB} + u$$
$$(14.39)\ (0.094)$$
Standard errors in brackets
$\bar{\rho} = 0.180$
$\bar{\sigma}^2 = 2.11 \qquad n = 61$

(b) Lagged response variable model
$$\hat{Y}_{MARCH} = 4.882 + 0.803Y_{FEB} + 0.117Wy_{MARCH}$$
$$(17.89)\ (0.095) \qquad (0.088)$$
Standard errors in brackets
$\bar{\sigma}^2 = 2.14 \qquad n = 61$

Display 8.7(a) when expanded implies additional space-time interaction that is

$$Y_{MARCH,i} = f(Y_{FEB,i}, \{Y_{MARCH,j}\}_{j\epsilon N(i)}, \{Y_{FEB,j}\}_{j\epsilon N(i)})$$

where $N(i)$ denotes the neighbours (as specified by W) of site i. In either event, however, the evidence does seem to support the existence of two elements underpinning spatial price variation: temporal continuity of prices at a site and local spatial competitive interaction.

Display 8.8. *Robust estimation of spatial error model with $V = V_{SAR}$*

(a) Robust estimation of β only
$$Y_{MARCH} = 64.559 + 0.518Y_{FEB} + u$$
$$(15.60)\ (0.102)$$
Standard errors in brackets
$\bar{\rho} = 0.200$
$n = 61$

(b) Robust estimation of β and ρ
$$Y_{MARCH} = 64.473 + 0.519Y_{FEB} + u$$
$$\bar{\rho}_R = 0.232$$

Figure 8.25 shows that there are some large residuals so it is worthwhile to assess the sensitivity of results to these observations. A robust analysis of the model in Display 8.7(*a*) was run using the Tukey bi-weight to downweight the influence of extreme residuals. The results are summarised in Display 8.8. First, only the regression parameter is estimated robustly, ρ is estimated by the usual maximum likelihood procedure. Part (*b*) of Display 8.8 reports the robust fitting of the same model this time using a robust procedure on the least squares estimator for ρ. Although least squares

Figure 8.25. Diagnostic plots for regression with spatial error model.

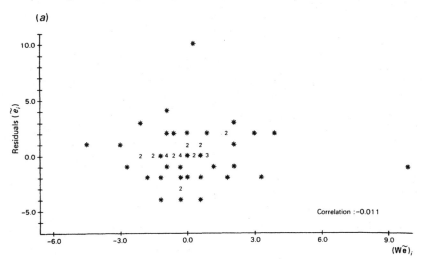

(*a*)

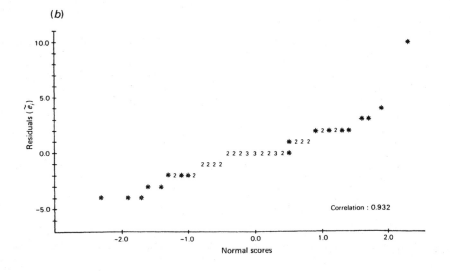

(*b*)

estimation is not generally to be recommended, in this case since ρ is small least squares should not introduce serious bias whilst the use of a robust weighting may improve the resistance properties of the estimator. The sites that have been strongly downweighted under the first analysis are shown in Figure 8.23(*a*) and are the sites that charged high prices in February and were least active in responding to the downward pressures on price. The estimate for the parameter associated with Y_{FEB} reflects this being considerably smaller than that obtained using non-robust estimation. The robust procedure seems therefore to provide a better estimate of competitive pressures (over time) amongst those sites actively competing in the market.

The two spatial regression models fitted assume spatial homogeneity. However, the retail market for petrol may be segmented with different demand (and/or supply) pressures across different groups or classes of retailers. Such spatial heterogeneity may extend to the different routeways leading into Sheffield with demand pressures reflecting differences in the socio-economic mix of customers that use them. Higher income groups may be far less sensitive to small differences in price than lower income groups, and they may also be 'tied-in' to certain retailers either through personal credit cards or business requirements thus making the retailers less responsive to local competitive pressures. On the other hand sales volumes may be higher in these areas, and given the importance corporations attach to preserving market share these might be the areas where competitive pressures are strongest.[7] A map of the large negative standardised residuals (< -1.0) from the model reported in Display 8.5 showed a large number of these in the southwest part of Sheffield along routeways 1, 2 and 4 (see Figure 6.5) which is a high income segment of the city. Such a block of residuals is consistent with a segmented-market hypothesis. Another potentially important distinction is between those retailers sited on the main routeways and those sited in the side-streets. There may be an important supply side distinction between these two groups of retailers with the latter often offering a broader range of automotive services but more crucially receiving less favourable terms from suppliers because they deal in lower volumes.

A series of models were fit in order to examine the segmented market hypothesis. The models together with some of the results are shown in Table 8.16. In all cases $n = 79$. The variable D_1 denotes the dummy variable measuring whether a site was on a main road ($D_1 = 0$) or not ($D_1 = 1$) and D_2 measures whether a site was on routeways 1, 2 or 4 ($D_2 = 0$) or not ($D_2 = 1$). Such segmentation effects are swamped whenever February prices are included in the model, but there is evidence of a main road/non-main road segmentation and rather weaker evidence of segmentation between

Table 8 16. *Testing for spatial heterogeneity in the petrol price data*

Model 1

$$Y_{MARCH} = \alpha + \beta D_1 + e$$

	Coefficient	St. Deviation	*t* value
$\hat{\alpha}$	143.81	0.58	247.51
$\tilde{\beta}$	4.65	1.21	3.83

$\hat{\sigma}^2 = 4.53$, $R^2 \times 100 = 16.0$, R^2(adjusted) $\times 100 = 14.9$.

Model 2

$$Y_{MARCH} = \alpha + \beta Y_{FEB} + \alpha\gamma D_1 + e$$

	Coefficient	St. Deviation	*t* value
$\hat{\alpha}$	4.91	13.05	0.38
$\hat{\beta}$	0.91	0.08	10.65
$\hat{\gamma}$	0.18	0.88	0.21

$\hat{\sigma}^2 = 2.89$, $R^2 \times 100 = 66.3$, R^2(adjusted) $\times 100 = 65.4$.

Model 3

$$Y_{MARCH} = \alpha + (\beta + \gamma D_1) Y_{FEB} + e$$

	Coefficient	St. Deviation	*t* value
$\hat{\alpha}$	5.13	13.31	0.39
$\hat{\beta}$	0.91	0.08	10.42
$\hat{\gamma}$	0.001	0.005	0.23

$\hat{\sigma}^2 = 2.89$, $R^2 \times 100 = 66.3$, R^2(adjusted) $\times 100 = 65.4$.

Model 4

$$Y_{MARCH} = \alpha + \beta D_2 + e$$

	Coefficient	St. Deviation	*t* value
$\hat{\alpha}$	143.30	1.01	141.86
$\hat{\beta}$	2.22	1.20	1.85

$\hat{\sigma}^2 = 4.84$, $R^2 \times 100 = 4.3$, R^2(adjusted) $\times 100 = 3.0$.

Model 5

$$Y_{\text{MARCH}} = \alpha + \beta Y_{\text{FEB}} + \gamma D_2 + e$$

	Coefficient	St. Deviation	t value
$\hat{\alpha}$	143.30	0.95	150.92
$\hat{\beta}$	4.35	1.30	3.34
$\hat{\gamma}$	0.82	1.20	0.68

$\hat{\sigma}^2 = 4.55$, $R^2 \times 100 = 16.5$, R^2(adjusted) $\times 100 = 14.3$.

Model 6

$$Y_{\text{MARCH}} = \alpha + (\beta + \gamma D_2) Y_{\text{FEB}} + e$$

	Coefficient	St. Deviation	t value
$\hat{\alpha}$	6.20	11.49	0.54
$\hat{\beta}$	0.90	0.07	11.92
$\hat{\gamma}$	0.007	0.004	1.60

$\hat{\sigma}^2 = 2.84$, $R^2 \times 100 = 67.4$, R^2(adjusted) $\times 100 = 66.5$.

high and low income areas. Interestingly when model 3 (Table 8.16) is fit to the 61 sites for which the matrix W was defined the GMC statistic as a standard normal deviate is 1.91. There seem to be several types of spatial effects present in this data. Although the statistical evidence is ambiguous it is suggestive of both interaction and segmentation effects.

The problems arising in the analysis of spatial heterogeneity in house price variation are discussed in Goodman (1981) and those arising in the case of urban population density data in Brueckner (1981) and Kau, Lee and Chen (1983). Stetzer (1982) proposes the use of jack-knifed parameters as an exploratory device for checking for spatial parameter variation.

NOTES

1. The expressions for the asymptotic variances in the case of these two models are complicated. Derivations are given in the cited literature. For the CAR model with parameters τ_1 and τ_2 the asymptotic variance increases more than exponentially with increases in the parameter values. A simple expression arises with the doubly geometric process. This is:

$$\lim_{n \to \infty} \text{Var}(r) = n^{-1}[(1 + \lambda_1 \lambda_2)/(1 - \lambda_1 \lambda_2)]^2$$

where λ_1 and λ_2 are the parameters of the two processes (Richardson and Hemon, 1982).

2.   We do not consider the problem of categorical data analysis with dependent spatial data. Developments in this area are discussed in Fingleton (1986).

3.   Briefly an added variable plot is constructed as follows. Suppose it is suspected that a new variable (X_{k+1}) should be added to the regression model

$$Y = \beta_0 + \beta_1 X_1 + \ldots + \beta_k X_k + e \qquad (A1)$$

First estimate (A.1) and obtain the residuals $(\hat{e})$. Now, regress X_{k+1} on the set $X_1, \ldots, X_k$ and obtain the residuals (e^*). This is the part of X_{k+1} not explained by $X_1, \ldots, X_k$. Now plot the elements of $\hat{e}$ against e^*. This is an added variable plot. If there is a strong linear trend this implies X_{k+1} should be added to (A.1), otherwise X_{k+1} can be left out. The plot is interpreted in a similar way to a scatter plot in regression. See Johnson and McCulloch (1987).

4.   For the model given by 8.12 with k deleted observations

$$\hat{\beta} = (\mathbf{X_o}^T \tilde{\mathbf{V}}_{oo}^{-1} \mathbf{X_o})^{-1} (\mathbf{X_o}^T \tilde{\mathbf{V}}_{oo}^{-1} \mathbf{y_o})$$
$$\hat{\sigma}^2 = (\mathbf{y_o} - \mathbf{X_o}\hat{\beta})^T \mathbf{V}_{oo}^{-1} (\mathbf{y_o} - \mathbf{X_o}\hat{\beta})/(n-k)$$

and the parameters of $\mathbf{V}$ are estimated by minimising

$$|\mathbf{V}_{oo}|^{1/(n-k)} ((\mathbf{y_o} - \mathbf{X_o}\tilde{\beta})^T \mathbf{V}_{oo}^{-1} (\mathbf{y_o} - \mathbf{X_o}\tilde{\beta}))$$

where $\mathbf{V}_{oo}$ is the submatrix of $\mathbf{V}$ for the subvector $\mathbf{y_o}$, the observed (or non-deleted) set of observations. $\mathbf{X_o}$ is the submatrix of $\mathbf{X}$ for the non-deleted observations (Martin, 1984). The description assumes that after deleting observations the labelling of areas is permutated so that areas 1 to $(n-k)$ are the observed set.

5.   Haggett, Cliff and Frey (1977, pp. 364–5) report the result of a regression of retail sales (y) on total personal income (x) for the counties of Ireland including and then excluding Cork and Dublin where levels of x are high. The purpose is to identify the sensitivity of the results to the high leverage effects of these two counties. They note that although the R^2 falls the residuals are no longer autocorrelated and conclude that 'the revised regression might therefore be regarded as more successful from a geographical point of view' (p. 365).

6.   The $\mathbf{W}$ matrix is rather sparse and this may have an effect on the validity of the inferential theory (see page 232).

7.   For a theoretical discussion of the origins of price heterogeneity arising from heterogeneity of demand associated with differences in income levels see Greenhut, Norman and Hung (1987, pp. 229–335).

Appendix: Robust estimation of the parameters of interaction schemes

In the case of a model such as (8.12) a robust estimator for β had been described in Chapter 2 assuming that $\mathbf{V}$ is known. Usually $\mathbf{V}$ is not known and if it is specified by an interaction model we might wish to estimate the unknown parameters of this part of the model robustly as well, since the same problems may infect this stage of estimation. Isolated large regression residuals $(\tilde{u})$ might give rise to outliers in the plot $(\{\tilde{u}_i, (\mathbf{W}\tilde{u})_i\})$ while a small cluster of large $\tilde{u}$ values might give rise to leverage problems. For reasons already discussed we only focus on robust estimation in the presence of outliers.

Consider the ordinary least squares estimator for the single parameter CAR or SAR scheme. In either case the spatial parameter is estimated by:

$$\tilde{u}^T W^T \tilde{u} / \tilde{u}^T W^T W \tilde{u}$$

which is the regression of $\tilde{u}$ on $W\tilde{u}$. A robust estimator is (Chapter 2):

$$\tilde{u}^T W^T W^* \tilde{u} / \tilde{u}^T W^T W^* W \tilde{u}$$

where $\tilde{u}$ are the residuals from the regression fit and W^* is a diagonal matrix with weights specified using the final set of (independent) residuals, $\tilde{e}$, from the previous cycle. This modified estimator is used in the iterative procedure. A robust estimate of σ^2 can be obtained either by substituting the robust estimate of the spatial parameter into the usual estimator (4.12, 4.17) or computing the MAD of the final set of independent residuals. This is also used in constructing confidence intervals.

In the case of the SAR model, particularly if ρ is large, maximum likelihood estimation is strongly to be preferred. This estimate is obtained by minimising a weighted least squares expression (see 4.13). It is again the term in σ^2 (4.12) that will be affected by extreme values. For the single parameter version an estimate $(\bar{\rho})$ is obtained by minimising with respect to ρ:

$$\{-2n^{-1}\ln|(I-\rho W)| + \ln\Sigma_{i=1,\ldots,n} f[(\bar{u}_i) - \rho\Sigma_{j=1,\ldots,n} w_{ij}\bar{u}_j]\}$$

where $\delta f(t)/\delta t = t w(t)$. Forms for $w(t)$ have been given in Table 2.4 and forms for $f(t)$ are given in Li (1985, table 8.2). In the case of the Tukey bi-weight

$$\begin{aligned} f(t) &= (B^2/6)\{1 - [1-(t/B)^2]^3\} & |t| \leqslant B \\ &= (B^2/6) & |t| > B \end{aligned}$$

It is likely that an estimator of this type could be very sensitive to the choice of starting point as well as choice of weighting function.

Other forms of resistant line fitting to the data $(\{\bar{u}_i, (W\bar{u})_i\})$ might also be considered (see Hoaglin *et al.*, 1985) providing the line is forced through the origin.

The merits (or otherwise) of these procedures remain to be examined.

Postscript

Chapter 1 defined three important objectives: the presentation of spatial data analysis as a part of general data analysis while alerting the reader to the special difficulties that spatial data may create; the presentation of a wide range of models for presenting spatial variation; the examination of the role of subject matter theory in spatial analysis. Good data analysis seeks a balance between being theoretically informed and letting the data speak. Balance is essential in order to avoid the twin problems of on the one hand using data analysis merely to confirm existing prejudices and on the other reporting ambiguous data patterns. We offer some observations to conclude the book that relate to these issues.

From a technical point of view some of the most serious difficulties facing the analyst of spatial data concern the wide range of possible data models and the need to implement awkward and laborious fitting procedures. The availability of specialist software that implements the fitting and evaluation of spatial models is desirable. But important though it is to develop specialist software for confirmatory spatial data analysis it should be evident that useful progress can be made by carrying out sensible exploratory analyses with standard software. Some pre-processing of the data may be necessary but thereafter simple graphical and resistant techniques of analysis can offer useful insights into the data, indicating the types of models that should be explored and hence determining the extent to which specialist software is needed in order to carry out confirmatory analyses. Such exploratory analyses may be particularly useful in situations (all too common) when more formal testing procedures are complicated, difficult to implement and of unknown reliability.

Not all the problems of spatial data analysis revolve around spatial correlation but where this attribute is present, and it is probably more prevalent in studies at small spatial scales than large, the analyst must give serious thought to its representation. Here we should distinguish between those problem areas such as mapping and interpolation where a good representation of spatial variation is the cornerstone of the work, from those problem areas where interest focuses on many attributes of the data and the analyst is more concerned to try to prevent spatial (auto) correlation from invalidating results or misleading the analyst or using it to reveal important data attributes. These last problems seem to be of particular importance in regression and correlation analysis in the social sciences. In either case however it is likely that the models described in Chapter 3 for representing spatial variation will often provide only crude approximations to the real patterns of variation encountered and the analyst needs to consider the extent to which results may prove sensitive to different but equally plausible representations of this variation.

An important concern has been to stress the need to carry out many analyses of the data and in particular to identify the sensitivity of model fits to:

(i)   modelling assumptions that derive from the substantive nature of the problem

(ii)   data characteristics, in particular extreme values and influential observations

(iii)   spatial attributes of the study area, in particular the distribution of observations, the nature of the areal partition, boundary influences, frame effects and the specification of inter-site relationships.

In some situations rather arbitrary decisions have to be taken about how to represent spatial relationships between observations or about the nature of boundary influences. The safest route is to consider alternatives and see to what extent results are stable, changing things one at a time (the list of explanatory variables, the weighting of observations, the inclusion of dependency effects, the allowance for error variance differences) to keep track of effects. Perhaps most important of all is the need to keep open the possibility that the set of plausible models for the data is always wider than has been currently imagined – the horizon of models as Leamer (1983) calls it. 'If you wish to make . . . discoveries, you will have to poke at the horizon and poke again' (Leamer, 1983, p. 40).

With so many possible analyses for a given set of data there is an obvious danger that matters can get out of hand; each data set yielding up a multitude of interpretations. With the increase in data, noted in Chapter 2, a future based on such a methodology might be viewed with some horror. But perhaps the crucial issue is not how many different analyses might be performed on any given set of data (all of which cannot be reported), but rather how informative is any given set of data. 'A data set yields useful information when the range of inferences is small enough to be useful and the corresponding family of models is broad enough to be believable' (Leamer and Leonard, 1983, p. 306). Data sets that yield only fragile models, the results of which are easily changed by small perturbations of the assumptions or the data should be clearly flagged. For spatial analysts confronted by the increase in spatially referenced data the criterion of informativeness may be a necessary one to apply.

Glossary

Section numbers in brackets.

Notation

^	Least squares estimator
~	Maximum likelihood estimator
$\rightarrow$	Approaches or goes to
$\mathbf{A}$	Matrix of trend surface coefficients
$C(r),C(s,t)$	Spatial covariance at distance r, lag (s,t)
$\hat{C}(r),\hat{C}(s,t)$	Estimate of $C(r),C(s,t)$ [4.1]
$C^*(r),C^*(s,t)$	Estimate of $C(r)$, $C(s,t)$ with edge correction [4.1]
$\mathbf{e}$	Vector of independent errors or disturbances
$\hat{\mathbf{e}},\tilde{\mathbf{e}}$	Vector of least square, maximum likelihood residuals
H_0,H_1	The null hypothesis, the alternative hypothesis
$MVN(\mathbf{\mu},\sigma^2\mathbf{V})$	Multivariate normal distribution with mean vector $\mathbf{\mu}$ and dispersion matrix $\sigma^2\mathbf{V}$
$N(\mu,\sigma^2)$	Normal distribution with mean μ and variance σ^2
R^2	Coefficient of multiple correlation
$R(r),R(s,t)$	Spatial correlation at distance r, lag (s,t)
$\hat{R}(r),\hat{R}(s,t)$	Estimate of $R(r),R(s,t)$ [4.1]
$R^*(r),R^*(s,t)$	Estimate of $R(r),R(s,t)$ with edge correction [4.1]
$\mathbf{u}$	Vector of spatially correlated errors or disturbances
$\hat{\mathbf{u}},\tilde{\mathbf{u}}$	Vector of least squares, maximum likelihood residuals
$\mathbf{W}$	Spatial weights or connectivity matrix [3.1.2]
$\mathbf{W}^+$	Diagonal weights matrix used in weighted least squares [2.3.3(h)]
$\mathbf{W}^*$	Diagonal weights matrix used in iterated weighted least squares. Elements of $\mathbf{W}$ are estimated from model residuals using a function (Table 2.3) [2.4.2(b)]
$\mathbf{X}$	Matrix of explanatory or predictor variable values used in regression
$\mathbf{y},\mathbf{Y}$	Vector of response values, corresponding random variables
$\mathbf{\beta}$	Vector of regression parameters
$\gamma(h),\hat{\gamma}(h)$	The semi-variogram, estimate of the semi-variogram at distance h [4.1]
$\mathbf{\theta}$	Vector of trend surface parameters
ν	Parameter of a moving average scheme
ρ	Parameter of a simultaneous autoregressive scheme
τ	Parameter of a conditional autoregressive scheme

Glossary

Selected statistical and mathematical terms

Akaikes information criterion: Used for statistical identification. It is an estimate of the measure of fit of any model to the data. Where the fits of several models are being compared, the best fitting model is the one which minimises this criterion. It is a mathematical formulation of the principle of parsimony. [4.2.5(a)]

Estimation terms

Maximum likelihood estimation: Method that selects that value of a parameter for which the probability (or probability density) of obtaining a given set of sample values is a maximum.

Least squares estimation: Method of curve fitting for estimating parameters where assumptions about underlying distributions cannot be made. The LS criterion demands that estimates be chosen so that the sum of the squares of the differences between the observed and predicted values is a minimum.

Estimate, Estimator: An estimate is a statistic used to provide an actual value for a parameter using given sample data. An estimator is the corresponding random variable which has a probability distribution under repeated sampling.

Unbiased estimator: Expected value of the estimator equals the parameter it is supposed to estimate.

Consistent estimator: Estimator where the probability that any estimate differs from the true value of the parameter by more than any arbitrary constant approaches zero as the size of the sample increases.

Relative efficiency: Given two unbiased estimators, one is said to be relatively more efficient than the other if its variance is less than the other.

Standard error: Square root of the sampling variance of an estimator.

Hypthesis testing terms

Lagrange multiplier test: A test of a hypothesis H_0 against an alternative H_1 which uses the likelihood function derived under H_0 and the set of restrictions on the parameters imposed on H_0 relative to H_1 [4.2.5(b)]

Likelihood ratio test: A test of a hypothesis H_0 against an alternative H_1 which uses the ratio of the two likelihood functions derived under H_0 and H_1 [4.2.5(a)]

Power of a test: The probability of rejecting H_0 when H_1 is true.

Type I error: Error committed when rejecting H_0 when it should be accepted.

Type II error: Error committed when accepting H_0 when it should be rejected.

Matrix terms

Eigenvalues: For a matrix $\mathbf{W}$, eigenvalues are the solutions (λ) to the equation $|\mathbf{W} - \lambda\mathbf{I}|$ where $\mathbf{I}$ is the identity matrix and $|\,.\,|$ denotes the determinant of a matrix.

Eigenvectors: The corresponding row vectors $\mathbf{a}$ or column vectors $\mathbf{b}$ for which $\mathbf{a}\mathbf{W} = \lambda\mathbf{a}$ or $\mathbf{W}\mathbf{b} = \lambda\mathbf{b}$.

Idempotent: A matrix $\mathbf{W}$ is idempotent if $\mathbf{W}^2 = \mathbf{W}$.

Median polish: Operation applied to a data table or matrix. Obtain the median for each row and subtract the corresponding value from every observation in the row. Now carry out this operation on the columns of the resulting matrix. Subtract column medians from each column. The operation is repeated until all rows and columns have zero medians at which stage the iterations stop. There are other stopping rules and the operation may start with columns rather than rows. [6.1.1(*b*)]

References

Abramowitz, M. and Stegun, I.A. (1984). *Handbook of Mathematical Functions.* New York: Dover.

Agterberg, F.P. (1970). Autocorrelation functions in geology. In *Geostatistics*, ed. D.F. Merriam, pp. 113–41. New York: Plenum.

(1984). Trend Surface Analysis. In *Spatial Statistics and Models*, ed. G.L. Gaile and C.J. Willmott, pp. 147–71. Dordrecht: Reidel.

Alexander, H.W. (1961). *Elements of Mathematical Statistics.* New York: Wiley.

Anas, A. and Eum, S.J. (1984). Hedonic analysis of a housing market in disequilibrium. *Journal of Urban Economics*, **15**, 87–106.

Anselin, L. (1986). Non nested tests on the weight structure in spatial autoregressive models: some Monte Carlo results. *Journal of Regional Sciences*, **26**, 267–84.

(1988a). Lagrange Multiplier test diagnostics for spatial dependence and spatial heterogeneity. *Geographical Analysis*, **20**, 1–17.

(1988b). *Spatial Econometrics: Methods and Models.* Dordrecht: Kluwer Academic Publishers.

Anselin, L. and Can, A. (1986). Model comparison and model validation issues in empirical work on urban density functions. *Geographical Analysis*, **18**, 179–97.

Arbia, G. (1986). The modifiable areal unit problem and the spatial autocorrelation problem: towards a joint approach. *Metron*, **44**, 325–41.

Armstrong, M. (1984). Improving the estimation and modelling of the variogram. In *Geostatistics for Natural Resources Characterization*, ed. G. Verly *et al.*, pp. 1–19. Dordrecht: Reidel.

Balling, R.C. (1984). Classification in climatology. In *Spatial Statistics and Models*, ed. G.L. Gaile and C.J. Willmott, pp. 81–108. Dordrecht: Reidel.

Barnes, R.J. and Johnson, Dr. Thys B. (1984). Positive kriging. In *Geostatistics for Natural Resources Characterization*, ed. G. Verly *et al.*, pp. 231–44. Dordrecht: Reidel.

Bartlett, M.S. (1935). Some aspects of the time correlation problem in regard to tests of significance. *Journal, Royal Statistical Society*, **98**, 536–43.

(1975). *The Statistical Analysis of Spatial Pattern.* London: Chapman and Hall.

Bellhouse, D.R. (1977). Some optimal designs for sampling in two-dimensions. *Biometrika*, **64**, 605–11.

Belsley, D.A., Kuh, E. and Welsch, R.E. (1980). *Regression Diagnostics.* New York: Wiley.

Bennett, R.J. (1979). *Spatial Time Series.* London: Pion.

Bennett, R.J., Haining, R.P. and Griffith, D.A. (1984). The problem of missing data on spatial surfaces. *Annals, Association American Geographers*, **74**, 138–56.

References

Bernstein, R., Lotspiech, J.B., Myers, H.J., Kolsky, H.G. and Lees, R.D. (1984). Analysis and processing of LANDSAT-4 sensor data using Advanced Image Processing Techniques and Technologies. *IEEE Transactions on Geoscience and Remote Sensing*, **GE-22**, 192–221.

Berry, B.J.L. (1967). *Geography of Market Centers and Retail Distribution*. Englewood Cliffs, New Jersey: Prentice Hall.

Berry, B.J.L. and Baker, A.M. (1968). Geographic sampling. In *Spatial Analysis*, ed. B.J.L. Berry and D.F. Marble, pp. 91–100. Englewood Cliffs, New Jersey: Prentice Hall.

Besag, J.E. (1974). Spatial interaction and the statistical analysis of lattice systems. *Journal, Royal Statistical Society*, **B36**, 192–236.

(1975). Statistical analysis of non-lattice data. *The Statistician*, **24**, 179–95.

(1977a). Efficiency of pseudo-likelihood estimators for simple Gaussian fields. *Biometrika*, **64**, 616–18.

(1977b). Errors in variables estimation for Gaussian lattice schemes. *Journal, Royal Statistical Society*, **B39**, 73–8.

(1978a). Some methods of statistical analysis for spatial data. *Bulletin of the International Statistical Institute*, **47**, 2, 77–92.

(1978b). Contribution to the discussion of a paper by M.S. Bartlett. *Journal, Royal Statistical Society*, **B40**, 165–6.

(1981a). On a system of two dimensional recurrence equations. *Journal, Royal Statistical Society*, **B43**, 302–9.

(1981b). On resistant techniques and statistical analysis. *Biometrika*, **68**, 463–9.

(1986). On the statistical analysis of dirty pictures. *Journal, Royal Statistical Society*, **B48**, 259–302.

Besag, J.E. and Diggle, P.J. (1977). Simple Monte Carlo tests for spatial pattern. *Applied Statistics*, **26**, 327–33.

Besag, J.E. and Moran, P.A.P. (1975). On the estimation and testing of spatial interaction in Gaussian lattice processes. *Biometrika*, **62**, 555–62.

Bivand, R.S. (1980). A Monte Carlo study of correlation coefficient estimation with spatially autocorrelated observations. *Quaestiones Geographicae*, **6**, 5–10.

(1984). Regression modelling with spatial dependence: an application to some class selection and estimation methods. *Geographical Analysis*, **16**, 25–37.

Boehm, B.W. (1967). Tabular representation of multivariate functions – with applications to topographic modelling. *Report RM-4636-PR*. Santa Monica, California: Rand Corporation.

Bolthausen, E. (1982). On the central limit theorem for stationary mixing random fields. *Annals of Probability*, **10**, 1047–50.

Box, G.E.P. (1979). Robustness in the strategy of scientific model building. In *Robustness in Statistics*, ed. R.L. Launer and G.N. Wilkinson, pp. 201–36. New York: Academic Press.

(1983). An apology for ecumenism in statistics. In *Scientific Inferences, Data Analysis and Robustness*, ed. G.E.P. Box, T. Leonard and C.-F. Wu, pp. 51–84. New York: Academic Press.

Brandsma, A.S. and Ketellapper, R.H. (1979a). A biparametric approach to spatial autocorrelation. *Environment and Planning*, **A11**, 51–8.

(1979b). Further evidence on alternative procedures for testing of spatial autocorrelation among regression disturbances. In *Exploratory and Explanatory Statistical Analysis of Spatial Data*, ed. C.P.A. Bartels and R.H. Ketellapper, pp. 113–36. Boston: Martinus Nijhoff.

Bras, R.L. and Rodriquez-Iturbe, I. (1976). Network design for the estimation of the areal mean of rainfall events. *Water Resources Research*, **12**, 1185–95.

Brueckner, J.K. (1981). Testing a vintage model of urban growth. *Journal of Regional Science*, **21**, 23–35.

Brush, J.E. (1953). The hierarchy of central places in southwestern Wisconsin. *Geographical Review*, **43**, 380–402.

Bunge, W. (1966). *Theoretical Geography*. Lund Studies in Geography. Lund: Gleerup.

Burgess, T.M. and Webster, R. (1980). Optimal interpolation and isorithmic mapping of soil properties. I. The semi variogram and punctual kriging. *Journal of Soil Science*, **31**, 315–31. II. Block kriging. *Journal of Soil Science*, **31**, 333–41.
   (1984). Optimal sampling strategies for mapping soil type. I. Distribution of boundary spacings. *Journal of Soil Science*, **35**, 641–54. II. Risk functions and sampling intervals. *Journal of Soil Science*, **35**, 655–65.

Burgess, T., Webster, M.R. and McBratney, A.B. (1981). Optimal interpolation and isorithmic mapping of soil properties, IV Sampling. *Journal of Soil Science*, **32**, 643–59.

Burridge, P. (1980). On the Cliff–Ord test for spatial correlation. *Journal, Royal Statistical Society*, **B42**, 107–8.
   (1981). Testing for a common factor in a spatial autoregression model. *Environment and Planning A*, **13**, 795–800.

Burrough, P.A. (1983). Multiscale sources of spatial variation in soil. I. The application of fractal concepts to nested levels of soil variation. *Journal of Soil Science*, **34**, 577–97. II. A non Brownian fractal model and its application in soil survey. *Journal of Soil Science*, **34**, 599–620.
   (1986). *Principles of Geographical information Systems for Land Resources Assessment*. Oxford Science Publications, Oxford: Clarendon Press.

Burrough, P.A., Bregt, A.K., de Heus, M.J. and Kloosterman, E.G. (1985). Complementary use of thermal imagery and spectral analysis of soil properties and wheat yields to reveal cyclic patterns in the Flevopolders. *Journal of Soil Science*, **36**, 141–52.

Campbell, J.B. (1981). Spatial correlation effects upon accuracy of supervised classification of land cover. *Photogrammetric Engineering and Remote Sensing*, **47**, 355–63.

Casetti, E. and Jones, J.P. III. (1987). Spatial aspects of the productivity slow down: an analysis of U.S. manufacturing data. *Annals, Association of American Geographers*, **77**, 76–88.

Chambers, J.M., Cleveland, W.S., Kleiner, B. and Tukey, P.A. (1983). *Gprahical Methods for Data Analysis*. London: Chapman Hall.

Chauvet, P. (1982). The variogram cloud. *17th APCOM Symposium*. Golden, Colorado: Colorado School of Mines.

Lord Chorley. (1987). *Handling Geographic Information*. Report of the Committee of Inquiry chaired by Lord Chorley. Department of the Environment, HMSO.

Chorley, R.J. and Haggett, P. (1965). Trend surface mapping in geographical research. *Transactions Institute of British Geographers*, **37**, 47–67.

Christakos, G. (1984). On the problem of permissable covariance and variogram models. *Water Resources Research*, **20**, 251–65.

Chung, C.F. (1984). Use of the Jackknife method to estimate auto-correlation functions (or variograms). In *Geostatistics for Natural Resources Characterization*, ed. G. Verly *et al.*, 55–69. Dordrecht: Reidel.

Cleveland, W.S. (1985). *The Elements of Graphing Data*. London: Chapman Hall.

Cleveland, W.S. and McGill, M.E. (1988). *Dynamic Graphics for Statistics*. London: Chapman Hall.

Cliff, A.D., Haggett, P. and Ord, J.K. (1985). *Spatial Aspects of Influenza Epidemics*. London: Pion.

References

Cliff, A.D., Haggett, P., Ord, J.K., Bassett, K. and Davies, R.B. (1975). *Elements of Spatial Structure: A Quantitative Approach.* Cambridge: Cambridge University Press.

Cliff, A.D., Haggett, P., Ord, J.K. and Versey, G.R. (1981). *Spatial Diffusion: An Historical Geography of Epidemics in an Island Community.* Cambridge: Cambridge University Press.

Cliff, A.D., Martin, R.L. and Ord, J.K. (1975). A test for spatial autocorrelation in choropleth maps based upon a modified χ^2 statistic. *Transactions, Institute of British Geographers,* **65,** 109–29.

Cliff, A.D. and Ord, J.K. (1975). The comparison of means when samples consist of spatially autocorrelated observations. *Environment and Planning A,* **7,** 725–34.
 (1981). *Spatial Processes: Models and Applications.* London: Pion.

Clifford, P. and Richardson, S. (1985). Testing the association between two spatial processes. *Statistics & Decisions,* **Suppl. No. 2,** 155–60.

Clifford, P., Richardson, S. and Hemon, D. (1989). Assessing the significance of the correlation between two spatial processes. *Biometrics,* **45,** 123–34.

Coale, A.J. and Stephan, F.F. (1962). The case of the Indians and the teenage widows. *Journal, American Statistical Association,* **57,** 338–47.

Cochran, W.G. (1963). *Sampling Techniques.* New York: Wiley.

Cook, D.G. and Pocock, S.J. (1983). Multiple regression in geographical mortality studies with allowance for spatially correlated errors. *Biometrics,* **39,** 361–71.

Cook, D.G., Pocock, S.J. and Shaper, A.G. (1982). Analysing geographic variation in cardio-vascular mortality: methods and results. *Journal Royal Statistical Society,* **A145,** 313–41.

Costanzo, C.M. (1985). Statistical inference in geography: modern approaches spell better times ahead. *Professional Geographer,* **35,** 158–64.

Costanzo, C.M., Hubert, L.M. and Golledge, R.G. (1983). A higher moment for spatial statistics. *Geographical Analysis,* **15,** 347–51.

Cox, N.J. (1983). On the estimation of spatial autocorrelation in geomorphology. *Earch Surface Processes and Landforms,* **8,** 89–93.

Cox, N.J. and Jones, K. (1981). Exploratory data analysis. In *Quantitative Geography,* ed. N. Wrigley and R.J. Bennett, pp. 135–43. London: Routledge & Kegan Paul.

Craig, R.G. (1979). Autocorrelation in LANDSAT data. *Proceedings of the 13th International Symposium on Remote Sensing of the Environment,* pp. 1517–24. Ann Arbor, Michigan.

Craig, R.G. and Labovitz, M.L. (1980). Sources of variation in LANDSAT autocorrelation. *Proceedings of the 14th International Symposium on Remote Sensing of the Environment,* pp. 1755–67. San Jose, Costa Rica.

Cressie, N. (1984). Towards resistant geostatistics. In *Geostatistics for Natural Resources Characterization,* ed. G. Verly *et al.,* pp. 21–44. Dordrecht: Reidel.
 (1985). Fitting variogram models by weighted least squares. *Journal International Association for Mathematical Geology,* **17,** 563–86.
 (1986). Kriging non stationary data. *Journal American Statistical Association,* **81,** 625–34.

Cressie, N. and Chan, N.H. (1989). Spatial modelling of regional variables. *Journal, American Statistical Association,* **84,** 393–401.

Cressie, N. and Hawkins, D.M. (1980). Robust estimation of the variogram. *International Journal of Mathematical Geology,* **12,** 115–26.

Cressie, N. and Read, T.R.C. (1989). Spatial data analysis of regional counts. *Biometrical Journal,* **6,** 699–719.

Cross, G.R. and Jain, A.K. (1983). Markov random field texture models. *IEEE Transactions on Pattern Analysis and Machine Intelligence,* **PAM 1–5,** 1, 25–39.

Curran, P.T. (1988). The semi variogram in remote sensing: an introduction. *Remote Sensing of Environment,* **24,** 493–507.

Curry, L.J. (1970). Applicability of space time moving average forecasting. In *Regional Forecasting; Proceedings of the 22nd Symposium of the Colston Research Society*, ed. M. Chisholm, A.E. Frey and P. Haggett. London: Butterworths.

(1977). Stochastic spatial distribution in equilibrium settlement theory. In *Man, Culture and Settlement*, ed. R.C. Eidt, chapter 12. New Delhi: Kalyan.

Dacey, M.F. (1965). A review of measures of contiguity for two and *k* colour maps. In *Spatial Analysis: A Reader in Statistical Geography*, pp. 479–95. Englewood Cliffs, New Jersey: Prentice Hall.

(1968). An empirical study of the areal distribution of houses in Puerto Rico. *Transactions, Institute of British Geographers*, **45**, 51–69.

Dalhaus, R. and Kunsch, H. (1987). Edge effects and efficient parameter estimation for stationary random fields. *Biometrika*, **74**, 877–82.

Davis, M.W.D. and David, M. (1978). Automatic kriging and contouring in the presence of trends. *Journal Canadian Petroleum Technology*, **17**, 90–9.

Delfiner, P. (1976). Linear estimation of non stationary spatial phenomena. In *Advanced Geostatistics in the Mining Industry*, ed. M. Guarascio, C.J. Huibregts and M. David, pp. 49–68. Dordrecht: Reidel.

Diaconis, P. (1985). Theories of data analysis: from magical thinking through classical statistics. In *Exploring Data Tables, Trends and Shapes*, ed. D.C. Hoaglin, F.M. Mosteller and J.W. Tukey, pp. 1–36. New York: Wiley.

Doreian, P. (1980). Linear models with spatially distributed data. *Sociological Methods and Research*, **9**, 29–60.

(1981). On the estimation of linear models with spatially distributed data. In *Sociological Methodology*, ed. S.S. Leinhardt, pp. 359–88. San Francisco, CA: Jossey Bass.

Doreian, P. and Hummon, N.P. (1976). *Modeling Social Processes*. New York: Elsevier.

Dougherty, E.L. and Smith, S.T. (1966). The use of linear programming to filter digitized map data. *Geophysics*, **31**, 253–9.

Dow, M.M., Burton, M.L. and White, D.R. (1982). Network autocorrelation: a simulation study of a foundational problem in regression and survey research. *Social Networks*, **4**, 169–200.

Dowd, P.A. (1984). The variogram and kriging: robust and resistant estimators. In *Geostatistics for Natural Resources Characterization*, ed. G. Verly et al., pp. 91–106. Dordrecht: Reidel.

Draper, N.R. and Guttman, I. (1980). Incorporating overlap effects from neighbouring units into response surface models. *Applied Statistics*, **29**, 128–34.

Edgington, E.S. (1969). Randomization tests. *Journal of Psychology*, **57**, 445–9.

Emerson, J.D. (1983). Mathematical aspects of transformation. In *Understanding Robust and Exploratory Data Analysis*, ed. D.C. Hoaglin, F.M. Mosteller and J.W. Tukey, pp. 247–82. New York: Wiley.

Emerson, J.D. and Hoaglin, D.C. (1983a). Resistant lines for y versus x. In *Understanding Robust and Exploratory Data Analysis*, ed. D.C. Hoaglin, F.M. Mosteller and J.W. Tukey, pp. 129–65. New York: Wiley.

(1983b). Analysis of two way tables by medians. In *Understanding Robust and Exploratory Data Analysis*, ed. D.C. Hoaglin, F.M. Mosteller and J.W. Tukey, pp. 166–210. New York: Wiley.

(1985). Resistant multiple regression, one variable at a time. In *Exploring Data Tables, Trends and Shapes*, ed. D.C. Hoaglin, F.M. Mosteller and J.W. Tukey, pp. 241–80. New York: Wiley.

Emerson, J.D. and Stoto, M.A. (1983). Transforming Data. In *Understanding Robust and Exploratory Data Analysis*, ed. D.C. Hoaglin, F.M. Mosteller and J.W. Tukey, pp. 97–128. New York: Wiley.

Emerson, J.D. and Wong, G.Y. (1985). Resistant non additive fits for two way tables. In *Exploring Data Tables, Trends and Shapes*, ed. D.C. Hoaglin, F.M. Mosteller and J.W. Tukey, pp. 67–124. New York: Wiley.

References

Evans, I.S. (1973). General geomorphometry, derivations of altitude and descriptive statics. In *Spatial Analysis in Geomorphology*, ed. R.J. Chorley, pp. 17–90. London: Methuen.

(1981). Census data handling. In *Quantitative Geography*, ed. N. Wrigley and R.J. Bennett, pp. 46–59. London: Routledge & Kegan Paul.

Fingleton, B. (1986). Analyzing cross-classified data with inherent spatial dependence. *Geographical Analysis*, **18**, 48–61.

Flowerdew, R. and Green, M. (1989). Statistical methods for inference between incompatible zonal systems. In *Accuracy of Spatial Databases*, ed. M. Goodchild and S. Gopal, pp. 239–47. London: Taylor and Francis.

Ford, E.D. (1976). The canopy of a Scots Pine forest: description of a surface of complex roughness. *Agricultural Meteorology*, **17**, 9–32.

Forster, B.C. (1980). Urban Residential ground cover using Landsat digital data. *Photogrammetric Engineering and Remote Sensing*, **46**, 547–58.

Freund, J.E. (1962). *Mathematical Statistics*. Englewood Cliffs: Prentice Hall.

Fuller, W.A. (1975). Regression analysis for sample survey. *Sankhya*, **37(c)**, Pt 3, 117–32.

Galtung, J. (1967). *Theory and Methods of Social Research*. New York: Columbia University Press.

Gatrell, A.C. (1977). Complexity and redundancy in binary maps. *Geographical Analysis*, **9**, 29–41.

(1979). Autocorrelation in spaces. *Environment and Planning A*, **II**, 507–16.

(1983). *Distance and Space: A Geographical Perspective*. Oxford: Clarendon Press.

Gentle, J.E. (1977). Least absolute values estimation: an introduction. *Communications in Statistics*, **B**, 313–28.

Ghosh, B. (1951). Random distances within a rectangle and between two rectangles. *Bulletin of the Calcutta Mathematical Society*, **43**, 17–24.

Glasbey, C.A. and O'Sullivan, M.F. (1988). Analysis of cone resistance data with missing observations below stones. *Journal of Soil Science*, **39**, 587–92.

Godambe, V.P. and Thompson, M.E. (1971). Bayes, fiducial and frequency aspects of statistical inference in regression analysis in survey sampling. *Journal, Royal Statistical Society*, **B33**, 361–90.

Good, I.J. (1983). The philosophy of exploratory data analysis. *Philosophy of Science*, **50**, 283–95.

Goodall, C. (1983). M-estimators of location: an outline of the theory. In *Understanding Robust and Exploratory Data Analysis*, ed. D.C. Hoaglin, F.M. Mosteller and J.W. Tukey, pp. 339–403. New York: Wiley.

Goodchild, M.F. (1984). Geocoding and geosampling. In *Spatial Statistics and Models*, ed. G.L. Gaile and C.J. Willmott, pp. 33–54. Dordrecht: Reidel.

(1986). *Spatial Autocorrelation*. CATMOG 47, Norwich, Geo Books.

Goodchild, M.F. and Gopal, S. (1989). *Accuracy of Spatial Databases*. London: Taylor and Francis.

Goodman, A.C. (1981). Housing submarkets within urban areas: definitions and evidence. *Journal of Regional Science*, **21**, 175–85.

Green, N.P., Finch, S. and Wiggins, J. (1985). The 'state of the art' in geographic information systems. *Area*, **17**, 295–301.

Green, P.J. and Sibson, R. (1978). Computing Dirichlet tessellations in the plane. *Computer Journal*, **21**, 168–73.

Greenhut, M.L., Norman, G. and Hung, C.-S. (1987). *The Economics of Imperfect Competition: A Spatial Approach*. Cambridge: Cambridge University Press.

Greig-Smith, P. (1964). *Quantitative Plant Ecology*, 2nd edn. London: Butterworths.

Griffith, D.A. (1981). Modeling urban population density in a multi-centred city. *Journal of Urban Economics*, **9**, 298–310.

(1982). Dynamic characteristics of spatial economic systems. *Economic Geography*, **58**, 177–196.

(1987). Spatial autocorrelation. *Resource Publications in Geography*, Association of American Geographers. State College, PA: Commercial Printing, Inc.

(1988). Estimating spatial autoregressive model parameters with commercial statistical packages. *Geographical Analysis*, **20**, 176–86.

Griffith, D.A., Bennett, R.J. and Haining, R.P. (1989). Statistical analysis of spatial data in the presence of missing observations: a methodological guide and an application to urban census data. *Environment and Planning A*, **21**, 1511–23.

Guptill, S.C. (1978). An evaluative technique for categorical maps. *Geographical Analysis*, **10**, 248–61.

Guyon, X. (1982). Parameter estimation for a stationary process on a d-dimensional lattice. *Biometrika*, **69**, 95–105.

Haggett, P. (1965). *Locational analysis in human geography*. London: Arnold.

(1976). Hybridizing alternative models of an epidemic diffusion process. *Economic Geography*, **52**, 136–46.

(1980). Geography and its boundaries. In *Die Bedeutung von Grenzen in der Geographie*, ed. H. Kishimoto, pp. 59–67. Zurich: Kummerley and Frey.

(1981). The edges of space. In *European Progress in Spatial Analysis*, ed. R.J. Bennett, pp. 51–70. London: Pion.

Haggett, P., Cliff, A.D. and Frey, A. (1977). *Locational Methods*. London: Arnold.

Haining, R.P. (1977). Model specification in stationary random fields. *Geographical Analysis*, **9**, 107–29.

(1978a). A spatial model for High Plains Agriculture. *Annals, Association of American Geographers*, **68**, 593–604.

(1978b). *Specification and Estimation Problems in Models of Spatial Dependence*. Evanston, IL: Northwestern University Press.

(1978c). Estimating spatial interaction models. *Environment and Planning A*, **10**, 305–20.

(1978d). The moving average model for spatial interaction. *Transactions, Institute of British Geographers*, N.S. **3**, 202–25.

(1979). Statistical tests and process generators for random field models. *Geographical Analysis*, **11**, 45–64.

(1980). Spatial autocorrelation problems. In *Geography and the Urban Environment*, ed. D.T. Herbert and R.J. Johnston, pp. 1–44. New York: Wiley.

(1981a). An approach to the statistical analysis of clustered map pattern. In *Dynamic Spatial Models*, ed. D. Griffith and R. Mackinnon, pp. 288–317. Rockville, Maryland, USA: Sijthoff & Noordhoff.

(1981b). Spatial interdependencies in population distributions: a study in univariate map analysis – I Rural population densities; II Urban population distributions. *Environment and Planning A*, **13**, 65–84, 85–96.

(1983a). Anatomy of a price war. *Nature*, **304**, 679–80.

(1983b). Modelling intra-urban price competition: an example of gasoline pricing. *Journal of Regional Science*, **23**, 517–28.

(1983c). Spatial and spatio-temporal interaction models and the analysis of patterns of diffusion. *Transactions, Institute of British Geographers*, NS8, 158–86.

(1984). Testing a spatial interacting-markets hypothesis. *Review of Economics and Statistics*, **66**, 576–83.

(1986). Intra urban retail price competition: corporate and neighbourhood aspects of spatial price variation. In *Spatial Pricing and Differentiated Markets*, ed. G. Norman, London Papers in Regional Science, vol. 16, pp. 144–64. London: Pion.

(1987a). Small area aggregate income models: theory and methods with an

397

application to urban and rural income data for Pennsylvania. *Regional Studies*, **21**, 519–30.

(1987b). Income diffusion and spatial econometric models. *Geographical Analysis*, **19**, 57–68.

(1987c). Trend surface analysis with regional and local scales of variation with an application to aerial survey data. *Technometrics*, **29**, 461–9.

(1988). Estimating spatial means with an application to remotely sensed data. *Communications in Statistics: Theory and Methods*, **17**, 573–97.

(1990). Models in human geography: problems in specifying, estimating and validating models for spatial data. In *Spatial Statistics: Past, Present and Future*, ed. D. A. Griffith, pp. 83–102. Ann Arbor: Michigan Document Services.

(1991a). Bivariate correlation with spatial data. *Geographical Analysis*, **23**, 210–27.

(1991b). Estimation with heteroscedastic and correlated errors: a spatial analysis of intra-urban mortality data. *Papers in Regional Science*, **70**, 223–41.

Haining, R.P., Griffith, D.A. and Bennett, R.J. (1983). Simulating two-dimensional autocorrelated surfaces. *Geographical Analysis*, **15**, 247–55.

(1984). A statistical approach to the problem of missing data using a first order markov model. *Professional Geographer*, **36**, 338–48.

(1989). Maximum likelihood estimation with missing spatial data and with an application to remotely sensed data. *Communications in Statistics: Theory and Methods*, **18**, 1875–94.

Hajrasulika, S., Baniabbassi, N., Metthey, J. and Nielsen, D.R. (1980). Spatial variability of soil sampling for salinity studies in southwest Iran. *Irrigation Science*, **1**, 197–208.

Hampel, F.R., Ronchetti, E.M., Rousseeuw, P.J. and Stahel, W.A. (1986). *Robust Statistics*. New York: Wiley.

Hawkins, D.M. and Cressie, N. (1984). Robust Kriging: a proposal. *Mathematical Geology*, **16**, 3–18.

Henley, S. (1981). *Non Parametric Geostatistics*. London: Applied Science Publications Ltd.

Hepple, L. (1976). A maximum likelihood model for econometric estimation with spatial series. In *Theory and Practice in Regional Science: London Papers in Regional Science*. Vol. 6, ed. I. Masser, pp. 90–104. London: Pion.

(1979). Bayesian analysis of the linear model with spatial dependence. In *Exploratory and Explanatory Statistical Analysis of Spatial Data*, ed. C.P.A. Bartels and R.H. Ketellapper, pp. 179–99. Boston: Martinus Nijhoff.

Hinkley, D. (1977). On a quick choice of power transformation. *Applied Statistics*, **26**, 67–9.

Hoaglin, D.C., Mosteller, F. and Tukey, J.W. (1983). *Understanding Robust and Exploratory Data Analysis*. New York: Wiley.

(1985). *Exploring Data Tables, Trends and Shapes*. New York: Wiley.

Hodder, I.R. and Orton, C. (1976). *Spatial Analysis in Archaeology*. Cambridge: Cambridge University Press.

Hoeksema, R.J. and Kitanidis, P.K. (1985). Analysis of the spatial structure of properties of selected acquifiers. *Water Resources Research*, **21**, 563–72.

Holmes, J.H. and Haggett, P. (1977). Graph theory interpretation of flow matrices: a note on maximization procedures for identifying significant links. *Geographical Analysis*, **9**, 388–99.

Horton, C.W., Hempkins, W.B. and Hoffman, A.A.J. (1964). A statistical analysis of some aeromagnetic maps from the Northwestern Canadian Shield. *Geophysics*, **4**, 582–601.

Huber, P.J. (1978). *Robust Statistical Procedures*. Philadelphia: SIAM.

Hubert, L. (1978). Non parametric tests for patterns in geographic variation: possible generalizations. *Geographical Analysis*, 10, 86–8.

Hubert, L.J. and Golledge, R.G. (1982). Measuring association between spatially defined variables: Tjostheim's index and some extensions. *Geographical Analysis*, 14, 273–8.

Hubert, L.J., Golledge, R.G. and Costanzo, C.M. (1981). Generalized procedures for evaluating spatial autocorrelation. *Geographical Analysis*, 13, 224–33.

(1985a). Tests of randomness: unidimensional and multidimensional. *Environment and Planning A*, 17, 373–85.

Hubert, L.J., Golledge, R.G., Costanzo, C.M. and Gale, N. (1985b). Measuring association between spatially defined variables: an alternative procedure. *Geographical Analysis*, 17, 36–46.

Hubert, L., Golledge, R., Costanzo, C., Gale, N. and Halperin, W. (1983). Non parametric tests for directional data. In *Recent Developments in Spatial Analysis: Methodology, Measurement Models*, ed. G. Bahrenberg, M. Fischer and P. Nijkamp, pp. 171–89. Aldershot: Gower.

Huff, D.L. and Black, W. (1978). A multivariate graphic display for regional analysis. In *Graphical Representation of Multivariate Data*, ed. P.C.C. Wang, pp. 199–218. New York: Academic Press.

Hughes, J.P. and Lettenmaier, D.P. (1981). Data requirements for kriging: Estimation and network design. *Water Resources Research*, 17, 1641–50.

Iglewicz, B. (1983). Robust scale estimators and confidence intervals for location. In *Understanding Robust and Exploratory Data Analysis*, ed. D.C. Hoaglin, F.M. Mosteller and J.W. Tukey, pp. 404–29. New York: Wiley.

Johnson, B.W. and NcCulloch, R.E. (1987). Added variable plots in linear regression. *Technometrics*, 29, 427–33.

Johnston, J. (1984). *Econometric Methods*, 3rd edn. London: McGraw Hill.

Johnston, R.J. (1984). Quantitative ecological analysis in human geography: an evaluation of four problem areas. In *Recent Developments in Spatial Data Analysis*, ed. G. Bahrenberg, M. Fischer and P. Nijkamp, pp. 131–41. Aldershot: Gower.

(1986). The neighbourhood effect revisited: spatial science or political regionalism. *Environment and Planning D*, 4, 41–55.

Jolliffe, I.T. (1987). Rotation of principal components: some comments. *Journal of Climatology*, 7, 507–10.

Jones, D.A., Gurney, R.J. and O'Connell, P.E. (1979). Network design using optimal estimation procedures. *Water Resources Research*, 15, 1801–12.

deJong, P., Sprenger, C. and Van Veen, F. (1984). On extreme values of Moran's *I* and Geary's *c*. *Geographical Analysis*, 16, 17–24.

Journel, A.G. and Huijbregts, C.J. (1978). *Mining Geostatistics*. London: Academic Press.

Kahneman, D., Slovic, P. and Tversky, A. (eds.) (1982). *Judgment under Uncertainty: Heuristics and Biases*. Cambridge: Cambridge University Press.

Kau, J.B., Lee, C.F. and Chen, R.C. (1983). Structural shifts in urban population density gradients: an empirical investigation. *Journal of Urban Economics*, 13, 364–77.

Kennedy, S. and Tobler, W.R. (1983). Geographic interpolation. *Geographical Analysis*, 15, 151–6.

Kershaw, K.A. (1973). *Quantitative and Dynamic Plant Ecology*, 2nd edn. London: Arnold.

Keyes, D.L., Basoglu, U., Kuhlmey, E.L. and Rhyner, M.L. (1976). Comparison of several sampling designs for geographical data. *Geographical Analysis*, 8, 295–304.

References

Kiefer, J. and Wynn, H.P. (1981). Optimum balanced block and Latin square designs for correlated observations. *Annals of Statistics*, **9**, 737–57.

(1983). Autocorrelation robust design of experiments. In *Scientific Inference, Data Analysis and Robustness*, ed. G.E.P. Box, T. Leonard and C.-F. Wu, pp. 279–99. New York: Academic Press.

Kindermann, R. and Snell, J.L. (1980). *Markov Random Fields and their Applications*. Providence, Rhode Island: American Mathematics Society.

Kitanidis, P.K. (1983). Statistical estimation of polynomial generalized covariance functions and hydrologic applications. *Water Resources Research*, **19**, 909–21.

Kooijman, S.A.L.M. (1976). Some remarks on the statistical analysis of grids especially with respect to ecology. *Annals of Systems Research*, **5**, 113–32.

Kramer, W. and Donninger, C. (1987). Spatial autocorrelation among errors and the relative efficiency of OLS in the linear regression model. *Journal American Statistical Association*, **82**, 577–9.

Krumbein, W.C. and Graybill, F.A. (1965). *An Introduction to Statistical Geology*. New York: McGraw Hill.

Krumbein, W.C. and Slack, H.A. (1956). Statistical analysis of low level radioactivity of Pennsylvanian black fissure shale in Illinois. *Bulletin, Geological Society of America*, **67**, 739–62.

Kunsch, H.R. (1983). Approximations to the maximum likelihood equations for some Gaussian random fields. *Scandinavian Journal of Statistics*, **10**, 239–46.

Labovitz, M.L. and Masuoka, E.J. (1984). The influence of autocorrelation in signature extraction: an example from a geobotanical investigation of Cotter Basin, Montana. *International Journal of Remote Sensing*, **5**, 315–32.

Lam, N.S. (1983). Spatial interpolation methods: a review. *American Cartographer*, **10**, 129–49.

Larimore, W.E. (1977). Statistical inference on stationary random fields. *Proceedings of the IEEE*, **65**, 961–70.

Lax, D.A. (1975). An interim report of a Monte Carlo study of robust estimators of width. *Technical Report No. 93*, (series 2). Department of Statistics, Princeton University.

Leamer, E.E. (1978). *Specification Searches: Ad Hoc Inference with Non Experimental Data*. New York: Wiley.

(1983). Lets take the con out of econometrics. *The American Economic Review*, **72**, 31–43.

Leamer, E.E. and Leonard, H. (1983). Reporting the fragility of regression estimates. *Review of Economics and Statistics*, **65**, 306–17.

Leonard, T. (1983). Some philosophies of inference and modelling. In *Scientific Inference, Data Analysis and Robustness*, ed. G.E.P., Box, T. Leonard and C.-F. Wu, pp. 9–23. New York: Academic Press.

Li, G. (1985). Robust regression. In *Exploring Data Tables, Trends and Shapes*, ed. D.C. Hoaglin, F.M. Mosteller and J.W. Tukey, pp. 281–343. New York: Wiley.

Lin, A.L. (1985). A note on testing for regional homogeneity of a parameter. *Journal of Regional Science*, **25**, 129–35.

Loftin, C. and Ward, S.K. (1983). A Spatial autocorrelation model of the effects of population density on fertility. *American Sociological Review*, **48**, 121–8.

Lovett, A.A., Bentham, C.G. and Flowerdew, R. (1986). Analysing geographic variations in mortality using Poisson regression: the example of ischaemic heart disease in England and Wales 1969–1973. *Social Science Medicine*, **23**, 936–45.

Madow, W.G. (1953). On the theory of systematic sampling III. Comparisons of centred and random start systematic sampling. *Annals, Mathematical Statistics*, **24**, 101–6.

Mantel, N. (1967). The detection of disease clustering and a generalized regression approach. *Cancer Research*, **27**, 209–20.

Mardia, K.V. and Marshall, R.J. (1984). Maximum likelihood estimation of models for residual covariance in spatial regression. *Biometrika*, 71, 135–46.

Martin, R.J. (1979). A subclass of lattice processes applied to a problem in plane sampling. *Biometrika*, 66, 209–17.

(1981). A note on the adequacy of fit of a spatial model to some agricultural uniformity data. *Biometrika*, 68, 336–8.

(1984). Exact maximum likelihood for incomplete data from a correlated Gaussian process. *Communications in Statistics: Theory and Methods*, 13, 1275–88.

(1987). Some comments on correction techniques for boundary effects and missing value techniques. *Geographical Analysis*, 19, 273–82.

(1989). Information loss for a spatial Gaussian process with incomplete data. *Communications in Statistics: Theory and Methods* (in press).

Matérn, B. (1960). Spatial variation. *Meddelander fran Statens Skogsforsknings institut*, 49, s., 1–144. Reprinted in *Springer-Verlag Lecture Notes in Statistics*, No. 36.

Matheron, G. (1971). *The Theory of Regionalized Variables*. Fontainebleau: Centre de Morphologie Mathématique.

(1973). The intrinsic random functions and their applications. *Advances in Applied Probability*, 5, 439–68.

Matula, D.W. and Sokal, R.R. (1980). Properties of Gabriel graphs relevant to geographic variation and the clustering of points in the plane. *Geographical Analysis*, 12, 205–22.

McBratney, A.B. and Webster, R. (1981). Detection of ridge and furrow pattern by spectral analysis of crop yield. *International Statistical Review*, 49, 45–52.

(1983a). How many observations are needed for regional estimation of soil properties. *Soil Science*, 135, 177–83.

(1983b). Optimal interpolation and isarithmic mapping of soil properties: V. Co-regionalisation and multiple sampling strategy. *Journal of Soil Science*, 34, 137–62.

(1986). Choosing functions for semi-variograms of soil properties and fitting them to sampling estimates. *Journal of Soil Science*, 37, 617–39.

McBratney, A.B., Webster, R. and Burgess, T.M. (1981). The design of optimal sampling schemes for local estimation and mapping of regionalized variables – I Theory and Method. *Computers and Geosciences*, 7, 331–4.

McCullagh, M.J. (1974). Estimation by kriging of the reliability of the Trent telemetry network. *Computer Applications*, 2, 357–74.

McGill, R., Tukey, J.W. and Larsen, W.A. (1978). Variations of box plots. *American Statistician*, 32, 12–16.

Mead, R. (1967). A mathematical model for the estimation of interplant competition. *Biometrics*, 23, 189–205.

Mejia, M.M. and Rodriguez-Iturbe, I. (1974). On the synthesis of random field sampling for the spectrum: an application to the generation of hydrologic spatial processes. *Water Resources Research*, 10, 705–11.

Mielke, P.W. (1978). Clarification and appropriate inferences for Mantel and Valand's nonparametric multivariate analysis technique. *Biometrics*, 34, 277–282.

Miesch, A.T. (1975). Variograms and variance components in Geochemistry and ore evaluation. In *Quantitative Studies in the Geological Sciences*, ed. E.H.T. Whitten, pp. 333–40. Colorado: Geological Society of America.

Milne, A. (1959). The centric systematic area sample treated as a random sample. *Biometrics*, 15, 270–97.

Moellering, H. (1984). Real maps, virtual maps and interactive cartography. In *Spatial Statistics and Models*, ed. G.L. Gaile and C.J. Willmott, pp. 109–32. Dordrecht: Reidel.

References

Moran, P.A.P. (1973a). A Gaussian markovian process on a square lattice. *Journal, Applied Probability,* 10, 54–62.

(1973b). Necessary conditions for Markovian processes on a lattice. *Journal, Applied Probability,* 10, 605–12.

Morrison, D.F. (1967). *Multivariate Statistical Methods.* New York: McGraw Hill.

Mulaik, S.A. (1985). Exploratory statistics and empiricism. *Philosophy of Science,* 52, 410–30.

Nisbett, R. and Ross, L. (1980). *Human Inference: Strategies and Shortcomings of Social Judgment.* Englewood Cliffs, New Jersey: Prentice Hall.

Nystuen, J.D. and Dacey, M.F. (1961). A graph theoretical interpretation of nodal regions, *Regional Science Association, Papers and Proceedings,* 7, 29–42.

O'Connell, P.E., Gurney, R.J., Jones, D.A., Miller, J.B., Nicholass, C.A. and Senior, M.R. (1979). A case study of rationalization of a rain gauge network in S.W. England. *Water Resources Research,* 15, 1813–22.

Oden, N.L. (1984). Assessing the significance of a spatial correlogram. *Geographical Analysis,* 16, 1–16.

Odland, J. (1988). *Spatial Autocorrelation.* California: Sage.

Oliver, M.A. and Webster, R. (1986). Semi variograms for modelling the spatial pattern of landform and soil properties. *Earth Surfaces Processes and Landforms,* 11, 491–504.

Omre, H. (1984). The variogram and its estimation. In *Geostatistics for Natural Resources Characterization,* ed. G. Verly *et al.,* pp. 107–25. Dordrecht: Reidel.

Openshaw, S. and Taylor, P.J. (1981). The modifiable areal unit problem. In *Quantitative Geography,* ed. N. Wrigley and R.J. Bennett, pp. 60–9. London: Routledge and Kegan Paul.

Orchard, R. and Woodbury, M. (1972). The missing information principle: theory and application. In *Proceedings of the Sixth Berkeley Symposium on Mathematical Statistics and Probability,* vol. I, ed. L. LeCam, J. Neyman and E. Scott, pp. 697–715. Berkeley: University of California Press.

Ord, J.K. (1975). Estimation methods for models of spatial interaction. *Journal, American Statistical Association,* 70, 120–6.

(1980). Tests of significance using non-normal data. *Geographical Analysis,* 12, 387–92.

(1981). Towards a theory of spatial statistics: a comment. *Geographical Analysis,* 13, 86–91.

Paelink, J.H.P. and Klaassen, L.H. (1979). *Spatial econometrics.* Farnborough, Hants.: Saxon House.

Pocock, S.J., Cook, D.G. and Beresford, S.A.A. (1981). Regression of area mortality rates on explanatory variables: what weighting is appropriate? *Applied Statistician,* 30, 286–96.

Quenouille, M.H. (1949). Problems in plane sampling. *Annals of Mathematical Statistics,* 20, 355–75.

Rao, C.R. (1971). Minimum variance quadratic unbiased estimation of variance components. *Journal of Multivariate Analysis,* 1, 445–56.

(1972). Estimation of variance and covariance components in linear models. *Journal, American Statistical Association,* 67, 112–15.

Rao, P.S.R.S. (1977). Theory of the MINQUE – a review. *Sankhya,* 39B, 3, 201–10.

Rayner, J.N. (1971). *An Introduction to Spectral Analysis.* London: Pion.

Rayner, J.N. and Golledge, R.G. (1972). Spectral analysis of settlement patterns in diverse physical and economic environments. *Environment and Planning A,* 4, 347–71.

(1973). The Spectrum of U.S. Route 40 re-examined. *Geographical Analysis,* 5, 338–50.

Rhind, D.W. (1981). Geographical information systems in Britain. In *Quantitative Geography*, ed. N. Wrigley and R.J. Bennett, pp. 17–35. London: Routledge and Kegan Paul.

Rhind, D. (1987). Recent developments in GIS in the U.K. *International Journal of Geographical Information Systems*, 1, 229–41.

Richardson, S. and Hemon, D. (1981). On the variance of the sample correlation between two independent lattice processes. *Journal of Applied Probability*, 18, 943–8.

(1982). Autocorrelation spatiale: ses conséquences sur la correlation empirique de deux processus spatiaux. *Revue de Statistique Appliquée*, 300, 41–51.

Richman, M.B. (1986). Rotation of principal components. *Journal of Climatology*, 6, 293–335.

Ripley, B.D. (1981). *Spatial Statistics*. New York: Wiley.

(1982). Edge effects in spatial stochastic processes. In *Statistics in Theory and Practice*, ed. B. Ranneby, pp. 247–62. Umea: Swedish University of Agricultural Sciences.

(1984). Spatial statistics: developments 1980–3. *International Statistical Review*, 52, 141–50.

(1988). *Statistical Inference for Spatial Processes*. Cambridge: Cambridge University Press.

Robinson, A.H. (1956). The necessity of weighting values in correlation analysis of areal data. *Annals of the Association of American Geographers*, 46, 233–6.

Rodriguez-Iturbe, I. and Mejia, J.M. (1974). The design of rainfall networks in time and space. *Water Resources Research*, 10, 713–28.

Rosenthal, R. (1978). How often are our numbers wrong? *American Psychologist*, 33, 1005–8.

Rossi, P.H., Wright, J.D. and Anderson, A.B. (1983). *Handbook of Survey Research*. New York: Academic Press.

Royaltey, H.H., Astrachan, E. and Sokal, R.R. (1975). Tests for pattern in geographic variation. *Geographical Analysis*, 7, 369–94.

Schulze, P. (1987). Once again: testing for regional homogeneity. *Journal of Regional Science*, 27, 129–33.

Semple, R.K. and Green, M.B. (1984). Classification in human geography. In *Spatial Statistics and Spatial Models*, ed. G.L. Gaile and C.J. Willmott, pp. 55–79. Dordrecht: Reidel.

Sen, A.K. (1976). Large sample size distribution of statistics used in testing for spatial correlation. *Geographical Analysis*, 8, 175–84.

Sen, A. and Soot, S. (1977). Rank tests for spatial correlation. *Environment and Planning A*, 9, 897–903.

Shepard, D.S. (1984). Computer mapping: the symap interpolation algorithm. In *Spatial Statistics and Models*, ed. G.L. Gaile and C.J. Willmott, pp. 133–45. Dordrecht: Reidel.

Siegal, S. (1956). *Nonparametric Statistics for the Behavioural Sciences*. New York: McGraw Hill.

Siemiatycki, J. (1978). Mantel's space time clustering statistic: computing higher moments and a comparison of various data transforms. *Journal of Statistical Computation and Simulation*, 7, 13–31.

Smith, T.R., Menon, S., Star, J.L. and Estes, J.E. (1987). Requirements and principles for the implementation and construction of large scale geographic information systems. *International Journal of Geographical Information Systems*, 1, 13–31.

Sokal, R.R. (1979). Ecological parameters inferred from spatial correlograms. In *Contemporary Quantitative Ecology and Related Ecometrics*, ed. G.P. Patil and M.L. Rosenzweig, pp. 167–96. Fairland, MA: International Co-operative Publishing House.

403

References

Sokal, R.R. and Oden, N.L. (1978). Spatial autocorrelation in biology: (1) Methodology, (2) Some biological applications of evolutionary and ecological interest. *Biological Journal of the Linnean Society*, 10, 199–228, 229–49.

Solow, A.R. (1984). The analysis of second order stationary processes: time series analysis, spectral analysis, harmonic analysis and geostatistics. In *Geostatistics for Natural Resources Characterization*, ed. G. Verly *et al.*, pp. 573–85. Dordrecht: Reidel.

Starks, T.H. and Fang, J.H. (1982). On the estimation of the generalized covariance function. *Journal of the International Association of Mathematical Geologists*, 14, 1, 57–64.

Stein, M.L. (1987). Minimum norm quadratic estimation of spatial variograms. *Journal, American Statistical Association*, 82, 765–72.

Stetzer, F. (1982). The analysis of spatial parameter variation with jack-knifed parameters. *Journal of Regional Science*, 22, 177–88.

Streitberg, B. (1979). Multivariate models of dependent spatial data. In *Exploratory and Explanatory Statistical Analysis of Spatial Data*, ed. C.P.A. Bartels and R.H. Ketellapper, pp. 139–77. Boston: Martinus Nijhoff.

Switzer, P. (1975). Estimation of the accuracy of qualitative maps. In *Display and Analysis of Spatial Data*, ed. J. Davis and M. McCullagh, pp. 1–13. New York: Wiley.

(1977). Estimation of spatial distributions from point sources with application to air pollution measurements. *Bulletin, International Statistical Institute*, XLVII, 2, 123–37.

(1980). Extensions of linear discriminant analysis for statistical classification of remotely sensed imagery. *Mathematical Geology*, 12, 367–76.

(1983). Some spatial statistics for the interpolation of satellite data. *Bulletin of the International Statistics Institute*, 50, 28.3, 962–72.

Tinkler, K.J. (1973). The topology of rural periodic market systems. *Geografiska Annaler B*, 55, 121–33.

Tjøstheim, D. (1978). A measure of association for spatial variables. *Biometrika*, 65, 109–14.

Tobler, W.R. (1969). Geographical filters and their inverses. *Geographical Analysis*, 1, 234–53.

(1979a). Cellular geography. In *Philosophy in Geography*, ed. S. Gale and G. Olsson, pp. 379–86. Dordrecht: Riley.

(1979b). Lattice tuning. *Geographical Analysis*, 11, 36–44.

(1979c). Smooth pycnophylactic interpolation for geographical regions. *Journal, American Statistical Association*, 74, 519–36.

Tobler, W.R. and Kennedy, S. (1985). Smooth multi-dimensional interpolation. *Geographical Analysis*, 17, 251–7.

Tobler, W. and Lau, J. (1978). Isopleth mapping using histosplines. *Geographical Analysis*, 10, 273–9.

Tubbs, J.D. and Cobberly, W.A. (1978). Spatial correlation and its effects upon classification results in LANDSAT. *Proceedings of the 12th International Symposium on Remote Sensing of the Environment*, pp. 775–81. Ann Arbor, Michigan.

Tukey, J.W. (1977). *Exploratory Data Analysis*. Reading, MA: Addison–Wesley.

Unwin, D.J. and Wrigley, N. (1987a). Control point distribution in trend surface modelling revisited: an application of the concept of leverage. *Transactions, Institute of British Geographers*, 12, 147–60.

(1987b). Towards a general theory of control point distribution effects in trend surface models. *Computers in Geosciences*, 13, 351–5.

Upton, G.J. (1985). Distance weighted geographic interpolations. *Environment and Planning A*, 17, 667–71.

404

Upton, G.J. and Fingleton, B. (1985). *Spatial Data Analysis by Example Volume 1: Point Pattern and Quantitative Data.* New York: Wiley.

Verly, G., David, M., Journel, A.G. and Marechal, A. (1984). *Geostatistics for Natural Resources Characterisation.* Dordrecht: Reidel.

Visvalingam, M. (1983). Operational definitions of area based social indicators. *Environment and Planning A,* **15,** 831–9.

(1988). Issues relating to basic spatial units. *Mapping Awareness,* **2,** 42–5.

Wahba, G. and Wendelberger, J. (1980). Some new mathematical methods for variational objective analysis using splines and cross validation. *Monthly Weather Review,* **8,** 1122–43.

Wartenberg, D. (1985). Multivariate spatial correlation: a method for exploratory geographical analysis. *Geographical Analysis,* **17,** 263–83.

Webster, R. (1985). Quantitative spatial analysis of soil in the field. *Advances in Soil Science,* **3,** 1–70.

Webster, R. and Burgess, T.M. (1981). Optimal Interpolation and isarithmic mapping of soil properties: III Changing drift and universal kriging. *Journal of Soil Sciences,* **32,** 505–24.

(1984). Sampling and bulking strategies for estimating soil properties in small regions. *Journal of Soil Science,* **35,** 127–40.

Weisberg, S. (1985). *Applied Linear Regression.* New York: Wiley.

Wetherill, G.B., Duncombe, P., Kenward, M., Köllerstrom, J., Paul, S.R. and Vowden, B.J. (1986). *Regression Analysis with Applications.* London: Chapman and Hall.

Whittle, P. (1954). On stationary processes in the plane. *Biometrika,* 41, 434–49.

(1962). Topographic correlation, power law covariance functions and diffusion. *Biometrika,* **49,** 305–14.

Willmott, C.J., Rowe, C.M. and Philpot, W.D. (1985). Small scale climatic maps: a sensitivity analysis of some common assumptions associated with grid point interpolation and contouring. *American Cartographer,* **12,** 5–16.

Woodcock, C.E. and Strahler, A.H. (1983). Characterising spatial patterns in remotely sensed data. *17th International Symposium on Remote Sensing of the Environment,* pp. 839–52. Ann Arbor, Michigan: University of Michigan.

Wrigley, N. (1977). Probability surface mapping: a new approach to trend surface mapping. *Transactions, Institute of British Geographers,* N.S. 2, 2, 129–40.

(1983). Quantitative methods: on data and diagnostics. *Progress in Human Geography,* 7, 567–77.

(1985). *Categorical Data Analysis.* London: Longman.

Yates, F. (1948). Systematic sampling. *Philosophical Transactions of the Royal Society of London,* **A241,** 345–77.

Youden, W.J. and A. Mehlich. (1937). Selection of efficient methods for soil sampling. *Contributions of the Boyce Thompson Institute of Plant Research,* 9, 59–70.

Yule, G.U. (1926). Why do we sometimes get nonsense correlations between two time series. *Journal Royal Statistical Society,* 89, 1–69.

Index

acquaintances, 74, 90

added variable plots, 332, 338, 355–7, 384

Akaike's information criterion, 143–5, 155, 390

anchoring, 11

areal partition
  effect on variances, 49–50, 129, 265, 272, 275, 352–3
  incompatibility of, 15, 309–10
  intra-area homogeneity, 47–9
  model sensitivity to, 47–9, 349
  properties of, 47–9

autocorrelation, spatial, 6, 27, 197, 228–39, 245, 253–8, 270–9, 280–2, 288, 325–6, 350–1, 355–6, 362, 364, 366, 377–9, 380, 384
  applications, 253–8, 270–9, 280–2, 350–1, 355–6, 362, 364, 366, 377–9, 380
  test: construction of, 231–4; interpretation of, 234–7; selection criteria, 237–9; statistics, 228–31

autocorrelations, partial, 145, 166, 236–7

automodels, spatial
  autobinomial, 100, 134
  autologistic, 99–100, 134, 160, 288–91
  autonormal, 86–90, 103–5, 108, 113, 127, 129, 130–3, 134–41, 142–7, 158, 161–7, 273–9, 306–7, 314–21, 356–65
  autopascal, 101, 291
  autopoisson, 101, 291
  for interpolation, 310

autoregressive models, spatial
  conditional, 86–90, 103–5, 108, 113, 127, 129, 130–3, 134–41, 142–7, 158, 161–7, 273–9, 306–7, 314–21, 356–65; and estimating spatial means, 161–7; applications of, 273–9, 306–7, 314–21, 356–65; estimators for parameters of: coding, 131–3; least

squares, 130–1; maximum likelihood, 127–9; hypothesis testing, 142–7; model definition, 86–90, 103–5, 108, 113; properties of estimators, 134–41; validation for, 158
  simultaneous, 81–3, 89–90, 104–8, 113, 124–6, 129, 130–3, 134–41, 142–7, 158, 161–6, 237, 253–9, 273–9, 280–2, 314–21, 356–65, 384–5; and estimating spatial means, 161–6; applications of, 253–9 273–9, 280–2, 314–21, 356–65; estimators for parameters of: coding, 131–3; least squares, 130–1; maximum likelihood, 124–6, 129; robust, 364, 384–5; hypothesis testing, 142–7; model definition, 81–3, 89–90, 104–8, 113; properties of estimators, 134–41; validation for, 158

autoregressive models, temporal, 65, 241, 314

bootstrapping, 47, 52

borders (of map), *see* boundary (of map)

boundary (of map), 38–40, 44–5, 48–9, 60, 65, 69, 101–10, 114, 118–19, 137–40, 178, 196, 252, 260–1, 279, 349, 360–1
  effects in estimation, 118–19, 137–40
  effects in modelling, 44–5, 48–9, 60, 252, 260–1, 279, 360–1
  effects in interpolation, 300–1
  models for, 101–10
  properties of, 38–40, 65, 69
  sampling, 178, 196

box plots, 201–13, 220–1, 224, 239–41, 281

Chauvet's cloud, 239–41

Chernoff faces, 225–7

Chorley report, 10, 17

classification, spatial, *see* regionalisation

cliques, 74, 87
conditional autoregressive models, *see*
    autoregressive models, spatial
confirmatory data analysis, 12–13, 50–2
connectivity matrix (W), 69–74, 82, 83,
    88–9, 110–13, 130–1, 146, 201,
    214, 228–39, 253–8, 265–79,
    341–4, 354–64, 377–83
    effect of on model properties, 110–13
    order relations, and, 69–74, 82–3, 88–9
    parameter estimation, 130–1
    pattern description, 201, 214, 228–39
    regression, 341–4, 354–64, 377–83
    trend surface analysis, 253–8, 265–79
contaminated normal data, 241, 245–6
correlation, bivariate, 8, 29, 313–23
correlation, spatial, 6, 10, 26, 40–3, 44,
    47, 48, 65, 90–4, 103–10, 119–23,
    180–3, 184, 185, 253–8, 260–4,
    314, 326, 330, 347, 351
    estimation 119–20, 171, 318
    models, 90–4, 103–10
correlograms, spatial, 235–6, 263
covariance, spatial, 6, 33, 42–3, 55–6, 65,
    80–4, 88–9, 90–4, 101–10, 111–13,
    119–20, 135, 150–8, 197, 244, 318,
    253–7
    estimation, 55–6, 119–20, 135, 197,
    244, 318
    fitting models, 150–8
    models, 90–4
cross-validation, 52, 155, 158

data adaptive modelling, 50–4
data
    accuracy, 14–15, 16, 53
    distribution of, 46, 200–13
    editing, 198, 280–1, 303–4
    extreme values, 46–7, 60, 153, 160–1;
    see also outliers
    missing, 14–15, 249–50, 291–3, 333
    pre-whitening, 318, 321
    sources, 13–15
    spatial configuration, 44–5, 243, 251–2,
    261–5, 294–5, 307–8
    spatial forms, 17–18
    storage, 18–21
    transformations, 49, 198, 222, 227–8,
    241, 272, 281–2, 332–3, 365
    types, 3, 34–5
    volume, 15–16
Delaunay triangulation, 70, 72
DFITS, 45, 270, 311, 364, 368, 376
Dirichlet partition, 70, 71, 116, 293,
    307–9
discontinuities, spatial, 38–40, 65, 114
discriminant analysis, 27
doubly geometric model, 106–9, 181
dummy variables (regional), 76, 80, 381–3

econometrics, spatial, 7, 11
edges
    links or joins (between sites), *see* order,
    spatial relationships; connectivity
    matrix (W)
    of map, *see* boundary (of map)
errors in variables model, 84, 127
exploratory data analysis, 4–7, 8, 10,
    12–13, 36–7, 50–2, 197–228

filter mapping, 250–1
frame effects, 48–9, 196, 252, 262–3, 372

Gabriel graph, 70, 72
Geary coefficient, 228–39, 246
generalised cross product statistic, 197,
    230–1, 232–44, 325–6
geographic information systems, 10, 15,
    20, 21
geostatistics, 6, 34, 197
graphics, computer, 9–10, 18–21, 225–7

heterogeneity, spatial, 22–3, 31–2, 43–4,
    65, 114–15, 334–8, 381–3
    of pattern, 22–3, 43, 65, 114–15
    of response, 22–3, 31–2, 43–4, 65,
    334–8, 381–3
heteroscedasticity, 332, 347, 365, 370–2
    see also size–variance relationships

inference frameworks, 32–7
    classical, 32–3, 35, 36, 37
    randomisation, 37
information, statistical, 40–3, 163–6
interaction models, spatial, *see* automodels,
    spatial; autoregressive models, spatial
interpolation, 8, 10, 27, 171, 172,
    249–50, 291–309
    cartographic, 293–6
    discrete, 310
    distance weighing, 294–6
    sequential, 297–304
    simultaneous, 304–7
interquartile range, 201
intrinsic random functions, 67–8, 94–8,
    150–1, 152–3, 157
    estimation, 150–1, 152–3, 157
    models, 67–8, 94–8
isotropic surfaces, 33, 66–7, 181–3,
    284–7

join-count statistics, 147, 228–39

Kriging, 8, 34, 155, 177, 186–8, 194,
    249, 297–303, 303–4, 304–6, 307,
    310
    block, 177, 187, 321
    resistant, 303–4
    unified universal, 304–6

Kriging (*cont.*)
  universal, 297–303
  variance, 298–301

Lagrange multiplier tests, 145–7, 347
land use studies, 193
leverage, 4, 44–5, 260, 262, 264, 275,
    278, 311, 333, 348–9, 360–4, 368,
    377, 384
likelihood ratio tests, 142–5, 146–7, 276
linearising transformations, 228

map transformations, 226–7
mapping, computer, 9–10, 18–21, 227,
    293–6, 310
Markov property, spatial, 86–8
mean, spatial
  estimation for area, 171–2, 177–86
  estimation for population, 123–4,
    127–9, 134–41, 161–6
  models for, 75–80
  see also trends, spatial; trend surface
    analysis; regression analysis
measurement errors
  attributes, 14–15, 21–2, 262
  locations, 15
median polish, 157, 215–20, 264, 284
minitab, xix, 197, 364
model specification, 7, 34–6, 118–23
  development, 50–4
  validation, 7, 45–6, 60, 158–61
modifiable areal units, *see* areal partition
Moran coefficient, 146–7, 228–39, 253–8,
    270–2, 275, 332, 334, 347, 366,
    377–9
mortality data, 199–206, 214, 265–79,
    365–75
moving average, spatial, 83–4, 105–6,
    108, 126, 133, 137, 273, 357–60
multicollinearity, 331, 333

non-stationarity, checks for, 222
normalising data transformations, 227–8,
    241, 281–2
nuggett variation, 29, 34, 155, 172, 287
numerical methods, problems, 127–9, 167,
    349–50

order relationships, spatial, 69–74
outliers, 4, 5, 13, 157, 198, 203, 210,
    212, 214–15, 217–19, 221, 222,
    224, 239–41, 264–5, 280–2, 303–4,
    333, 348–9, 352
  spatial, 5, 198, 214–15, 217–19, 221,
    222, 224, 264–5, 280–2
  see also data, extreme values

periodicity, spatial, 115, 178, 183, 286–7
petrol data, 201, 207–13, 372–83

pocket plot, 241
Poisson regression, 368–75
pollution, monitoring, 29, 171, 216–20,
    253–8, 280–2
polynomial generalised covariance
    functions, 96–7, 147–58, 167, 282–7
  estimators, 150–4
  properties of estimators, 154–8
precipitation, monitoring, 22, 171, 173,
    194–5
principal components analysis, 61, 225,
    321, 330

Rankit plots, 201–13, 272, 280, 282,
    332, 353, 368–70, 376, 380
raster data, 19
regionalisation, 22–3, 43–4, 172, 223
regionalised variables, 6, 10, 193
regression, correlated errors, 31, 43,
    123–4, 134–5, 164–6, 339, 341,
    347, 356–65, 379–81, 384–5
  applications, 356–65, 379–81
  estimation, 43, 123–4, 134–5, 164–6,
    347, 384–5
  model, 31, 339, 341
regression, lagged explanatory, 8, 30–1,
    339–40, 341, 344–7, 354–60
  applications, 354–60
  estimation, 344–7
  model, 30–1, 339–40, 341
regression, lagged response, 8, 30–1,
    340–1, 344–7, 354–60, 379–81
  applications, 354–60, 379–81
  estimation, 344–7
  model, 30–1, 340–1
regression, space–time, 372–83
regression, standard, 8, 11, 16, 30, 56–60,
    76, 123–4, 164–6, 330–50, 350–4,
    354–5, 365–6
  applications, 350–4, 354–5, 365–6
  estimation, 56–60, 123–4, 164–6
  model, 16, 30, 76, 330–50
regression, weighted, 49–50, 129, 332,
    334, 352, 368–72
remote sensing, 17–18, 21–2, 26–7, 172,
    216–20, 249–50, 252, 283
replication, 33
residuals, 45, 128, 146–7, 157, 158–61,
    167, 270–8, 280–2, 311, 334–8,
    347, 351, 368–71, 376–83
  model validation, 158–61, 167, 270–8,
    280–2, 334–8, 351, 368–71, 376–83
  test for spatial correlation, 146–7
resistant data analysis, 4–5, 12–13, 51–2,
    54–5, 197–228, 303–4
  see also robust data analysis
robust data analysis, 4–5, 12–13, 51–2,
    54–60, 197–8, 239–44, 364–5,
    368–71, 384–5

of centre, 55–6
of covariances, 244
of regression, 56–60, 364–5, 368–71
of semi-variogram, 241–4
of spatial parameters, 364–5, 384–5
see also resistant data analysis

sampling, spatial, 26–7, 171–96
  aligned 176, 181–3
  forms of: area, 177; point, 175–6
  hierarchical, 190–1
  random, 27, 175, 177–96
  stratified, 176, 177–96
  systematic, 27, 176, 177–96
semi-variogram, 34, 55–6, 68, 94–7, 114,
    121–3, 151–2, 154–8, 167, 171,
    175, 186–8, 191, 197, 218–19,
    239–44, 246, 299–304, 310, 321
  estimation of, 121–3, 171, 175, 218–19
  model fitting, 151–2, 154–8, 167,
    282–7
  robust estimation, 55–6, 157, 197,
    239–44
  theoretical models, 34, 94–7, 122
signed chi-square, 201
simulation, 116–17, 160, 245–6
  contaminated data, 245–6
  spatial, 116–17, 160
simultaneous autoregressive models, *see*
    autoregressive models, spatial
size–variance relationships, 49–50, 129,
    265, 272, 275, 352–3
soil studies, 22, 189, 193–4, 283, 284–7
splines, 293
space–time processes, 24, 113, 259–63
spatial dependency, 40–3
spatial processes, 24–6

Spearman rank correlation coefficient, 314,
    321, 322–3, 326–30
spectral function, 120–1
square root differences cloud, 239–41
stationarity, 33, 38, 66–7, 69, 82, 94–7,
    102–4
  deviations from mean, of, 33, 67
  increments, of, 33, 67–8
  local, 66, 82, 102–4
  strong, 66–7
  weak, 33, 66–7
stem and leaf plots, 201–13, 281, 282
stochastic process, spatial, 66, 67

Tjostheim's index, 29–30, 324–30
trends, spatial, 29, 178, 182–3, 197, 198,
    215–22, 244, 287, 314
trend surface analysis, 32, 44–5, 49,
    75–6, 77–9, 123–4, 127–9, 134–5,
    164–6, 215, 251–82, 331, 366–73
  Correlated errors, applications, 253–82;
    estimation, 123–4, 127–9, 134–5,
    164–6; weighted, 276–8
  Problems with: border effects, 252,
    260–1; frame effects, 252, 262;
    irregular areas, 265–80; outliers,
    280–2; unequal coverage, 252,
    261–4
  standard model, 75–6, 123, 251–2

variance stabilising transformations, 228,
    272, 365
variation, spatial, 37–40
  contrast with time, 37–8
vector data, 18–19, 20

Winsorising, 303–4

Printed in the United Kingdom
by Lightning Source UK Ltd.
1965